Fields Institute Communications

VOLUME 65

The Fields Institute for Research in Mathematical Sciences

The Fields Institute is a centre for research in the mathematical sciences, located in Toronto, Canada. The Institute's mission is to advance global mathematical activity in the areas of research, education and innovation. The Fields Institute is supported by the Ontario Ministry of Training, Colleges and Universities, the Natural Sciences and Engineering Research Council of Canada, and seven Principal Sponsoring Universities in Ontario (Carleton, McMaster, Ottawa, Toronto, Waterloo, Western and York), as well as by a growing list of Affiliate Universities in Canada, the U.S. and Europe, and several commercial and industrial partners.

For further volumes:
www.springer.com/series/10503

Javad Mashreghi • Emmanuel Fricain
Editors

Blaschke Products and Their Applications

The Fields Institute for Research in the Mathematical Sciences

FIELDS

Springer

Editors
Javad Mashreghi
Département de Mathématiques
et de Statistique
Université Laval
Québec, QC
Canada

Emmanuel Fricain
Laboratoire Paul Painlevé
Université des Sciences et Technologies
Lille 1
Villeneuve d'Ascq cedex
France

ISSN 1069-5265 ISSN 2194-1564 (electronic)
Fields Institute Communications
ISBN 978-1-4614-5340-6 ISBN 978-1-4614-5341-3 (eBook)
DOI 10.1007/978-1-4614-5341-3
Springer New York Heidelberg Dordrecht London

Library of Congress Control Number: 2012950139

Mathematics Subject Classification: 30D50, 30D40, 30D55, 30E20, 30E25, 32A36

Printed on acid-free paper

Springer is part of Springer Science+Business Media (www.springer.com)

Conference on *Blaschke Products and Their Applications*
Fields Institute
Toronto
July 25 to 29, 2011

Preface

Infinite Blaschke products were introduced by Blaschke in 1915. However, finite Blaschke products, as a subclass of rational functions, has existed long before without being specifically addressed as finite Blaschke products. In 1929, R. Nevanlinna introduced the class of bounded analytic functions with almost everywhere unimodular boundary values. Then the term inner function was coined much later by A. Beurling in his seminal study of the invariant subspaces of the shift operator. The first extensive study of the properties of inner functions was made by W. Blaschke, W. Seidel and O. Frostman. The Riesz technique in extracting the zeros of a function in a Hardy space is considered as the first step of the full canonical factorization of such elements. The disposition of zeros of an inner function is intimately connected with the existence of radial limits of the inner function and its derivatives.

For almost a century, Blaschke products have been studied and exploited by mathematicians. Their boundary behaviour, the asymptotic growth of various integral means of Blaschke products and their derivatives, their applications in several branches of mathematics in particular as solutions to extremal problems, their membership in different function spaces and their dynamics are examples from a long list of active research domains in which they show their face.

With the exclusive help of Fields Institute, we held a conference on Blaschke Products and their Application from July 25 to 29, 2011, at the University of Toronto. The purpose of the conference was to bring together a wide spectrum of mathematicians in this area. With more than 50 specialists and young researchers from around the globe, we had 36 talks. There were 28 one-hour talks and 8 thirty-minute talks. Besides discussing Blaschke products, or more generally inner functions, and their properties, their applications in other domains were also extensively discussed. In particular, the following topics were of primary attention:

i. Approximation theory (L. Baratchart, A. Boivin, P. Gorkin, V. Prokhorov),
ii. Boundary values (W. Ross),
iii. Conformal metrics (O. Roth),
iv. Critical points (S. Favorov, D. Kraus),
v. Differential equations (J. Benbourenane, J. Heittokangas),
vi. Dynamical systems (O. Ivrii),

vii. Geometry (U. Daepp),
viii. Harmonic analysis (M. Pap),
ix. Hyperbolic geometry (L. Baribeau),
x. Integral means (D. Vukotic),
xi. Inner functions (A. Nicolau),
xii. Interpolation (P. Gorkin, G. Semmler),
xiii. Morse theory (L. Baratchart),
xiv. Operator theory (H. Bommier, S. Charpentier, D. Drissi),
xv. Pluripotential theory (A. Edigarian, W. Zwonek),
xvi. Riemann-Hilbert problem (C. Glader),
xvii. Ritt's theory (P. Tuen-Wai Ng),
xviii. Spectral theory of Toeplitz operators (E. Shargorodsky),
xix. Theory of analytic functions (I. Chyzhykov, R. Fournier, Q. Rahman),
xx. Theory of computation (T. McNicholl),
xxi. Truncated Toeplitz operators (J. Cima, W. Ross).

These talks were highly appreciated by the participants. It also confirms the fact that Blaschke products impressively appear in a large number of various fields and this conference allowed us to bring together a wide spectrum of prominent mathematicians of different domains.

This proceedings is the outcome of the conference. It contains 15 research-survey papers which are presented in alphabetical order of their titles. We would like to thank all the participants, the authors for their valuable contributions, and the Fields Institute for its unique and generous support of this event.

Javad Mashreghi
Emmanuel Fricain

Contents

Applications of Blaschke Products to the Spectral Theory of Toeplitz Operators

Sergei Grudsky and Eugene Shargorodsky

Abstract The chapter is a survey of some applications of Blaschke products to the spectral theory of Toeplitz operators. Topics discussed include Toeplitz operators with bounded measurable symbols, factorisation with an infinite index, compositions with Blaschke products, representation of functions with a given asymptotic behaviour of the argument in a neighbourhood of a discontinuity in the form of a composition of a continuous function with a Blaschke product, and applications to the KdV equation.

Keywords Toeplitz operators · Spectral theory · Discontinuous symbols · Blaschke products

Mathematics Subject Classification Primary 47B35 · 30J10 · Secondary 47A10 · 30H10

1 Introduction

Let $\mathbb{T} = \{\zeta \in \mathbb{C} : \ |\zeta| = 1\}$ be the unit circle and let $H^p(\mathbb{T})$, $1 \le p \le \infty$ denote the Hardy space, that is $H^p(\mathbb{T}) := \{f \in L^p(\mathbb{T}) : \ f_n = 0 \text{ for } n < 0\}$, where f_n is the nth Fourier coefficient of f. Let $T(a) : H^p(\mathbb{T}) \to H^p(\mathbb{T})$, $1 < p < \infty$ denote the Toeplitz operator generated by a function $a \in L^\infty(\mathbb{T})$, i.e. $T(a)f = P(af)$, $f \in H^p(\mathbb{T})$, where P is the Riesz projection:

$$Pg(\zeta) = \frac{1}{2} g(\zeta) + \frac{1}{2\pi i} \int_{\mathbb{T}} \frac{g(w)}{w - \zeta} \, dw, \quad \zeta \in \mathbb{T}. \tag{1}$$

S. Grudsky
Departamento de Matematicas, CINVESTAV del I.P.N., Apartado Postal 14-740, 07000, Mexico, D.F., Mexico
e-mail: grudsky@math.cinvestav.mx

E. Shargorodsky (✉)
Department of Mathematics, King's College London, Strand, London WC2R 2LS, UK
e-mail: eugene.shargorodsky@kcl.ac.uk

J. Mashreghi, E. Fricain (eds.), *Blaschke Products and Their Applications*,
Fields Institute Communications 65, DOI 10.1007/978-1-4614-5341-3_1,

$P : L^p(\mathbb{T}) \to H^p(\mathbb{T})$, $1 < p < \infty$ is a bounded projection and

$$P\left(\sum_{n=-\infty}^{+\infty} g_n \zeta^n\right) = \sum_{n=0}^{+\infty} g_n \zeta^n.$$

Toeplitz operators on the real line are defined similarly: let

$$Pf(x) = \frac{1}{2} f(x) + \frac{1}{2\pi i} \int_{\mathbb{R}} \frac{f(\tau)}{\tau - x} \, d\tau, \quad x \in \mathbb{R}. \tag{2}$$

Then $P : L^p(\mathbb{R}) \to L^p(\mathbb{R})$, $1 < p < \infty$ is a bounded projection and its range $H^p(\mathbb{R}) := PL^p(\mathbb{R})$ is the Hardy space corresponding to the upper half plane. The Toeplitz operator generated by a function (symbol) $a \in L^\infty(\mathbb{R})$ is defined as follows

$$T(a)f := P(af), \qquad T(a) : H^p(\mathbb{R}) \to H^p(\mathbb{R}).$$

Linear fractional transformations usually allow one to switch between Toeplitz operators on $\mathbb{R}$ and those on $\mathbb{T}$ without difficulty. Most of the present paper deals with the case of $\mathbb{T}$, although we pass to Toeplitz operators on $\mathbb{R}$ when discussing symbols with discontinuities of the (semi-)almost periodic type.

Toeplitz operators are closely related to the Riemann-Hilbert problem. They represent a universal and a most powerful tool that has been applied to a wide variety of problems in elasticity theory, fluid dynamics, physics, geometry, combinatorics, integrable systems, orthogonal polynomials, random matrices, probability and stochastic processes, information and control theory, and in many other fields. Toeplitz operators constitute one of the most important classes of non-self-adjoint operators with a very rich spectral theory, which utilizes methods of operator theory, function theory and the theory of Banach algebras. Their spectral properties are well understood in the case of piece-wise continuous, almost periodic or semi-almost periodic symbols (see the next section for more information and references). Unfortunately much less is known about properties of Toeplitz operators with general bounded measurable symbols.

The aim of the present survey is to describe an approach to the study of spectral properties of Toeplitz operators with symbols having "bad" discontinuities. This approach is based on a generalisation of the Wiener-Hopf factorisation that involves inner functions (Sect. 4) and on results on representation of functions with a given asymptotic behaviour of the argument in a neighbourhood of the discontinuity in the form of a Blaschke product or, more generally, in the form of a composition of a continuous function with a Blaschke product (Sect. 5). When dealing with compositions involving Blaschke products in the context of Toeplitz operators, one needs to study compositions of Muckenhoupt weights with Blaschke products. The corresponding results are described in Sect. 3. Section 2 is a brief introduction to the spectral theory of Toeplitz operators. Section 6 is devoted to applications of Blaschke products to the KdV equation. The final Sect. 7 contains a list of some open problems.

In order to keep the presentation simple, we do not consider Toeplitz operators on weighted Hardy spaces and block Toeplitz operators, i.e. Toeplitz operators with matrix symbols ($a \in L^\infty_{N\times N}$).

2 Spectra of Toeplitz Operators

A bounded linear operator A on a Banach space X is said to be normally solvable if its range $\operatorname{Ran} A$ is closed. We put $\operatorname{Ker} A = \{f \in X : Af = 0\}$ and $\operatorname{Coker} A := X/\operatorname{Ran} A$. If A is normally solvable and $\dim \operatorname{Ker} A < \infty$, then A is called a Φ_+-operator. If $\dim \operatorname{Coker} A < \infty$, then A is normally solvable and is called a Φ_--operator. A Fredholm operator is an operator that is both Φ_- and Φ_+. The index of a Fredholm operator A is the integer $\operatorname{Ind} A := \dim \operatorname{Ker} A - \dim \operatorname{Coker} A$. The operator A is right (left) invertible if there is a bounded linear operator B on X such that $AB = I$ $(BA = I)$, where I is identity operator on X, and the operator A is invertible if there is a bounded operator B on X such that $AB = BA = I$. It is easy to see that if A is left (right) invertible, then A is a $\Phi_+(\Phi_-)$-operator.

The spectrum and the essential spectrum of A are defined as follows:

$$\operatorname{Spec}(A) := \{\lambda \in \mathbb{C} : \ A - \lambda I \text{ is not invertible}\},$$

$$\operatorname{Spec}_e(A) := \{\lambda \in \mathbb{C} : \ A - \lambda I \text{ is not Fredholm}\}.$$

For any algebra $\mathfrak{A}$, we denote by $G\mathfrak{A}$ the group of invertible elements of $\mathfrak{A}$.

Theorem 1 ([46]) *The spectrum of* $T(a) : H^p(\mathbb{T}) \to H^p(\mathbb{T})$ *is connected.*

Theorem 2 ([9, 42], see also [17, Chap. 7, Theorem 5.1] or [5, Theorem 2.38]) *Let* $a \in L^\infty(\mathbb{T})$, $a \neq 0$. *Then* $T(a) : H^p(\mathbb{T}) \to H^p(\mathbb{T})$ *has a trivial kernel or a dense range.*

This theorem implies that a nonzero Toeplitz operator $T(a) : H^p(\mathbb{T}) \to H^p(\mathbb{T})$ is normally solvable if and only if it is Φ_- or Φ_+.

Theorem 3 ([22, 42], see also [17, Chap. 7, Theorem 4.1] or [5, Theorem 2.30]) *Let* $a \in L^\infty(\mathbb{T})$, $a \neq 0$. *If* $T(a) : H^p(\mathbb{T}) \to H^p(\mathbb{T})$ *is normally solvable then* $a \in GL^\infty(\mathbb{T})$, *i.e.*

$$\operatorname*{ess\,inf}_{t \in \mathbb{T}} \left|a(t)\right| > 0.$$

Theorem 4 ([10, 44], see also [5, Proposition 2.32]) *Let* $a \in L^\infty(\mathbb{T})$. *Then* $T(a) : H^p(\mathbb{T}) \to H^p(\mathbb{T})$ *is invertible (Fredholm,* Φ_- *or* Φ_+*) if and only if* $a \in GL^\infty(\mathbb{T})$ *and* $T(a/|a|) : H^p(\mathbb{T}) \to H^p(\mathbb{T})$ *is invertible (Fredholm,* Φ_- *or* Φ_+ *respectively). Moreover, if* $a \in GL^\infty(\mathbb{T})$, *then*

$$\dim \operatorname{Ker} T(a) = \dim \operatorname{Ker} T(a/|a|), \qquad \dim \operatorname{Coker} T(a) = \dim \operatorname{Coker} T(a/|a|).$$

Let $C(\mathbb{T})$ be the space of all continuous functions on the unit circle $\mathbb{T}$. Suppose $b \in C(\mathbb{T})$ and $b(t) \neq 0$, $\forall t \in \mathbb{T}$. Then the winding number of b is defined as follows

$$\operatorname{wind} b := \frac{1}{2\pi}[\arg b]_{\mathbb{T}},$$

where $[\arg b]_{\mathbb{T}}$ denotes the total increment of $\arg b(t)$ as the variable t travels around $\mathbb{T}$ in the counterclockwise direction.

Theorem 5 ([15], see also [17, Chap. 3, Theorem 7.1] or [5, Theorem 2.42]) *Let $a \in C(\mathbb{T})$. Then $\mathrm{Spec_e}(T(a)) = a(\mathbb{T})$ and*

$$\mathrm{Ind}\big(T(a) - \lambda I\big) = -\,\mathrm{wind}(a - \lambda), \quad \forall \lambda \in \mathbb{C} \setminus a(\mathbb{T}).$$

Theorem 6 ([16], see also [17, Chap. 9, Theorem 3.1] or [5, Proposition 5.39]) *Let $a \in L^\infty(\mathbb{T})$ be piecewise continuous and let*

$$\mathrm{Arc}_p(a;t) := \left\{ \zeta \in \mathbb{C} \,\middle|\, \arg \frac{a(t-0) - \zeta}{a(t+0) - \zeta} = \frac{2\pi}{p} \right\}$$

if $a(t-0) \neq a(t+0)$. Then

$$\mathrm{Spec_e}\big(T(a)\big) = \left(\bigcup_{t \in \mathbb{T}} \big\{ a(t \pm 0) \big\} \right) \cup \left(\bigcup_{a(t-0) \neq a(t+0)} \mathrm{Arc}_p(a;t) \right).$$

Let $H^\infty(\mathbb{T}) + C(\mathbb{T})$ be the Banach algebra of all functions of the form $h + f$ with $h \in H^\infty(\mathbb{T})$ and $f \in C(\mathbb{T})$ (see [34, 35]). An element a is invertible in $H^\infty(\mathbb{T}) + C(\mathbb{T})$ if and only if its harmonic extension to the unit disk is bounded away from zero in some annulus $1 - \delta < |z| < 1$ ([11], [12, 7.36], see also [5, Theorem 2.62]).

Theorem 7 ([11], [12, 7.36], see also [5, Theorem 2.65] and [13, Theorem 2.7]) *Suppose $a \in H^\infty(\mathbb{T}) + C(\mathbb{T})$ and $\operatorname{ess\,inf}_{t \in \mathbb{T}} |a(t)| > 0$.*

(1) *$T(a) : H^p(\mathbb{T}) \to H^p(\mathbb{T})$ is Fredholm if and only if $1/a \in H^\infty(\mathbb{T}) + C(\mathbb{T})$, in which case $\mathrm{Ind}(T(a)) = -\,\mathrm{wind}(a_r)$, where $r \in (0,1)$ is sufficiently close to 1, $a_r(e^{i\theta}) := \widehat{a}(re^{i\theta})$ and $\widehat{a}$ is the harmonic extension of a to the unit disk.*
(2) *If $1/a \notin H^\infty(\mathbb{T}) + C(\mathbb{T})$, then $T(a)$ is left invertible and $T(1/a) : H^p(\mathbb{T}) \to H^p(\mathbb{T})$ is its left inverse.*

A number $c \in \mathbb{C}$ is called a (*left, right*) *cluster value* of a measurable function $a : \mathbb{T} \to \mathbb{C}$ at a point $t \in \mathbb{T}$ if $1/(a - c) \notin L^\infty(W)$ for every neighbourhood (left semi-neighbourhood or right semi-neighbourhood respectively) $W \subset \mathbb{T}$ of t. Cluster values are invariant under changes of the function on measure zero sets. We denote the set of all left (right) cluster values of a at t by $a(t-0)$ (by $a(t+0)$), and use also the following notation $a(t) = a(t-0) \cup a(t+0)$, $a(\mathbb{T}) = \bigcup_{t \in \mathbb{T}} a(t)$. It is easy to see that $a(t-0)$, $a(t+0)$, $a(t)$ and $a(\mathbb{T})$ are closed sets. Hence they are all compact if $a \in L^\infty(\mathbb{T})$.

It follows from Theorem 3 that

$$a(\mathbb{T}) \subseteq \mathrm{Spec_e}\big(T(a)\big). \tag{3}$$

Suppose that for each $t \in \mathbb{T}$ the set $a(t)$ consists of two points

$$a_1(t), a_2(t) \in \mathbb{C}$$

(which may coincide). We say that $t \in \mathbb{T}_{\mathrm{I}}$ if $a_1(t) \neq a_2(t)$ and each of the sets $a(t-0)$ and $a(t+0)$ consists of one point, i.e. if a has a left and a right limits at t and they do not coincide. We say that $t \in \mathbb{T}_{\mathrm{II}}$ if at least one of the sets $a(t-0)$, $a(t+0)$ consists of two points, i.e. if a does not have a left or a right limit at t.

Let

$$\mathcal{R}_p(a;t) := \left\{ \zeta \in \mathbb{C} \,\middle|\, \frac{2\pi}{\max\{p,p'\}} \leq \arg \frac{a_1(t)-\zeta}{a_2(t)-\zeta} \leq \frac{2\pi}{\min\{p,p'\}} \right\}, \tag{4}$$

where $p' = p/(p-1)$.

Theorem 8 ([7, 8, 43], see also [5, 5.50–5.58]) *Suppose $a \in L^\infty(\mathbb{T})$ and for each $t \in \mathbb{T}$ the set $a(t)$ consists of at most two points. Then*

$$\mathrm{Spec}_{\mathrm{e}}\big(T(a)\big) = a(\mathbb{T}) \cup \Big(\bigcup_{t\in\mathbb{T}_{\mathrm{I}}} \mathrm{Arc}_p(a;t) \Big) \cup \Big(\bigcup_{t\in\mathbb{T}_{\mathrm{II}}} \mathcal{R}_p(a;t) \Big).$$

A complete description of the (essential) spectrum of $T(a)$ in terms of $a(t \pm 0)$, $t \in \mathbb{T}$ is no longer possible if $a(t)$ is allowed to contain more than two points (see [5, 4.71–4.78] and [38]). We return to this topic in Sect. 4. Here, we continue with a general result on factorisation.

Definition 1 Let $1 < p < \infty$. We say that a function $a \in GL^\infty(\mathbb{T})$ admits a p-factorisation if it can be represented in the form

$$a(t) = a_-(t) t^\kappa a_+(t), \quad t \in \mathbb{T}, \tag{5}$$

where κ is an integer, called the index of factorisation, and the functions $a^\pm$ satisfy the following conditions:

(1) $\overline{a_-} \in H^p(\mathbb{T})$, $\overline{a_-^{-1}} \in H^{p'}(\mathbb{T})$, $a_+ \in H^{p'}(\mathbb{T})$, $a_+^{-1} \in H^p(\mathbb{T})$, $p' = p/(p-1)$;
(2) the operator $(1/a_+)Pa_+I$ is bounded on $L^p(\mathbb{T})$.

It is not difficult to see that a p-factorisation is unique up to a constant factor. The set of all functions $a \in GL^\infty(\mathbb{T})$ that admit a p-factorisation will be denoted by $\mathrm{fact}(p)$.

Theorem 9 ([42, 44, 45], see also [17, Chap. 8, Theorems 4.1 and 4.2] or [5, Theorem 5.5]) *Let $a \in GL^\infty(\mathbb{T})$. The Toeplitz operator $T(a) : H^p(\mathbb{T}) \to H^p(\mathbb{T})$, $1 < p < \infty$ is Fredholm if and only if $a \in \mathrm{fact}(p)$. If representation* (5) *holds, then* $\mathrm{Ind}\, T(a) = -\kappa$, *and for $\kappa = 0$ ($\kappa > 0$ or $\kappa < 0$) the operator $T(a)$ is invertible (left invertible or right invertible respectively); moreover,*

$$\big[T(a)\big]^{-1} = P \frac{1}{t^\kappa a_+} P \frac{1}{a_-} I \tag{6}$$

is the corresponding inverse operator. Further, for $\kappa < 0$ *we have*

$$\dim \operatorname{Ker} T(a) = |\kappa| \quad \textit{and} \quad \operatorname{Ker} T(a) = \operatorname{span}\left\{ \frac{t^{j-1}}{a^+},\ j = 1, 2, \ldots, |\kappa| \right\}, \tag{7}$$

while for $\kappa > 0$ *we have* $\dim \operatorname{Coker} T(a) = \kappa$, *and* $f \in \operatorname{Ran} T(a)$ *if and only if the following orthogonality conditions are satisfied*:

$$\int_{\mathbb{T}} f(t) \frac{1}{t^j a_-(t)}\, dt = 0, \quad j = 1, 2, \ldots, \kappa. \tag{8}$$

It is not always easy to check whether or not $a \in \text{fact}(p)$. The following result describes a rather broad subclass of fact(p). A function $a \in GL^\infty(\mathbb{T})$ is called *locally p-sectorial* if for every $t \in \mathbb{T}$ there exist an open arc $\ell(t) \subset \mathbb{T}$ containing t and functions $g_\pm^{(t)} \in GH^\infty(\mathbb{T})$ such that

$$\bigcup_{\tau \in \ell(t)} \left(\overline{g_-^{(t)}}\, a g_+^{(t)}\right)(\tau) \subset \left\{ z = re^{i\theta} \in \mathbb{C} :\ r > 0,\ |\theta| < \frac{\pi}{\max\{p, p'\}} \right\}.$$

It is easy to see that $a \in GL^\infty(\mathbb{T})$ is locally p-sectorial if $a(t)$ lies in an open sector with the vertex at the origin and an angular opening not exceeding $2\pi/\max\{p, p'\}$ for every $t \in \mathbb{T}$.

Theorem 10 ([41, 42], see also [17, Chap. 12] or [5, 5.12–5.21]) *Let* $a \in GL^\infty(\mathbb{T})$ *be locally p-sectorial. Then* $T(a) : H^p(\mathbb{T}) \to H^p(\mathbb{T})$, $1 < p < \infty$ *is Fredholm.*

In the case of the space $L^p(\mathbb{R})$ the notion of a p-factorisation takes the following form. We say that a function $a \in GL^\infty(\mathbb{R})$ admits a *p-factorisation with respect to the real line* $\mathbb{R}$ if it can be represented in the form

$$a(x) = a_-(x) \left(\frac{x - i}{x + i} \right)^\kappa a_+(x), \quad x \in \mathbb{R}, \tag{9}$$

where κ is an integer, called the *index of factorisation*, and the functions $a_\pm$ satisfy the following conditions:

(1) $\frac{a_-(x)}{x-i} \in \overline{H^p(\mathbb{R})}$, $\frac{1}{a_-(x)(x-i)} \in \overline{H^{p'}(\mathbb{R})}$, $\frac{a_+(x)}{x+i} \in H^{p'}(\mathbb{R})$, $\frac{1}{a_+(x)(x+i)} \in H^p(\mathbb{R})$, $p' = p/(p-1)$;
(2) the operator $(1/a_+)Pa_+I$ is bounded in $L^p(\mathbb{R})$.

The algebra $AP(\mathbb{R})$ of almost periodic functions is defined as the smallest closed subalgebra of $L^\infty(\mathbb{R})$ that contains the set $\{e_\lambda : \lambda \in \mathbb{R}\}$, where $e_\lambda(x) = e^{i\lambda x}$. We denote by $C(\overline{\mathbb{R}})$ the set of all continuous functions f on $\mathbb{R}$ that have finite limits $f(-\infty)$ and $f(+\infty)$ at $\pm\infty$, and by $C(\dot{\mathbb{R}})$ the subspace of $C(\overline{\mathbb{R}})$ consisting of functions continuous at infinity, i.e. such that $f(-\infty) = f(+\infty)$. Finally, the smallest closed subalgebra of $L^\infty(\mathbb{R})$ that contains $AP(\mathbb{R}) \cup C(\overline{\mathbb{R}})$ is denoted by $SAP(\mathbb{R})$ and is called the algebra of semi-almost periodic functions. Every function $b \in SAP(\mathbb{R})$ can be represented in the form

$$b(x) = \big(1 - w(x)\big)b_l(x) + w(x)b_r(x) + c_0(x), \tag{10}$$

where $b_l, b_r \in AP(\mathbb{R})$, $c_0 \in C(\dot{\mathbb{R}})$ with $c_0(\infty) = 0$, and w is a function from $C(\overline{\mathbb{R}})$ such that

$$w(-\infty) = 0 \quad \text{and} \quad w(+\infty) = 1 \tag{11}$$

(see [36]). The functions b_l and b_r are uniquely determined and independent of the choice of w. They are called the left and the right almost periodic representatives of b.

According to H. Bohr's theorem, every function $b \in GAP(\mathbb{R})$ can be written in the form

$$b(x) = e^{i\mu(b)x + c(x)}, \quad x \in \mathbb{R} \tag{12}$$

with $\mu(b) \in \mathbb{R}$ and $c \in AP(\mathbb{R})$. The number $\mu(b)$ is called the mean motion of b and it is given by the formula

$$\mu(b) = \lim_{T\to\infty} \frac{1}{2T} \arg b(x)\Big|_{x=-T}^{T},$$

where $\arg b$ is any continuous branch of the argument of b. If $\mu(b) = 0$, the geometric mean $\lambda(b)$ is defined by

$$\lambda(b) = e^{M(c)}, \tag{13}$$

where $M(c)$ is the mean value of c,

$$M(c) = \lim_{T\to\infty} \frac{1}{2T} \int_{-T}^{T} c(x)dx.$$

Note that the function $c \in AP(\mathbb{R})$ in (12) is unique up to an additive constant in $2\pi i\mathbb{Z}$. Hence definition (13) does not depend on a particular choice of c.

For a $b \in SAP(\mathbb{R})$, set

$$\mu_-(b) := \mu(b_l), \qquad \mu_+(b) := \mu(b_r)$$

(see (10)). If $\mu_\pm(b) = 0$, set

$$\lambda_-(b) := \lambda(b_l), \qquad \lambda_+(b) := \lambda(b_r).$$

Theorem 11 ([31–33, 36], see also [13, Theorem 4.24]) *Let $a \in SAP(\mathbb{R})$. If $T(a) : H^p(\mathbb{R}) \to H^p(\mathbb{R})$, $1 < p < \infty$ is normally solvable, then*

$$\inf_{x\in\mathbb{R}} \big|a(x)\big| > 0.$$

Suppose this condition is satisfied.

(1) *If* $\mu_\pm(a) = 0$, *then* $T(a)$ *is Fredholm if and only if*

$$\frac{1}{2\pi} \arg \frac{\lambda_+(a)}{\lambda_-(a)} - \frac{1}{p} \notin \mathbb{Z}.$$

If this condition is not satisfied, then the $T(a) : H^p(\mathbb{R}) \to H^p(\mathbb{R})$ *is not normally solvable.*

(2) *If* $\mu_\pm(a) \geq 0$ *and* $\mu_+^2(a) + \mu_-^2(a) \neq 0$, *then* $T(a)$ *is left invertible and* $\dim \operatorname{Coker} T(a) = \infty$.

(3) *If* $\mu_\pm(a) \leq 0$ *and* $\mu_+^2(a) + \mu_-^2(a) \neq 0$, *then* $T(a)$ *is right invertible and* $\dim \operatorname{Ker} T(a) = \infty$.

(4) *If* $\mu_+(a)\mu_-(a) < 0$, *then* $T(a)$ *is not normally solvable in any of the spaces* $H^p(\mathbb{R})$, $1 < p < \infty$ *and* $\dim \operatorname{Ker} T(a) = \dim \operatorname{Coker} T(a) = 0$.

3 Compositions with Blaschke Products and the A_p Condition

The results in Sect. 2 give an explicit description of the (essential) spectrum of $T(a)$ if $a(t)$ consists of at most two points for every t or if a is semi-almost periodic. Both cases include piecewise continuous symbols treated in Theorem 6. Suppose now $a \in L^\infty(\mathbb{T})$ has a "bad" discontinuity at $t = 1$ or at any other point of $\mathbb{T}$. Then one cannot, in general, tell whether or not $T(a) : H^p(\mathbb{T}) \to H^p(\mathbb{T})$ is Fredholm. A possible way of approaching this problem is to try representing a in the form $a = a_0 \circ v$, where a_0 is a simple, e.g., piecewise continuous function and $v : \mathbb{T} \to \mathbb{T}$ is a suitable measurable transformation. If $v(1) = \mathbb{T}$, then $a = a_0 \circ v \in L^\infty(\mathbb{T})$ has a bad discontinuity at $t = 1$, namely $a(1) = a_0(\mathbb{T})$.

Suppose $T(a_0) : H^p(\mathbb{T}) \to H^p(\mathbb{T})$ is Fredholm. Then a_0 admits a factorisation of the form (5)

$$a_0(t) = a_-(t) t^\kappa a_+(t), \quad t \in \mathbb{T} \ (\kappa \in \mathbb{Z})$$

(see Theorem 9). Hence

$$a(t) = a_-\big(v(t)\big) v^\kappa(t) a_+\big(v(t)\big), \quad t \in \mathbb{T}. \tag{14}$$

Since we would like to have

$$\begin{aligned} &\overline{a_- \circ v} \in H^p(\mathbb{T}), && \overline{(a_- \circ v)^{-1}} \in H^{p'}(\mathbb{T}), \\ &a_+ \circ v \in H^{p'}(\mathbb{T}), && (a_+ \circ v)^{-1} \in H^p(\mathbb{T}), \end{aligned} \tag{15}$$

we need v to have an analytic extension to the unit disk. Given that $|v| = 1$ on $\mathbb{T}$, it is natural to assume that v is a nonconstant inner function. Since $v(1) = \mathbb{T}$, natural choices for v are the singular inner function

$$v(\zeta) = \exp\left(\sigma \frac{\zeta + 1}{\zeta - 1}\right), \quad \sigma = \text{const} > 0 \tag{16}$$

and infinite Blaschke products with zeroes converging to $t = 1$.

Suppose v is an inner function. Then the following variant of Littlewood's subordination principle shows that (15) does indeed hold.

Theorem 12 ([28], [37, Sect. 1.3] and [13, Theorem 5.5]) *Let v be a nonconstant inner function and let γ_v be defined by*

$$(\gamma_v f)(t) = f\big(v(t)\big), \quad t \in \mathbb{T}.$$

(1) *The mapping γ_v is a bounded linear operator on the space $L^p(\mathbb{T})$, $1 \le p < \infty$. The subspace $H^p(\mathbb{T})$ is invariant under γ_v.*
(2) *The mapping γ_v is an automorphism of the algebra $L^\infty(\mathbb{T})$. The subalgebra $H^\infty(\mathbb{T})$ is invariant under γ_v.*
(3) *For any $f \in L^p(\mathbb{T})$, $1 \le p \le \infty$,*

$$\left(\frac{1-|v(0)|}{1+|v(0)|}\right)^{1/p} \|f\|_p \le \|\gamma_v f\|_p \le \left(\frac{1+|v(0)|}{1-|v(0)|}\right)^{1/p} \|f\|_p. \tag{17}$$

The middle factor in the factorisation (5) is the finite Blaschke product t^κ and the index of the corresponding Toeplitz operator is $-\kappa$. If $\kappa \neq 0$ in (14) and if v is an inner function which is not a finite Blaschke product, then one would expect $T(a): H^p(\mathbb{T}) \to H^p(\mathbb{T})$ to be semi-Fredholm with an infinite index. This is indeed the case under natural conditions on the first and the third factors, and the corresponding representation is called a generalised factorisation with an infinite index. The function $a = a_0 \circ v$ is called v-periodic. These notions are discussed in Sect. 4 (see Theorems 16 and 17).

Finally, we need to find out whether or not the factorisation (14) satisfies condition (2) of Definition 1, i.e. whether or not the operator

$$\frac{1}{a_+ \circ v} P(a_+ \circ v)I: \quad L^p(\mathbb{T}) \to L^p(\mathbb{T})$$

is bounded.

Let $\rho: \mathbb{T} \to [0, +\infty]$ be a measurable function. According to the Hunt–Muckenhoupt–Wheeden theorem ([24]), the operator $(1/\rho)P\rho I$ is bounded on $L^p(\mathbb{T})$, $1 < p < \infty$ if and only if ρ satisfies the A_p condition:

$$\sup_I \left(\frac{1}{|I|}\int_I \rho^p(t)|dt|\right)^{\frac{1}{p}} \left(\frac{1}{|I|}\int_I \rho^{-p'}(t)|dt|\right)^{\frac{1}{p'}} = C_p < \infty, \tag{18}$$

where $I \subset \mathbb{T}$ is an arbitrary arc, $|I|$ denotes its length, and $p' = p/(p-1)$.

Hence we arrive at the following question:

$$\text{does } \rho \in A_p \text{ imply } \rho \circ v \in A_p \text{ for an arbitrary inner function } v? \tag{19}$$

Note by the way that if $v(0) = 0$, then $v : \mathbb{T} \to \mathbb{T}$ is measure preserving, i.e. $|v^{-1}(E)| = |E|$ for any measurable $E \subset \mathbb{T}$ (see, e.g., [28] or take f equal to the indicator function of E in (17) with $p = 1$).

Using Theorem 9 one can easily show (see [2, Sect. 1]) that (19) is equivalent to the following question: does the invertibility of $T(b) : H^p(\mathbb{T}) \to H^p(\mathbb{T})$ imply that of $T(b \circ v) : H^p(\mathbb{T}) \to H^p(\mathbb{T})$?

The answer is positive in the case $p = 2$ (see, e.g., [2, Theorem 3]). This follows, e.g., from the Helson–Szegö theorem ([23], see also [14, Chap. IV, Theorem 3.4]):

$$\rho \in A_2 \iff \rho = \exp(f + \widetilde{g}), \qquad f, g \in L^\infty(\mathbb{T}, \mathbb{R}), \qquad \|g\|_\infty < \pi/4,$$

where $\widetilde{g}$ is the harmonic conjugate of g.

Similarly, a theorem by N.Ya. Krupnik ([25, 26], see also [17, Sect. 12.5]) says that

$$\rho \in A_p \cap A_{p'} \iff \rho = \exp(f + \widetilde{g}), \qquad f, g \in L^\infty(\mathbb{T}, \mathbb{R}),$$
$$\|g\|_\infty < \frac{\pi}{2\max\{p, p'\}}, \qquad p' = \frac{p}{p-1},$$

and it is not difficult to show that

$$\rho \in A_p \cap A_{p'} \implies \rho \circ v \in A_p \cap A_{p'}$$

(see [2, Theorem 4]).

One can also prove that the reverse of the implication in (19) is true.

Theorem 13 ([2]) *Let $1 < p < \infty$, $p' = p/(p-1)$ and let v be an inner function.*

(1) *Suppose ρ is a weight such that $\rho \in L^p$ and $\rho^{-1} \in L^{p'}$. If $\rho \circ v \in A_p$ then $\rho \in A_p$.*
(2) *Suppose $a \in L^\infty(\mathbb{T})$. If $T(a \circ v) : H^p(\mathbb{T}) \to H^p(\mathbb{T})$ is invertible then $T(a) : H^p(\mathbb{T}) \to H^p(\mathbb{T})$ is invertible.*

In spite of all the above results, the answer to (19) turns out to be negative.

Let $\{z_k\}_{k=-\infty}^{\infty}$ be a sequence of nonzero points in the open unit disk satisfying

$$\lim_{k \to \pm\infty} z_k = 1 \quad \text{and} \quad \sum_{k=-\infty}^{\infty} \left(1 - |z_k|\right) < \infty. \tag{20}$$

The first condition in (20) guarantees that the Blaschke product

$$B(t) := \prod_{k=-\infty}^{\infty} \frac{|z_k|}{z_k} \frac{z_k - t}{1 - \overline{z_k} t}, \qquad t \in \mathbb{T} \tag{21}$$

extends to an analytic function on $\mathbb{C} \setminus (\bigcup_k \{\overline{z}_k^{-1}\} \cup \{1\})$. In particular, B is continuous on $\mathbb{T} \setminus \{1\}$.

Write $z_k = r_k e^{i\theta_k}$ with $0 < r_k < 1$ and $-\pi < \theta_k \le \pi$. Put

$$\theta_k := \begin{cases} (\operatorname{sign} k)e^{-|k|} & \text{for } k \neq 0, \\ -1 & \text{for } k = 0, \end{cases}$$

$$\Delta_k := \begin{cases} \theta_k - \theta_{k+1} & \text{for } k = 1, 2, 3, \dots, \\ \theta_{k-1} - \theta_k & \text{for } k = 0, -1, -2, \dots, \end{cases} \tag{22}$$

$$\delta_k := \min\left\{ \left(\frac{\Delta_k}{\log \Delta_k} \right)^2, \left(\frac{\Delta_{k-1}}{\log \Delta_k} \right)^2 \right\},$$

choose a number $M > 1$, and define $r_k \in (0, 1)$ by

$$r_k := (1 - \delta_k/M)/(1 + \delta_k/M). \tag{23}$$

Theorem 14 ([2]) *Let $p \in (1, 2) \cup (2, \infty)$, $1/p + 1/p' = 1$ and let σ be any number in the interval $(1/p', 1/p)$ if $1 < p < 2$ and any number in the interval $(-1/p', -1/p)$ if $2 < p < \infty$. Then*

$$w(t) := |t - 1|^{-\sigma} \tag{24}$$

is a weight in A_p, but if $M > 1$ is sufficiently large and $B_M = B$ is the Blaschke product (21) *with the zeroes given by* (22)–(23), *then*

$$w\big(B_M(t)\big) = \big|B_M(t) - 1\big|^{-\sigma} \tag{25}$$

is not a weight in A_p.

Theorem 14 shows that there exists a Blaschke product for which the implication in (19) does not hold. We now describe a class of Blaschke products for which this implication does hold.

Consider the Blaschke product

$$B\big(e^{i\theta}\big) = \prod_{k=1}^{\infty} \frac{r_k - e^{i\theta}}{1 - r_k e^{i\theta}}, \qquad \theta \in [-\pi, \pi], \tag{26}$$

where $r_k \in (0, 1)$ and $\sum_{k=1}^{\infty} (1 - r_k) < \infty$.

Theorem 15 ([21]) *Suppose $r_1 \le r_2 \le \cdots \le r_n \le \cdots$, and*

$$\inf_{k \ge 1} \frac{1 - r_{k+1}}{1 - r_k} > 0. \tag{27}$$

If ρ satisfies the A_p condition, then $\rho \circ B$ also satisfies the A_p condition.

Corollary ([21]) *Let $1 < p < \infty$, $a \in L^\infty(\mathbb{T})$, and let a Blaschke product B satisfy the conditions of Theorem* 15. *Then $T(a) : H^p(\mathbb{T}) \to H^p(\mathbb{T})$ is invertible if and only if $T(a \circ B) : H^p(\mathbb{T}) \to H^p(\mathbb{T})$ is invertible.*

Proof The invertibility of $T(a \circ B)$ implies that of $T(a)$ according to Theorem 13 (see [2, Theorem 12]). The opposite implication follows from Theorem 15 (see [2, Sect. 1]). □

Theorem 15 and its Corollary remain true if the Blaschke product (26), (27) is substituted with the singular inner function (16) (see [19, 21]).

4 More on the Spectra of Toeplitz Operators

We start with extending Definition 1 and Theorem 9 to the case of $\Phi_\pm$ operators. We say that a function $a \in L^\infty(\mathbb{T})$ *admits a generalised factorisation with an infinite index in the space* $L^p(\mathbb{T})$, $1 < p < \infty$ if it admits a representation

$$a = bh \quad \text{or} \quad a = bh^{-1}, \tag{28}$$

where

(1) $b \in \text{fact}(p)$ (see Definition 1);
(2) $h \in H^\infty(\mathbb{T}),\ 1/h \in L^\infty(\mathbb{T})$;
(3) $q/h \notin H^\infty(\mathbb{T})$ for any polynomial q.

In this case, we also say that *a admits an* (h, p)*-factorisation.*

The class of functions admitting a generalized factorisation with an infinite index in $L^p(\mathbb{T})$ will be denoted by $\text{fact}(\infty, p)$.

Let $Q := I - P$, where P is the projection defined by (1).

Theorem 16 ([13, Theorem 2.6]) *Assume* $a \in \text{fact}(\infty, p)$ *and* $\operatorname{ind} b = 0$. *If* $a = bh^{-1}$, *then the operator* $T(a) : H^p(\mathbb{T}) \to H^p(\mathbb{T})$ *is right invertible*, $\dim \operatorname{Ker} T(a) = \infty$, *and the operator*

$$\left[T(a)\right]^{-1} = \frac{h}{b^+} P \frac{1}{b^-} I, \tag{29}$$

where $b = b^+ b^-$ *is the* p*-factorisation of the function* b, *is a right inverse of* $T(a)$. *For a function* φ *to belong to* $\operatorname{Ker} T(a)$ *it is necessary and sufficient that*

$$\varphi = \frac{h}{b^+} Q \frac{b^+}{h} \psi, \quad \textit{where } \psi \in \operatorname{Ker} T\left(h^{-1}\right). \tag{30}$$

If $a = bh$, *then* $T(a) : H^p(\mathbb{T}) \to H^p(\mathbb{T})$ *is left invertible*, $\dim \operatorname{Coker} T(a) = \infty$, *and the operator*

$$\left[T(a)\right]^{-1} = P \frac{1}{b^+ h} P \frac{1}{b^-} I \tag{31}$$

is a left inverse for $T(a)$. *For a function* f *to belong to* $\operatorname{Ran} T(a)$ *it is necessary and sufficient that*

$$\int_\Gamma \psi_j(t) f(t) dt = 0, \quad j = 1, 2, \ldots, \tag{32}$$

where

$$\psi_j = \frac{1}{b^-} Q \frac{(t-z_0)^{-j}}{hb^+} \in L_-^{p'}(\mathbb{T}) := QL^{p'}(\mathbb{T}), \quad p' = p/(p-1),$$

and $z_0 \in \mathbb{C}$ *is a fixed point such that* $|z_0| > 1$.

Functions admitting a generalized factorisation with an infinite index often arise as compositions with inner functions (cf. (14)). A function $a \in L^\infty(\mathbb{T})$ is called *u-periodic* if it admits a representation

$$a(t) = g\big(u(t)\big), \tag{33}$$

where $g \in L^\infty(\mathbb{T})$ and u is an inner function.

Theorem 17 ([18, Theorem 5.2]) *Let* $g \in C(\mathbb{T})$ *and suppose* $g(t) \neq 0, \forall t \in \mathbb{T}$ *and* $\operatorname{wind} g = \kappa$. *Then for every* $1 < p < \infty$ *and every inner function* $u \in H^\infty(\mathbb{T})$ *the* *u-periodic function* (33) *admits a* $(u^{|\kappa|}, p)$*-factorisation*

$$a(t) = g_-\big(u(t)\big)u^\kappa(t)g_+\big(u(t)\big),$$

where $g(t) = g_-(t)t^\kappa g_+(t)$ *is a factorisation of the type* (5). *Moreover, if* g *is a rational function, then*

$$(g_+ \circ u)^{\pm 1} \in H^\infty(\mathbb{T}), \qquad (g_- \circ u)^{\pm 1} \in \overline{H^\infty(\mathbb{T})}.$$

Remark 1 Theorem 17 cannot be extended to arbitrary symbols $g \in \operatorname{fact}(p)$ due to the difficulty described by Theorem 14. However, it can be extended to all locally p-sectorial symbols g (see [13, Theorem 5.8]). It also holds for all $g \in \operatorname{fact}(p)$ if one restricts the class of inner functions u to those for which the conclusions of Theorem 15 and its Corollary hold.

Similarly to the situation with Theorem 9, it is not always easy to check whether or not $a \in \operatorname{fact}(\infty, p)$. A broad subclass of $\operatorname{fact}(\infty, p)$ is desribed in Sect. 5 in terms of the asymptotic behaviour of the argument in a neighbourhood of a discontinuity.

Let us now consider compositions with homeomorphisms $\alpha : \mathbb{R} \to \mathbb{R}$ in the context of Toeplitz operators with (semi-)almost periodic symbols on $\mathbb{R}$. We will confine ourselves to the $H^2(\mathbb{R})$ setting to avoid difficulties related to Theorem 14. We start with a negative result.

Theorem 18 ([3]) *There exist* $b \in GAP(\mathbb{R})$ *and an orientation preserving homeomorphism* $\alpha : \mathbb{R} \to \mathbb{R}$ *such that* $T(b) : H^2(\mathbb{R}) \to H^2(\mathbb{R})$ *is Fredholm while* $T(a)$ *with* $a(x) = b(\alpha(x))$ *is not.*

In order to obtain positive results, one needs to restrict the class of homeomorphisms $\alpha : \mathbb{R} \to \mathbb{R}$. Let, similarly to the case of $\mathbb{T}$ considered in Sect. 2, $H^\infty(\mathbb{R}) + C(\dot{\mathbb{R}})$ be the Banach algebra of all functions of the form $h + f$ with $h \in H^\infty(\mathbb{R})$ and $f \in C(\dot{\mathbb{R}})$ (see [34, 35]).

Theorem 19 ([3]) *Let $b \in AP(\mathbb{R})$ and suppose*

$$e^{i\lambda\alpha} \in H^{\infty}(\mathbb{R}) + C(\dot{\mathbb{R}}), \quad \forall \lambda > 0. \tag{34}$$

Put $a(x) = b(\alpha(x))$. We then have the following.

(i) *If $T(b) : H^2(\mathbb{R}) \to H^2(\mathbb{R})$ is invertible, then $T(a)$ is a Φ-operator.*
(ii) *If $T(b) : H^2(\mathbb{R}) \to H^2(\mathbb{R})$ is left invertible, then $T(a)$ is a Φ_+-operator.*
(iii) *If $T(b) : H^2(\mathbb{R}) \to H^2(\mathbb{R})$ is right invertible, then $T(a)$ is a Φ_--operator.*

Theorem 36 provides sufficient conditions for (34) to hold (see also Theorem 37).

The following result extends Theorem 19 to semi-almost periodic symbols and it is natural to substitute condition (34) with the following one

$$(1 - w)e^{i\lambda\alpha}, \; we^{i\lambda\alpha} \in H^{\infty}(\mathbb{R}) + C(\dot{\mathbb{R}}) \quad \text{for all } \lambda > 0, \tag{35}$$

where $w \in C(\overline{\mathbb{R}})$ is a fixed function subject to (11).

Theorem 20 ([3]) *Let the homeomorphism α satisfy condition* (35) *and let $b \in SAP(\mathbb{R})$. Put $a(x) = b(\alpha(x))$. If $T(b) : H^2(\mathbb{R}) \to H^2(\mathbb{R})$ is a Φ-operator, then $T(a)$ is also a Φ-operator.*

Let us now return to the comment made after Theorem 8. Consider, for example, $a \in L^{\infty}(\mathbb{T})$ such that $a(1)$ consists of three points, $a(1 \pm 0) = a(\mathbb{T}) = \{c_1, c_2, c_3\} \subset \mathbb{C}$ and the closed triangle $\Delta(c_1, c_2, c_3)$ with the vertices c_1, c_2, c_3 is non-degenerate. Then the (essential) spectrum of $T(a) : H^2(\mathbb{T}) \to H^2(\mathbb{T})$ is a connected set which contains $\{c_1, c_2, c_3\}$ and is contained in $\Delta(c_1, c_2, c_3)$ ([12, Theorem 7.45], [6, 22], see also Theorems 1, 3, 10 above). It turns out however that this set is not determined solely by c_1, c_2, c_3. A. Böttcher has constructed examples where the spectrum of $T(a) : H^2(\mathbb{T}) \to H^2(\mathbb{T})$

(i) does not contain any points of the boundary of the triangle $\Delta(c_1, c_2, c_3)$ other than c_1, c_2, c_3;
(ii) contains a side of $\Delta(c_1, c_2, c_3)$ and no other point of the boundary apart from c_1, c_2, c_3;
(iii) coincides with the union of two sides of $\Delta(c_1, c_2, c_3)$;
(iv) coincides with the boundary of $\Delta(c_1, c_2, c_3)$;
(v) coincides with $\Delta(c_1, c_2, c_3)$

(see [5, 4.71–4.78]). These striking examples and the results obtained in [38, 39] imply that if $a(t)$ is not required to contain at most two points for every $t \in \mathbb{T}$, then it is no longer possible to describe the (essential) spectrum of $T(a)$ in terms of the cluster values of a. In other words, it is no longer sufficient to know the values of a, it is rather important to know "how these values are attained" by a. This field seems to be wide open at present.

Since a complete description of the essential spectrum of $T(a)$ in terms of the cluster values of $a \in L^{\infty}(\mathbb{T})$ is impossible, it is natural to try finding "optimal" sufficient conditions for a point λ to belong to the essential spectrum.

We need the following notation. Let $K \subset \mathbb{C}$ be an arbitrary compact set and $\lambda \in \mathbb{C} \setminus K$. Then the set

$$\sigma(K;\lambda) = \left\{ \frac{w-\lambda}{|w-\lambda|} \;\middle|\; w \in K \right\} \subseteq \mathbb{T}$$

is compact as a continuous image of a compact set. Hence the set $\Delta_\lambda(K) := \mathbb{T} \setminus \sigma(K;\lambda)$ is open in $\mathbb{T}$. So, $\Delta_\lambda(K)$ is the union of an at most countable family of open arcs.

We call an open arc of $\mathbb{T}$ p-large if its length is greater than or equal to $2\pi/\max\{p,p'\}$, where $p' = p/(p-1)$, $1 < p < \infty$.

We know that $a(\mathbb{T}) \subseteq \mathrm{Spec}_e(T(a))$ (see (3)). Böttcher's examples mentioned above show that no point from $\mathbb{C} \setminus a(\mathbb{T})$ will always belong to the (essential) spectrum of $T(a): H^2(\mathbb{T}) \to H^2(\mathbb{T})$, unless $a(\mathbb{T})$ lies on a straight line. The following result shows that the situation is somewhat different for $p \neq 2$.

Theorem 21 ([40]) *Let $1 < p < \infty$, $a \in L^\infty(\mathbb{T})$, $\lambda \in \mathbb{C} \setminus a(\mathbb{T})$ and suppose that, for some $t \in \mathbb{T}$,*

(i) *$\Delta_\lambda(a(t-0))$ (or $\Delta_\lambda(a(t+0))$) contains at least two p-large arcs,*
(ii) *$\Delta_\lambda(a(t+0))$ (or $\Delta_\lambda(a(t-0))$ respectively) contains at least one p-large arc.*

Then λ belongs to the essential spectrum of $T(a): H^p(\mathbb{T}) \to H^p(\mathbb{T})$.

Suppose $a(t)$ consists of two points. Then condition (ii) in the above theorem is automatically satisfied, while condition (i) means that a does not have a left limit at t (or a right limit at t respectively) and that λ belongs to the set (4). Hence, Theorem 21 is in a sense an extension of Theorem 8.

Condition (i) is optimal in the following sense.

Theorem 22 ([39]) *Let $t \in \mathbb{T}$, $K \subset \mathbb{C}$ be a compact set, $\lambda \in \mathbb{C} \setminus K$, and suppose $\Delta_\lambda(K)$ contains at most one p-large arc. Then there exists $a \in L_\infty(\mathbb{T})$ such that*

$$a(t \pm 0) = a(t) = a(\mathbb{T}) = K$$

and

$$T(a) - \lambda I: \quad H^r(\mathbb{T}) \to H^r(\mathbb{T})$$

is invertible for any $r \in [\min\{p,p'\}, \max\{p,p'\}]$.

While condition (i) is the main reason why λ belongs to $\mathrm{Spec}_e(T(a))$, the rôle of (ii) is to make sure that the behaviour of $a(\tau)$ as τ approaches t from the other side does not counterbalance the effect of (i). It turns out that condition (ii) cannot be dropped.

Theorem 23 ([21]) *There exists $a \in L^\infty(\mathbb{T})$ such that $a(1-0) = \{\pm 1\}$, $|a| \equiv 1$, $T(a): H^p(\mathbb{T}) \to H^p(\mathbb{T})$ is invertible for any $p \in (1,2)$, and $T(1/a): H^p(\mathbb{T}) \to H^p(\mathbb{T})$ is invertible for any $p \in (2,+\infty)$.*

The proof of this theorem relies on the Corollary of Theorem 15 and on Theorem 34.

5 Modelling of Monotone Functions with the Help of Blaschke Products

Suppose $a \in GL^\infty(\mathbb{T})$. Then Theorem 4 allows one to reduce the study of the operator $T(a) : H^p(\mathbb{T}) \to H^p(\mathbb{T})$, $1 < p < \infty$ to that of $T(a/|a|) : H^p(\mathbb{T}) \to H^p(\mathbb{T})$. We can therefore assume without loss of generality that $|a| = 1$, i.e. that

$$a\big(\exp(i\theta)\big) = \exp\big(2\pi i f(\theta)\big), \qquad \theta \in (-\pi, \pi], \tag{36}$$

where f is a measurable real-valued function. Suppose a has a discontinuity at $t = 1$. We aim at finding conditions on the behaviour of f in a neighbourhood of $t = 1$ under which a can be represented in terms of Blaschke products in such a way that one can then apply Theorems 16 and 17. Although our motivation comes from the theory of Toeplitz operators, we believe that the results presented in this section may be of some interest in their own right.

Since the discontinuity is at $t = 1$, it is natural to consider Blaschke products with zeroes converging to 1. Our first result is about the argument of such a Blaschke product. Let

$$B\big(e^{i\theta}\big) = \prod_{k=1}^{\infty} \frac{\overline{z_k}}{|z_k|} \frac{z_k - e^{i\theta}}{1 - \overline{z_k}\, e^{i\theta}}, \qquad \theta \in (-\pi, \pi], \tag{37}$$

where $z_k = r_k \exp(i\theta_k) \neq 0$, $\theta_k \in (-\pi, \pi]$, $r_k = |z_k| < 1$, $\sum_{k=1}^{\infty}(1 - r_k) < \infty$.

Theorem 24 ([13, Theorem 2.8]) *Suppose B has the form* (37) *and*

$$\lim_{k\to\infty} z_k = 1.$$

Then one can choose a branch of $\arg B(e^{i\tau})$ *which is continuous and increasing on* $(0, 2\pi)$, *and which satisfies the following conditions*

$$\lim_{\tau\to 0+0} \arg B\big(e^{i\tau}\big) =: A_+ < 0, \qquad \lim_{\tau\to 2\pi-0} \arg B\big(e^{i\tau}\big) =: A_- > 0.$$

Moreover, at least one of these limits is infinite and

$$\arg B\big(e^{i\theta}\big) = \begin{cases} -2(\sum_{\theta_k \ge \theta}(\pi + \varphi_k(\theta)) + \sum_{\theta_k < \theta} \varphi_k(\theta)), & \theta \in (0, \pi], \\ 2(\sum_{\theta_k \le \theta}(\pi - \varphi_k(\theta)) - \sum_{\theta_k > \theta} \varphi_k(\theta)), & \theta \in [-\pi, 0), \end{cases} \tag{38}$$

where

$$\varphi_k(\theta) = \arctan\left(\varepsilon_k \cot \frac{\theta - \theta_k}{2}\right), \qquad \varepsilon_k = \frac{1 - r_k}{1 + r_k}. \tag{39}$$

The next result shows that the argument of a Blaschke product may grow arbitrarily slowly or arbitrarily fast as $t \to 1$ and that the growth on the left from 1 may be different from that on the right.

Theorem 25 ([13, Theorem 2.9]) *Suppose that a real-valued function f is continuous and increasing on $(-\pi, 0)$ and $(0, \pi)$, and that at least one of the limits $\lim_{\theta \to 0 \pm 0} f(\theta)$ is infinite. Then there exists a Blaschke product B of the form* (37) *such that*

$$\left|\arg B\left(e^{i\theta}\right) - f(\theta)\right| \le \text{const}, \quad \theta \in (-\pi, \pi) \setminus \{0\}. \tag{40}$$

Theorem 25 allows one to factor out a Blaschke product from the symbol of a Toeplitz operator in such a way that the resulting Toeplitz operator has a symbol with a bounded and continuous argument on $\mathbb{T} \setminus \{1\}$. Unfortunately not much is known about such operators, so the above theorem is not sufficient for our purposes.

Suppose a has the form (36), where the function f is continuous and monotonically increasing on the intervals $(-\pi, 0)$ and $(0, \pi)$, and satisfies

$$\lim_{\theta \to 0 \pm 0} f(\theta) = \mp\infty. \tag{41}$$

With no loss of generality we can take $f(-\pi + 0) = f(\pi - 0) = 0$. Let

$$\vartheta(x) := f^{-1}(-x), \quad x \in \mathbb{R} \setminus \{0\}. \tag{42}$$

Then ϑ is monotonically decreasing on $(-\infty, 0)$ and $(0, \infty)$, and

$$\vartheta(\pm\infty) = 0, \qquad \vartheta(0 \pm 0) = \pm\pi.$$

Further, let

$$\Delta(n) = \begin{cases} \vartheta(n) - \vartheta(n+1), & n = +0, 1, 2, \ldots, \\ \vartheta(n-1) - \vartheta(n), & n = -0, -1, -2, \ldots. \end{cases}$$

Consider the sequence of functions

$$\psi_n(s) = \frac{\vartheta(n) - \vartheta(n+s)}{\Delta(n)}, \quad s \in I := [-1/2, 1/2].$$

We assume that this sequence converges monotonically on I and that

$$\begin{aligned} \lim_{n \to +\infty} \psi_n(s) &= \psi(s), \\ \lim_{n \to -\infty} \psi_n(s) &= -\psi(-s), \end{aligned} \tag{43}$$

where the function ψ is monotonically increasing and continuous on I. Finally, we put

$$\xi(n) = \begin{cases} \vartheta(n+1)/\vartheta(n), & n = +0, 1, 2, \ldots, \\ \vartheta(n-1)/\vartheta(n), & n = -0, -1, -2, \ldots, \end{cases}$$

$$\alpha(n) = 1 - \xi(n).$$

We will need the following technical result.

Theorem 26 ([13, Proposition 5.6]) *Suppose the function* ϑ *has the form* (42) *and satisfies condition* (43). *Then*

$$\lim_{n\to\pm\infty} \frac{\Delta(n\pm 1)}{\Delta(n)} = d, \quad 0 \le d \le 1. \tag{44}$$

Moreover,

$$\begin{aligned} &\lim_{n\to\infty} \xi(n) = d, \qquad \lim_{n\to\infty} \alpha(n) = 1 - d, \\ &\lim_{n\to\infty} \frac{\alpha(n+1)}{\alpha(n)} = 1, \end{aligned} \tag{45}$$

and

$$\psi(1/2) - d\psi(-1/2) = 1.$$

We will need the following two auxiliary functions

$$A(\theta) = \sum_{\vartheta(j)>\theta} \arctan\bigl(\alpha(j)\bigr) - \sum_{\vartheta(j)<-\theta} \arctan\bigl(\alpha(j)\bigr), \quad \theta \in (0,\pi), \tag{46}$$

and

$$C(n) = \sum_{j=m(n)}^{n-\sigma} \arctan \frac{\Delta(j)}{\vartheta(j) - \vartheta(n)} - \sum_{j=n+\sigma}^{M(n)} \arctan \frac{\Delta(j)}{\vartheta(n) - \vartheta(j)}, \tag{47}$$

where $\sigma = \operatorname{sign} n$, the number $m = m(n)$ is the j of the smallest modulus for which $|\vartheta(j)| \le \frac{3}{2}|\vartheta(n)|$, while $M = M(n)$ is the j of the largest modulus for which $|\vartheta(j)| \ge \frac{1}{2}|\vartheta(n)|$.

The quantity $A(\theta)$ relates the behaviour of $\vartheta(x)$ as $x \to +\infty$ to its behaviour as $x \to -\infty$; in other words, it connects the behaviour of f in a right semi-neighbourhood of zero to its behaviour in a left semi-neighbourhood (see (42)). The quantity $C(n)$ characterises the behaviour of $\vartheta(x)$ near the point $x = n$.

Theorem 27 ([13, Theorem 5.10]) *Suppose the function* $a \in GL^\infty(\mathbb{T})$ *is continuous on* $\mathbb{T} \setminus \{1\}$ *and has the form* (36) *with a function* f *that is monotonically increasing on* $(-\pi, 0)$ *and* $(0, \pi)$ *and satisfies condition* (41). *In addition, assume that condition* (43) *is satisfied, that* $d = 1$ *in* (44), *and that the limits*

$$\lim_{\theta\to 0\pm 0} A(\theta) = a, \quad a \in \mathbb{R}, \tag{48}$$

$$\lim_{n\to\pm\infty} C(n) = 0 \tag{49}$$

exist, where $A(\theta)$ *and* $C(n)$ *are defined by* (46) *and* (47).

Then a *admits the representation*

$$a(t) = B(t)g\big(B(t)\big)d(t), \tag{50}$$

which is a (B, p)*-factorisation, with* $g, d \in C(\mathbb{T})$. *Moreover, the winding number of the function* g *is equal to zero, the Blaschke product* B *is constructed from the zeroes* $z_j = r_j \exp\{i\vartheta(j)\}$, *where* $r_j = (1 - \Delta(j)/2)/(1 + \Delta(j)/2)$, $j = \pm 1, \pm 2, \ldots$, *and the product*

$$b(t) := g\big(B(t)\big)d(t)$$

admits a p*-factorisation of the form* (5) *for any* $1 < p < \infty$.

Theorem 28 ([13, Theorem 5.12]) *Suppose the function* f *satisfies all the conditions of Theorem* 27 *and that in condition* (43)

$$\psi(s) = s. \tag{51}$$

Then the function a *given by* (36) *belongs to* $H^\infty(\mathbb{T}) + C(\mathbb{T})$ *and admits the representation*

$$a(t) = U(t)c(t), \tag{52}$$

where c *is a continuous function on* $\mathbb{T}$ *and the inner function* U *has the form*

$$U(t) = \frac{r_0 + B(t)}{1 + r_0 B(t)}, \qquad r_0 = e^{-2},$$

with the same Blaschke product B *as in* (50).

Conditions (43), (48), (49), under which Theorems 27 and 28 hold, cover a very large class of symbols with arguments that increase in a neighbourhood of the discontinuity. However, they are not always easy to verify. The following theorems provide more convenient sufficient conditions. We assume as above that f is monotonically increasing on $[-\pi, 0)$ and $(0, \pi]$, and satisfies (41).

Theorem 29 ([13, Proposition 5.8]) *Let* f *be twice continuously differentiable on* $[-\pi, \pi] \setminus \{0\}$ *and let* f' *be monotonically decreasing (increasing) on* $(0, \pi)$ *(on* $(-\pi, 0)$ *respectively) and satisfy*

$$\lim_{\theta \to 0} \frac{f''(\theta)}{(f'(\theta))^2} = 0. \tag{53}$$

Then (43) *holds with the function* $\psi(s) \equiv s$.

It is not difficult to see that (53) implies

$$\lim_{\theta \to 0} \frac{1}{\theta f'(\theta)} = 0. \tag{54}$$

Theorem 30 ([13, Proposition 5.9]) *Suppose that f is twice continuously differentiable on $[-\pi, \pi] \setminus \{0\}$ and that f' is monotonically nonincreasing (nondecreasing) on $(0, \pi)$ (on $(-\pi, 0)$ respectively) and satisfies (54) and*

$$\lim_{\theta \to 0} \frac{f''(\theta)|\theta|^{1/2}}{(f'(\theta))^{3/2}} = 0. \tag{55}$$

Then (49) holds.

Theorem 31 ([13, Proposition 5.10]) *Suppose that all the assumptions of Theorem 29 are satisfied and that*

$$\left| \frac{f'(\theta)}{f'(\frac{3}{2}\theta)} \right| \le g, \tag{56}$$

where $g > 1$ does not depend on $\theta \in (-\pi, \pi) \setminus \{0\}$. Then (49) holds.

Theorem 32 ([13, Proposition 5.11]) *Suppose the function f is odd: $f(-\theta) = -f(\theta)$. Then (48) holds.*

Theorem 33 ([13, Proposition 5.12]) *Suppose f is continuously differentiable on $(-\pi, \pi) \setminus \{0\}$ and the function $\psi(\theta) = (\theta f'(\theta))^{-1}$ tends monotonically to zero as $\theta \to 0$. Then (48) holds whenever one of the following three conditions is satisfied:*

$$\int_0^\pi \left[f'(s) \arctan \frac{1}{sf'(s)} - f'(-s) \arctan \frac{1}{sf'(-s)} \right] ds < \infty; \tag{57}$$

$$\int_0^\pi \left| \frac{1}{(sf'(s))^2} - \frac{1}{(sf'(-s))^2} \right| \frac{ds}{s} < \infty; \tag{58}$$

$$\int_0^\pi \frac{1}{(sf'(s))^2} \frac{ds}{s} < \infty \quad \textit{and} \quad \int_0^\pi \frac{1}{(sf'(-s))^2} \frac{ds}{s} < \infty. \tag{59}$$

Below are several examples where the conditions of Theorem 27 are satisfied (see [13, Sect. 5.6]).

Example 1 Power whirls.

Consider the function

$$f(\theta) = \begin{cases} -c_+ \theta^{-\lambda_+}, & \theta > 0, \\ c_- |\theta|^{-\lambda_-}, & \theta < 0, \end{cases}$$

where $c_\pm > 0$ and $\lambda_\pm \in (0, \infty)$. It obviously satisfies the conditions of Theorems 29 and 30 (and of Theorem 31), as well as condition (59), and consequently all the conclusions of Theorem 27 are valid for f.

One can consider a more general case that often arises in the theory of the Riemann-Hilbert problem with an infinite index

$$f(\theta)=\begin{cases}-c_+(\theta)\theta^{-\lambda_+}, & \theta>0,\\ c_-(\theta)|\theta|^{-\lambda_-}, & \theta<0,\end{cases} \tag{60}$$

where $\lambda_\pm>0$, and the functions $c_\pm$ are continuous on $[0,\pi]$ and $[-\pi,0]$ respectively. Let us assume that $c_\pm(\theta)$ are twice continuously differentiable on $[-\pi,0)$ and $(0,\pi]$ and that

$$\lim_{\theta\to 0\pm 0} c'_\pm(\theta)\theta=0, \qquad \lim_{\theta\to 0\pm 0} c''_\pm(\theta)\theta^2=0. \tag{61}$$

The conditions of Theorem 27 can be verified with the help of Theorems 29, 30, and 33.

Example 2 Power-logarithmic whirls. Now let

$$f(\theta)=\begin{cases}-c_+\theta^{-\lambda_+}(\log|\theta|^{-1})^{\beta_+}, & \theta>0,\\ c_-|\theta|^{-\lambda_-}(\log|\theta|^{-1})^{\beta_-}, & \theta<0,\end{cases}$$

where $c_\pm>0$ and $\lambda_\pm\in(0,\infty)$, $\beta_\pm\in\mathbb{R}$. The applicability of Theorem 27 in this case is verified in the same way as in Example 1.

Example 3 Exponential and superexponential growth of the argument. Let

$$f(\theta)=\begin{cases}-c_+\exp\{d_+\theta^{-\lambda_+}\}, & \theta>0,\\ c_-\exp\{d_-|\theta|^{-\lambda_-}\}, & \theta<0,\end{cases}$$

where $c_\pm>0$, $d_\pm>0$ and $\lambda_\pm\in(0,\infty)$, or let

$$f(\theta)=\begin{cases}-c_+\exp\{g_+\exp(d_+\theta^{-\lambda_+})\}, & \theta>0,\\ c_-\exp\{g_-\exp(d_-|\theta|^{-\lambda_-})\}, & \theta<0,\end{cases} \tag{62}$$

where $c_\pm>0$, $d_\pm>0$, $g_\pm>0$, and $\lambda_\pm\in(0,\infty)$. The conditions of Theorem 27 are verified as in the preceding cases. Let us mention only that Theorem 31 does not apply here, while Theorem 30 does.

These examples show that the conditions of Theorem 27 are well suited to rapidly growing arguments $f(\theta)$. In particular, it is easy to see that a function f constructed via a composition of a finite number of exponentials similarly to (62) also satisfies (53), (55) and (59).

Let us now consider the case of slowly growing arguments $f(\theta)$.

Example 4 Logarithmic whirls.

Let

$$f(\theta)=\begin{cases}-c(\log\theta^{-1})^{\beta}, & \theta>0,\\ c(\log|\theta|^{-1})^{\beta}, & \theta<0,\end{cases} \tag{63}$$

where $\beta>0$, $c>0$. If $\beta>1$, then f satisfies the conditions of Theorem 27, which can be verified by evaluating the limits (53), (55) and applying Theorem 32. On the other hand, if $\beta\in(0,1]$, then f fails to satisfy the condition $d=1$ in Theorem 27. The critical case $\beta=1$ is the most important for us and we will consider it below (see Theorems 34, 35).

Similarly to Example 1, one can replace the constants $c_{\pm}$ in Examples 2–4 with continuous functions.

Example 5 Asymmetric whirls.

In Examples 1–3, condition (49) can be verified separately for left and right semi-neighbourhoods of the point $\theta=0$ with the help of (59). This allows one to construct new examples that satisfy the conditions of Theorem 27 from the ones mentioned above by combining different types of whirling to the left and to the right. For instance, one can take

$$f(\theta)=\begin{cases}-c_{+}\exp(g\exp(\theta^{-\lambda})), & \theta>0,\\ c_{-}\log^{\beta}(|\theta|^{-1}), & \theta<0,\end{cases} \tag{64}$$

where $c_{\pm}>0$, $g>0$, $\lambda>0$, and $\beta>3/2$. The corresponding function (36) combines very fast oscillations to the right of the point $\theta=0$ with very slow oscillations to the left of it.

It was mentioned in Example 4 that Theorem 27 does not cover the case of slow oscillations and that the natural boundary of its domain of applicability is the case of pure logarithmic whirls. In this case, we have the following result which was a key ingredient in the proof of Theorem 23.

Theorem 34 (See [13, Theorem 2.10 and the end of the proof of Theorem 5.9]) *Suppose $a\in GL^{\infty}(\mathbb{T})$ is continuous on $\mathbb{T}\setminus\{1\}$ and has the form* (36) *with a function f satisfying the condition*

$$\lim_{\theta\to\pm0}\left(f(\theta)\pm\frac{1}{2}\log|\theta|^{-1}\right)=0.$$

Then a admits the representation

$$a(t)=B(t)g\big(B(t)\big)d(t), \tag{65}$$

where $g, d \in C(\mathbb{T})$, the winding number of g is 0, and B is the infinite Blaschke product with the zeroes

$$z_k = \frac{2 - \exp(-k + 1/2)}{2 + \exp(-k + 1/2)}.$$

Note by the way that B in the above theorem is an interpolating Blaschke product (see the end of Sect. 5.4 in [13]).

The following result is a generalisation of Theorem 34 (see [13, Sect. 5.7]).

Theorem 35 ([13, Theorem 5.11]) *Let a function $a \in GL^\infty(\mathbb{T})$ be continuous on $\mathbb{T} \setminus \{1\}$ and have the form* (36) *with a function f that is monotonically increasing on $(-\pi, 0) \cup (0, \pi)$ and satisfies the condition* (41). *Assume, in addition, that condition* (43) *is satisfied, that* (45) *holds with some number $0 < d < 1$, and that*

$$\lim_{n \to +\infty} \left(-\frac{\vartheta(n)}{\vartheta(-n)} \right) = 1.$$

Then the function a admits the following representation, which is a (B, p)-factorisation simultaneously for all $1 < p < \infty$:

$$a(t) = B(t) g\big(B(t)\big) d(t),$$

where $g, d \in C(\mathbb{T})$. Moreover, the winding number of the function g is equal to zero and the Blaschke product B is constructed from the zeroes $z_j = r_j \exp\{i\vartheta(j)\}$, where

$$r_j = \big(1 - \Delta(j)/2\big) / \big(1 + \Delta(j)/2\big), \quad j = \pm 1, \pm 2, \ldots.$$

Using a linear fractional transformation, one can easily transplant the above results from $\mathbb{T}$ to $\mathbb{R}$. The following analogue of a special case of Theorem 28 is of a direct relevance to Theorem 19.

Theorem 36 ([3, 4]) *Let $\alpha : \mathbb{R} \to \mathbb{R}$ be an orientation preserving homeomorphism that is twice continuously differentiable for all sufficiently large values of $x > 0$ and is such that*

$$\liminf_{x \to +\infty} \frac{x\alpha''(x)}{\alpha'(x)} > -2, \tag{66}$$

$$\lim_{x \to +\infty} \frac{\alpha''(x)}{(\alpha'(x))^2} = 0, \tag{67}$$

$$\lim_{x \to +\infty} x^{1/2} \frac{\alpha''(x)}{(\alpha'(x))^{3/2}} = 0, \tag{68}$$

$$\lim_{x \to +\infty} \big(\alpha(x) + \alpha(-x)\big) = 0. \tag{69}$$

Then

$$e^{i\lambda\alpha} \in H^\infty(\mathbb{R}) + C(\dot{\mathbb{R}}), \quad \forall\lambda > 0.$$

Moreover the following representation holds

$$e^{i\lambda\alpha(x)} = B_\lambda(x)\, C_\lambda(x), \tag{70}$$

where B_λ *is a Blaschke product with an infinite number of zeroes accumulating at infinity and* C_λ *is a unimodular function belonging to* $C(\dot{\mathbb{R}})$.

Condition (66) is equivalent to the requirement that $x^2\alpha'(x)$ is strictly increasing for large values of x. Conditions (66)–(68) are satisfied for large classes of functions. Here are some examples:

$$\alpha(x) = c\,x^\gamma, \quad \gamma > 0,$$
$$\alpha(x) = c\ln^\delta(x+1), \quad \delta > 1,$$
$$\alpha(x) = c\,x^\gamma \ln^\delta(x+1), \quad \gamma > 0,\ \delta \in (-\infty, \infty),$$
$$\alpha(x) = c_1 \exp\bigl(c_2 x^\gamma\bigr), \quad \gamma > 0$$

with some positive constants c, c_1, c_2 (cf. Examples 1–4).

On the other hand, there are plenty of orientation preserving homeomorphisms $\alpha : \mathbb{R} \to \mathbb{R}$ for which $e^{i\alpha} \notin H^\infty(\mathbb{R}) + C(\dot{\mathbb{R}})$. This is a consequence of the following result.

Theorem 37 ([1, 3]) *Given any orientation preserving homeomorphism* $\eta : \mathbb{R} \to \mathbb{R}$, *there exists an orientation preserving homeomorphism* $\alpha : \mathbb{R} \to \mathbb{R}$ *such that* $\alpha - \eta \in L^\infty(\mathbb{R})$ *and* $e^{i\alpha} \notin H^\infty(\mathbb{R}) + C(\dot{\mathbb{R}})$.

Theorem 36 implies the following sufficient condition for (35) to hold.

Theorem 38 ([3]) *Suppose there exists* $\delta > 1$ *such that* $\beta(x) := \alpha(x) - (\log x)^\delta$ *is strictly increasing and twice continuously differentiable for all sufficiently large values of* $x > 0$, *and suppose* β *satisfies* (66)–(68) (*with* β *in place of* α). *Then* $we^{i\lambda\alpha} \in H^\infty(\mathbb{R}) + C(\dot{\mathbb{R}})$ *for all* $\lambda > 0$, *where* w *is the same as in* (10)–(11).

The final topic of this section is motivated by applications to the KdV equation (see Sect. 6). We are interested in conditions under which the argument of the quotient of two Blaschke products with purely imaginary zeroes in the upper half-plane is continuous on the real line. Consider the Blaschke product

$$B(z) = \prod_{k=1}^{\infty} \frac{z - ix_k}{z + ix_k}, \quad z \in \mathbb{C}_+ := \{z \in \mathbb{C} \,|\, \operatorname{Im} z > 0\} \tag{71}$$

with purely imaginary zeroes such that

$$x_1 > \cdots > x_k > x_{k+1} > \cdots > 0 \quad \text{and} \quad \lim_{k\to\infty} x_k = 0. \tag{72}$$

In this case, the standard Blaschke condition (see, e.g., [14, Chap. II, (2.3)] or [27, (13.13)]) reads

$$\sum_{k=1}^{\infty} x_k < \infty. \tag{73}$$

Theorem 24 takes the following simple form.

Theorem 39 *Let* $\arg B$ *denote the branch of the argument of the Blaschke product* (71)–(73) *which is continuous on* $\mathbb{R}\setminus\{0\}$ *and satisfies* $\lim_{x\to\pm\infty}\arg B(x)=0$, *and let the branch of* arctan *be chosen so that* $\arctan x \in (-\frac{\pi}{2}, \frac{\pi}{2})$. *Then* $\arg B$ *is increasing on* $\mathbb{R}\setminus\{0\}$,

$$\arg B(x) = -\arg B(-x), \quad x\in\mathbb{R}, \tag{74}$$

$$\lim_{x\to\pm 0} \arg B(x) = \mp\infty, \tag{75}$$

and

$$\arg B(x) = -2\sum_{k=1}^{\infty} \arctan\frac{x_k}{x}, \quad x\neq 0. \tag{76}$$

Let f_B be a continuous and decreasing on $(0,+\infty)$ function such that

$$f_B(k) = x_k.$$

Let $\Delta_k = x_k - x_{k+1}$ and $\Delta_k^{(2)}(s) = f_B(k+s) - f_B(k) + s(x_k - x_{k+1})$, $s\in[-1/2, 1/2]$.

Theorem 40 ([20]) *Suppose the sequence* $\{x_k\}$ *is such that*

$$\lim_{k\to\infty}\frac{x_k - x_{k+1}}{x_k} = 0 \tag{77}$$

and

$$\lim_{k\to\infty}\sup_{s\in[-1/2,1/2]}\left(\frac{|\Delta_k^{(2)}(s)|}{\Delta_k}\right) = 0. \tag{78}$$

Then

$$\arg B(x) = -2\int_{1/2}^{\infty}\arctan\frac{f_B(u)}{x}du + O(x), \tag{79}$$

where $\lim_{x\to 0} O(x) = 0$.

Theorem 41 ([20]) *Let a function f_B be continuously differentiable on $(0,+\infty)$ and satisfy all the conditions of Theorem 40. Then, for $x>0$*

$$\arg B(x) = -2x \int_0^1 \frac{\varphi_B(y)}{x^2+y^2}dy + \frac{\pi}{2} + O_1(x), \tag{80}$$

where $\varphi_B(y) := f_B^{-1}(y)$ is the inverse function of f_B and

$$\lim_{x\to 0} O_1(x) = 0.$$

Let now $R(x) = B_1(x)/B_2(x)$, where $B_1(x)$ and $B_2(x)$ are Blaschke products with the zeroes $if_{B_j}(k)$, $j=1,2$, where the functions f_{B_j} satisfy the conditions of Theorem 41. Introduce the function

$$r(y) := \varphi_{B_1}(y) - \varphi_{B_2}(y),$$

where $\varphi_{B_j}(y) := f_{B_j}^{-1}(y)$, $j=1,2$.

Theorem 42 ([20]) *Suppose at least one of following two conditions holds:*

(i)

$$r(y) = r_0 + O_2(y)$$

with some $r_0 \in \mathbb{R}$, $\lim_{y\to 0} O_2(y) = 0$;

(ii)

$$\int_0^y r(s)ds = r_1 y + O_3(y)$$

with some $r_1 \in \mathbb{R}$, $\lim_{y\to 0}(\frac{O_3(y)}{y}) = 0$.

Then

$$\arg R(x) = r_2 + O_4(x),$$

with some $r_2 \in \mathbb{R}$, $\lim_{x\to 0} O_4(x) = 0$.

The following corollary of Theorem 42 together with Theorem 36 play an important rôle in the proof of Theorem 43.

Corollary ([20]) *Let*

$$B_j(z) = \prod_{k=1}^{\infty} \frac{z - ix_k^{(j)}}{z + ix_k^{(j)}}, \quad j = 1,2$$

be two Blaschke products with interlacing ($x_k^{(1)} > x_k^{(2)} > x_{k+1}^{(1)}$) imaginary zeroes accumulating at 0, and let f be a real continuously differentiable function such that

$f(2x)$ and $f(2x-1)$ satisfy the conditions of Theorem 41 *and*

$$f(k)=\begin{cases} x^{(1)}_{\frac{k+1}{2}}, & k \text{ is odd,} \\ x^{(2)}_{\frac{k}{2}}, & k \text{ is even.} \end{cases}$$

Then $\arg B_1/B_2$ *is continuous on the real line.*

6 Applications to the KdV Equation

Let P be the projection defined by (2), $Q := I - P$ and let

$$(Jf)(x) = f(-x) : L^2(\mathbb{R}) \to L^2(\mathbb{R}) \tag{81}$$

be the reflection operator. The Hankel operator with the symbol $a \in L^\infty(\mathbb{R})$ is defined by the formula

$$\big(\mathbb{H}(a)f\big)(x) := (JQaf)(x)\colon \quad H^2(\mathbb{R}) \to H^2(\mathbb{R}). \tag{82}$$

The symbol

$$\phi(x) = e^{i(tx^3+cx)}d(x), \quad t>0,\ c\in\mathbb{R} \tag{83}$$

arises in the inverse scattering transform method for the Korteweg-de Vries (KdV) equation (see [29, 30]). The form of the unimodular function $d(x)$ depends on the properties of the initial data for the KdV equation. In certain important cases the function d has the form

$$d(x) = \frac{B_1(x)}{B_2(x)} I(x), \tag{84}$$

where B_1, B_2 are Blaschke products with zeroes converging to 0 along the imaginary axis and I is an inner function ($I \in H^\infty(\mathbb{R})$ and $|I(x)| = 1$ a.e. on the real line).

The proof of the following result relies on Theorems 7, 36 and 42.

Theorem 43 ([20]) *Let* $\phi(x) = e^{i(tx^3+cx)}\frac{B_1(x)}{B_2(x)}I(x)$, $t>0$, $c\in\mathbb{R}$, *where* B_j, $j = 1,2$ *are Blaschke products with zeroes* $\{if_{B_j}(k)\}$ *and the real-valued functions* f_{B_j}, $j=1,2$ *satisfy the conditions of Theorems* 40–42. *Then the Toeplitz operator*

$$T(\phi)\colon \quad H^2(\mathbb{R}) \to H^2(\mathbb{R})$$

is left invertible, the Hankel operator

$$\mathbb{H}(\phi)\colon \quad H^2(\mathbb{R}) \to H^2(\mathbb{R})$$

is compact and the operator

$$I + \mathbb{H}(\phi)\colon \quad H^2(\mathbb{R}) \to H^2(\mathbb{R})$$

is invertible.

Theorem 43 plays a crucial rôle in the proof of case 3 in the following theorem. Consider the Cauchy problem for the Korteweg-de Vries equation

$$\frac{\partial u(x,t)}{\partial t} - 6u(x,t)\frac{\partial u(x,t)}{\partial x} + \frac{\partial^3 u(x,t)}{\partial x^3} = 0, \quad t \geq 0,\ x \in \mathbb{R}, \tag{85}$$

$$u(x,0) = q(x), \tag{86}$$

and the Schrödinger operator $H_q = -d^2/dx^2 + q(x)$ on $L^2(\mathbb{R})$. Let $H_q^D = -d^2/dx^2 + q(x)$ be the corresponding operator on $L^2(-\infty, 0)$ with the Dirichlet boundary condition $u(0) = 0$.

Theorem 44 ([20]) *Assume that the initial profile $q(x)$ in* (86) *is real-valued, locally integrable, supported in $(-\infty, 0)$ and such that*

$$\inf \mathrm{Spec}(H_q) = -a^2 > -\infty. \tag{87}$$

Then the Cauchy problem for the KdV equation (85)–(86) *has a unique solution $u(x,t)$ which is a meromorphic function in x on the whole complex plane with no real poles for any $t > 0$ if at least one of the following conditions holds*:

1. *The operator H_q^D has a non-empty absolutely continuous spectrum*;
2. $\mathrm{Spec}(H_q^D) \cap \mathbb{R}_-$ *is a set of uniqueness of an $H^\infty(\mathbb{R})$ function*;
3. $\{\mathrm{Spec}(H_q^D) \cup \mathrm{Spec}(H_q)\} \cap \mathbb{R}_-$ *is a discrete set $\{-x_n^2\}_{n\geq 1}$ such that the sequence $\{x_n\}_{n\geq 1}$ satisfies the conditions of the Corollary of Theorem* 42.

7 Some Open Problems

There are of course many open problems in the spectral theory of Toeplitz operators. Here we list just a few of them, mainly those that are directly related to the topics discussed above.

1. Describe inner functions/Blaschke products υ for which $\rho \in A_p \Longrightarrow \rho \circ \upsilon \in A_p$ (cf. Theorems 14 and 15). In particular, is the condition (27) necessary for this implication to hold in the case of Blaschke products with positive zeroes? Perhaps one should try to describe pairs (ρ, υ), where $\rho \in A_p$ and υ is an inner function, such that $\rho \circ \upsilon \in A_p$.
2. Find conditions on an orientation preserving homeomorphism $\alpha : \mathbb{R} \to \mathbb{R}$ that are necessary and sufficient for

   $$e^{i\lambda\alpha} \in H^\infty(\mathbb{R}) + C(\dot{\mathbb{R}}), \quad \forall \lambda > 0$$

   to hold (cf. Theorem 36).
3. According to Theorem 16, $a \in \mathrm{fact}(\infty, p)$ is a sufficient condition for the right/left invertibility of $T(a) : H^p(\mathbb{T}) \to H^p(\mathbb{T})$, $1 < p < \infty$. Is this also a necessary condition for the right/left invertibility or even for the $\Phi_\pm$ property of $T(a)$? The answer is positive for $p = 2$ (see [13, Sect. 2.7]).

4. Study spectral properties of $T(a) : H^p(\mathbb{T}) \to H^p(\mathbb{T})$ when a belongs to a Douglas algebra, i.e. to a closed subalgebra of $L^\infty(\mathbb{T})$ containing $H^\infty(\mathbb{T})$ (cf. Theorem 7). According to the Chang–Marshall theorem, every such algebra is generated by $H^\infty(\mathbb{T})$ and the complex conjugates of some inner functions (see, e.g., [14, Chap. IX]).

Finally, we would like to reiterate that very little is known about the (essential) spectrum of $T(a) : H^p(\mathbb{T}) \to H^p(\mathbb{T})$, $1 < p < \infty$ for a general $a \in L^\infty(\mathbb{T})$.

Acknowledgements The first author was supported by PROMEP (México) via "Proyecto de Redes" and by CONACYT grant 102800.

References

1. Böttcher, A., Grudsky, S.: Toeplitz operators with discontinuous symbols: phenomena beyond piecewise continuity. Oper. Theory, Adv. Appl. **90**, 55–118 (1996)
2. Böttcher, A., Grudsky, S.: On the composition of Muckenhoupt weights and inner functions. J. Lond. Math. Soc., II. Ser. **58**(1), 172–184 (1998)
3. Böttcher, A., Grudsky, S.M., Spitkovsky, I.M.: Toeplitz operators with frequency modulated semi-almost periodic symbols. J. Fourier Anal. Appl. **7**, 523–535 (2001)
4. Böttcher, A., Grudsky, S.M., Spitkovsky, I.M.: Block Toeplitz operators with frequency-modulated semi-almost periodic symbols. Int. J. Math. Math. Sci. **34**, 2157–2176 (2003)
5. Böttcher, A., Silbermann, B.: Analysis of Toeplitz Operators. Springer, Berlin (2006)
6. Brown, A., Halmos, P.R.: Algebraic properties of Toeplitz operators. J. Reine Angew. Math. **213**, 89–102 (1963)
7. Clancey, K.F.: One dimensional singular integral operators on L^p. J. Math. Anal. Appl. **54**, 522–529 (1976)
8. Clancey, K.F.: Corrigendum for the article "One dimensional singular integral operators on L^p". J. Math. Anal. Appl. **99**, 527–529 (1984)
9. Coburn, L.A.: Weyl's theorem for nonnormal operators. Mich. Math. J. **13**, 285–288 (1966)
10. Devinatz, A.: Toeplitz operators on H^2 spaces. Trans. Am. Math. Soc. **112**(2), 304–317 (1964)
11. Douglas, R.G.: Toeplitz and Wiener-Hopf operators in $H^\infty + C$. Bull. Am. Math. Soc. **74**, 895–899 (1968)
12. Douglas, R.G.: Banach Algebra Techniques in Operator Theory. Springer, New York (1998)
13. Dybin, V., Grudsky, S.M.: Introduction to the Theory of Toeplitz Operators with Infinite Index. Birkhäuser, Basel (2002)
14. Garnett, J.B.: Bounded Analytic Functions. Academic Press, New York (1981)
15. Gohberg, I.: On an application of the theory of normed rings to singular integral equations. Usp. Mat. Nauk **7:2**(48), 149–156 (1952) (Russian)
16. Gokhberg, I., Krupnik, N.: Algebra generated by one-dimensional singular integral operators with piecewise continuous coefficients. Funct. Anal. Appl. **4**, 193–201 (1970) (translation from Funkts. Anal. Prilozh. **4**(3), 26–36 (1970))
17. Gohberg, I., Krupnik, N.: One-Dimensional Linear Singular Integral Equations I & II. Birkhäuser, Basel (1992)
18. Grudsky, S.M.: Toeplitz operators and the modelling of oscillating discontinuities with the help of Blaschke products. In: Elschner, J., et al. (eds.) Problems and Methods in Mathematical Physics. Oper. Theory Adv. Appl., vol. 121, pp. 162–193. Birkhäuser, Basel (2001)
19. Grudskij, S.M., Khevelev, A.B.: On invertibility in $L^2(R)$ of singular integral operators with periodic coefficients and a shift. Sov. Math. Dokl. **27**, 486–489 (1983) (translated from Dokl. Akad. Nauk SSSR **269**, 1303–1306 (1983))

20. Grudsky, S.M., Rybkin, A.: Quotient of Blaschke products and compactness and invertibility of Hankel and Toeplitz operators. Oper. Theory, Adv. Appl., Springer Basel AG **228**, 127–150 (2013)
21. Grudsky, S.M., Shargorodsky, E.: Spectra of Toeplitz operators and compositions of Muckenhoupt weights with Blaschke products. Integral Equ. Oper. Theory **61**, 63–75 (2008)
22. Hartman, P., Wintner, A.: The spectra of Toeplitz's matrices. Am. J. Math. **76**(4), 867–882 (1954)
23. Helson, H., Szegö, G.: A problem in prediction theory. Ann. Mat. Pura Appl. **51**(1), 107–138 (1960)
24. Hunt, R., Muckenhoupt, B., Wheeden, R.: Weighted norm inequalities for the conjugate function and Hilbert transform. Trans. Am. Math. Soc. **176**, 227–251 (1973)
25. Krupnik, N.Ya.: Some consequences of the Hunt–Muckenhoupt–Wheeden theorem. Mat. Issled. **47**, 64–70 (1978) (Russian)
26. Krupnik, N.Ya.: Banach Algebras with Symbol and Singular Integral Operators. Birkhäuser, Basel (1987)
27. Mashreghi, J.: Representation Theorems in Hardy Spaces. Cambridge University Press, Cambridge (2009)
28. Nordgren, E.A.: Composition operators. Can. J. Math. **20**, 442–449 (1968)
29. Rybkin, A.: Meromorphic solutions to the KdV equation with non-decaying initial data supported on a left half line. Nonlinearity **23**(5), 1143–1167 (2010)
30. Rybkin, A.: The Hirota τ-function and well-posedness of the KdV equation with an arbitrary step-like initial profile decaying on the right half line. Nonlinearity **24**(10), 2953–2990 (2011)
31. Saginashvili, A.I.: Singular integral operators with coefficients having semi-almost-periodic type discontinuities. Soobshch. Akad. Nauk Gruz. SSR **94**, 289–291 (1979) (Russian)
32. Saginashvili, A.I.: Singular integral operators with semialmost periodic discontinuities at the coefficients. Soobshch. Akad. Nauk Gruz. SSR **95**, 541–543 (1979) (Russian)
33. Saginashvili, A.I.: Singular integral equations with coefficients having discontinuities of semi-almost-periodic type. Transl., Ser. 2, Am. Math. Soc. **127**, 49–59 (1985) (translated from Tr. Tbil.. Mat. Inst. Razmadze **66**, 84–95 (1980))
34. Sarason, D.: Generalized interpolation in H^∞. Trans. Am. Math. Soc. **127**, 179–203 (1967)
35. Sarason, D.: Algebras of functions on the unit circle. Bull. Am. Math. Soc. **79**(2), 286–299 (1973)
36. Sarason, D.: Toeplitz operators with semi-almost periodic symbols. Duke Math. J. **44**, 354–364 (1977)
37. Shapiro, J.H.: Composition Operators and Classical Function Theory. Springer, Berlin (1993)
38. Shargorodsky, E.: On singular integral operators with coefficients from $P_n\mathbb{C}$. Tr. Tbil. Mat. Inst. Razmadze **93**, 52–66 (1990) (Russian)
39. Shargorodsky, E.: On some geometric conditions of Fredholmity of one-dimensional singular integral operators. Integral Equ. Oper. Theory **20**(1), 119–123 (1994)
40. Shargorodsky, E.: A remark on the essential spectra of Toeplitz operators with bounded measurable coefficients. Integral Equ. Oper. Theory **57**, 127–132 (2007)
41. Simonenko, I.B.: The Riemann boundary value problem for n pairs of functions with measurable coefficients and its application to the investigation of singular integrals in the spaces L_p with weight. Izv. Akad. Nauk SSSR, Ser. Mat. **28**, 277–306 (1964) (Russian)
42. Simonenko, I.B.: Some general questions in the theory of the Riemann boundary problem. Math. USSR, Izv. **2**, 1091–1099 (1968) (translated from Izv. Akad. Nauk SSSR Ser. Mat. **32**(5), 1138–1146 (1968))
43. Spitkovskij, I.M.: Factorization of matrix-functions belonging to the classes $\tilde{A}_n(p)$ and TL. Ukr. Math. J. **35**, 383–388 (1983) (translation from Ukr. Mat. Zh. **35**(4), 455–460 (1983))
44. Widom, H.: Inversion of Toeplitz matrices II. Ill. J. Math. **4**(1), 88–99 (1960)
45. Widom, H.: Singular integral equations in L_p. Trans. Am. Math. Soc. **97**(1), 131–160 (1960)
46. Widom, H.: Toeplitz operators on H_p. Pac. J. Math. **19**(3), 573–582 (1966)

Approximating the Riemann Zeta-Function by Strongly Recurrent Functions

P.M. Gauthier

Abstract Bhaskar Bagchi has shown that the Riemann hypothesis holds if and only if the Riemann zeta-function $\zeta(z)$ is strongly recurrent in the strip $1/2 < \Re z < 1$. In this note we show that $\zeta(z)$ can be approximated by strongly recurrent functions sharing important properties with $\zeta(z)$.

Keywords Riemann hypothesis · Strong recurrence

Mathematics Subject Classification Primary 11M26 · Secondary 30E10

1 Introduction

In 1982, Bagchi ([1, 2]) showed a surprising equivalence between the Riemann hypothesis and a statement in topological dynamics, a subject which has its origins in the motion of particles. One is reminded of the amazing similarity between quantum dynamical systems and zeros of the Riemann zeta-function discovered by the chance encounter of Freeman Dyson and Hugh Montgomery in 1972.

For $-\infty \leq \alpha < \beta \leq +\infty$, we denote by $S = S(\alpha, \beta)$ the strip $\alpha < \Re s < \beta$ and by $\mathcal{O}(S)$ the set of functions holomorphic in S. We denote by $meas(E)$ the Lebesque measure of a Borel subset E of $\mathbb{R}$. For a Borel set $E \subset \mathbb{R}$, we denote by $\overline{d}_{\mathbb{R}}(E)$ and $\underline{d}_{\mathbb{R}}(E)$ respectively the upper and lower densities of E in $\mathbb{R}$, defined as follows.

$$\overline{d}_{\mathbb{R}}(E) := \limsup_{T \to +\infty} \frac{meas(E \cap [-T, T])}{2T}$$

$$\underline{d}_{\mathbb{R}}(E) := \liminf_{T \to +\infty} \frac{meas(E \cap [-T, T])}{2T}.$$

A function $f \in \mathcal{O}(S)$ is said to be *strongly recurrent* if, for each compact set $K \subset S$, and for each $\epsilon > 0$, the set of $t \in \mathbb{R}$ such that $\max_K |f(z) - f(z + it)| < \epsilon$ is of positive upper density.

P.M. Gauthier (✉)
Département de mathématiques et de statistique, Université de Montréal, Montréal, Canada H3C 3J7
e-mail: gauthier@dms.umontreal.ca

J. Mashreghi, E. Fricain (eds.), *Blaschke Products and Their Applications*, Fields Institute Communications 65, DOI 10.1007/978-1-4614-5341-3_2,

Theorem 1 (Bagchi) *The Riemann hypothesis holds if and only if the Riemann zeta-function is strongly recurrent in the strip* $1/2 < \Re z < 1$.

We shall also consider a discrete form of strong recurrence. If E is a finite subset of $\mathbb{Z}$, we denote by $\#E$ the number of elements of E. For a subset E of $\mathbb{Z}$, we denote by $\overline{d}_{\mathbb{Z}}(E)$ and $\underline{d}_{\mathbb{Z}}(E)$ respectively the upper and lower densities of E in $\mathbb{Z}$, defined as follows.

$$\overline{d}_{\mathbb{Z}}(E) := \limsup_{N\to+\infty} \frac{\#(E \cap \{-N,\ldots,N\})}{2N+1}$$

$$\underline{d}_{\mathbb{Z}}(E) := \liminf_{N\to+\infty} \frac{\#(E \cap \{-N,\ldots,N\})}{2N+1}.$$

For $\Delta \in \mathbb{R}$, $\Delta \neq 0$ we say that a function $f \in \mathcal{O}(S)$ is *strongly recurrent modulo* Δ, if for each compact set $K \subset S$, and for each $\epsilon > 0$, the set of $k \in \mathbb{Z}$ such that $\max_K |f(z) - f(z + i(k\Delta))| < \epsilon$ is of positive upper density in $\mathbb{Z}$.

The following particular case of a theorem of Walter H. Gottschalk and Gustav A. Hedlund [8] allows us to pass between continuous and discrete dynamics.

Theorem 2 (Inheritance Theorem) *Let* $f \in \mathcal{O}(S)$ *and* $\Delta \in \mathbb{R}, \Delta \neq 0$. *Then,* f *is strongly recurrent if and only if* f *is strongly recurrent modulo* Δ.

Our main results are the following.

Theorem 3 *For each* $\Delta \in \mathbb{R}$ *different from* 0, *there exists a sequence of functions* φ_n *meromorphic on* $\mathbb{C}$, *each of which is strongly recurrent in* $1/2 < \Re z < 1$ *modulo* Δ *and hence strongly recurrent. Moreover:*

(1) $\varphi_n \to \zeta$ *uniformly on compact subsets of* $\mathbb{C}$;
(2) φ_n *has only a simple pole at* $z = 1$ *with residue* 1;
(3) $\varphi_n(x) \in \mathbb{R}, \forall x \in \mathbb{R} \setminus \{1\}$.

Theorem 4 *Let* S *be the fundamental strip* $0 < \Re z < 1$. *There is a function* $h \in \mathcal{O}(S)$ *which is strongly recurrent and which satisfies the functional equation of the Riemann zeta-function.*

The functional equation for the Riemann zeta-function is the following:

$$\zeta(z)\pi^{-z/2}\Gamma\left(\frac{z}{2}\right) = \zeta(1-z)\pi^{-(1-z)/2}\Gamma\left(\frac{1-z}{2}\right)$$

and we say that a function h satisfies the functional equation for the Riemann zeta-function if the previous equation is satisfied, when ζ is replaced by h. Regarding this functional equation, we recall the famous theorem of Hans L. Hamburger [10], which states that the only function which satisfies this functional equation and has the same general character as ζ is ζ itself.

Theorem 5 (Hamburger) *If f is a Dirichlet series convergent in $\Re z > 1$ with a meromorphic continuation to $\mathbb{C}$ as a function of finite order with only finitely many poles and, if f satisfies the functional equation for the Riemann zeta-function ζ, then $f = \zeta$.*

Theorem 3 was presented at the 2011 Winter Meeting of the Canadian Mathematical Society, in the session on Composition Operators, organized by Javad Mashreghi and Nina Zorboska. We shall give the proofs after introducing some preparatory material.

2 Symmetric Function Theory

Let $\overline{\mathbb{C}}$ denote the closed complex plane $\overline{\mathbb{C}} = \mathbb{C} \cup \{\infty\}$. For a subset $E \subset \overline{\mathbb{C}}$, denote $\overline{E} = \{z : \overline{z} \in E\}$ and $\overline{\infty} = \infty$. Let us say that E is *real-symmetric* if $\overline{E} = E$. For a function f defined on a set E, we denote by $\overline{f}$ the function defined on $\overline{E}$ by the formula $\overline{f}(z) = \overline{f(\overline{z})}$. We shall say that a function f defined on a real-symmetric set E is real-symmetric if $\overline{f} = f$. A meromorphic function on a real-symmetric domain meeting the real-axis is real-symmetric if and only if $f(z) \in \mathbb{R} \cup \{\infty\}$ for $z \in \mathbb{R} \cup \{\infty\}$. Some theorems in function theory remain true in the real-symmetric category.

A function is said to be meromorphic (holomorphic) on a set E if it is meromorphic (holomorphic) in an open neigborhood of E. We denote by $\mathcal{M}(E)$ and $\mathcal{O}(E)$ the family of meromorphic respectively holomorphic functions on E. If E is real-symmetric we define $\mathcal{M}_{\mathbb{R}}(E)$ to be the family of functions f meromorphic on E and real-symmetric. We denote by $\mathcal{O}_{\mathbb{R}}(E)$ the subfamily of functions in $\mathcal{M}_{\mathbb{R}}(E)$ which are holomorphic on E.

Theorem 6 (Symmetric Mittag-Leffler) *Let P be a discrete subset of $\mathbb{C}$ with $\overline{P} = P$. For each $p \in P$, let Q_p be a non-constant polynomial and suppose $\overline{Q}_p = Q_{\overline{p}}$. Then, there is a function $h \in \mathcal{M}_{\mathbb{R}}(\mathbb{C})$ whose poles are precisely the points of P and with principal parts $Q_p(1/(z-p))$, $p \in P$.*

Proof Let h_+ be a meromorphic function on $\mathbb{C}$ whose poles are precisely the points $p \in P$ with $\Im p > 0$ and with the prescribed principal parts. Set $h_- = \overline{h}_+$. Let ϕ be a meromorphic function on $\mathbb{C}$ whose poles are precisely the real points of P and with the prescribed principal parts. Set $h_o = (\phi + \overline{\phi})/2$. Then, the function $h = h_+ + h_- + h_o$ has the required properties. □

For a closed set E, we denote by $A(E)$, as usual, the family of functions continuous on E and holomorphic on the interior E^o. We shall say that a closed set $E \subset \mathbb{C}$ is a *set of uniform approximation* if, for each $f \in \mathcal{A}(E)$, and each $\epsilon > 0$, there is an entire function g such that $|f(z) - g(z)| < \epsilon$, for each $z \in E$. The following result is due to Norair U. Arakelian (see [6]).

Theorem 7 *A closed set $E \subset \mathbb{C}$ is a set of uniform approximation if and only if $\overline{\mathbb{C}} \setminus E$ is connected and locally connected.*

We may also state a real-symmetric theorem on uniform approximation.

Theorem 8 (Symmetric Approximation) *If E is a set of uniform approximation, E is real-symmetric and $f \in \mathcal{M}_{\mathbb{R}}(E)$, then, for each $\epsilon > 0$, there is a function $g \in \mathcal{M}_{\mathbb{R}}(\mathbb{C})$, whose poles in $\mathbb{C}$ are the same as those of f on E and with same principal parts, such that*

$$\left|f(z) - g(z)\right| < \epsilon \exp\left(-|z|^{1/4}\right), \quad \textit{for all } z \in E.$$

Proof By Theorem 6, there is an $h \in \mathcal{M}_{\mathbb{R}}(\mathbb{C})$ with $f - h \in \mathcal{O}_{\mathbb{R}}(E)$. By another theorem of Arakelian (see [6]), there is a $\Phi \in \mathcal{O}(\mathbb{C})$ which approximates $f - h$ as required. Then, $\phi = (\Phi + \overline{\Phi})/2$ also approximates $f - h$. Set $g = \phi + h$. Then, g has the required properties. □

We shall require a stronger approximation than uniform approximation. A closed set $E \subset \mathbb{C}$ is a *set of tangential approximation* if, for each $f \in A(E)$, and each positive continuous function ϵ, there is an entire function g such that $|f(z) - g(z)| < \epsilon(z)$, for each $z \in E$. Of course, a set of tangential approximation is *a fortiori* a set of uniform approximation. Let us say that a family E_α of subsets of $\mathbb{C}$ has no *long islands* if, for each compact $K \subset \mathbb{C}$ there is a (larger) compact set Q such that each E_α which intersects K is contained in Q. The following theorem gives a condition which characterizes sets of tangential approximation. I discovered this condition and proved the necessity. The sufficiency was established by Ashot H. Nerssesian (see [6]).

Theorem 9 *A set $E \subset \mathbb{C}$ of uniform approximation is a set of tangential approximation if and only if the family of components of the interior has no long islands.*

Just as for uniform approximation, there is a real-symmetric version of this theorem. Let us say that a real-symmetric set E of uniform approximation is a set of real-symmetric tangential approximation if for each real-symmetric $f \in A(E)$ and each real-symmetric positive continuous function ϵ, there is a real-symmetric entire function g such that $|f(z) - g(z)| < \epsilon(z)$, for each $z \in E$. Just as for uniform approximation, and with a similar proof, we have the following symmetric tangential approximation theorem.

Theorem 10 *A real-symmetric set $E \subset \mathbb{C}$ of uniform approximation is a set of real-symmetric tangential approximation if and only if the family of components of the interior has no long islands.*

3 Frequent Hypercyclicity

Let $T : X \to X$ be an operator from a linear space X into itself. The (forward) orbit of a vector $x \in X$ under the action of T is the set of vectors $O(x) = \{Tx, T^2x, \ldots\}$, where $T^n x$ is defined inductively as $T(T^{n-1}x)$. A vector $x \in X$ is said to be a cyclic vector for the operator T if the subspace generated by the orbit $O(x)$ is dense in X and the operator T is said to be a cyclic operator if it has a cyclic vector. A vector $x \in X$ is said to be hypercyclic for T if the orbit itself $O(x)$ is dense in X. We shall say that x is hypercyclic for T in a subset $Y \subset X$ if for each $y \in Y$, there is a sequence $\{n_k\}$ in $\mathbb{N}$ such that $T^{n_k}x \to y$. For an excellent overview of hypercyclicity, see the survey [9] by Carl-Goswin Grosse-Erdmann.

For $a \in \mathbb{C}$, let $\varphi_a : \mathbb{C} \to \mathbb{C}$ denote the translation $\varphi_a(z) = z + a$ and let C_a denote the composition operator on the space of complex-valued functions on $\mathbb{C}$, defined as $C_a f(z) = (f \circ \varphi_a)(z) = f(z+a)$. George D. Birkhoff showed that, for each $a \neq 0$, the composition operator C_a is hypercyclic on the space of entire functions. That is, there exists a (hypercyclic) entire function f. This means that the translates of f are dense in the space of all entire functions. More precisely, for each entire function g, there is a sequence of natural numbers $\{n_k\}$ such that $f(z + n_k a) \to g(z)$ uniformly on compact. Such a hypercyclic function f is also said to be a universal function.

My advisor, Wladimir Seidel, and Joseph L. Walsh [15] established an analog of Birkhoff's theorem in the disc, replacing translation by non-euclidian translation. There is no difficulty in extending the results of Birkhoff, Seidel and Walsh to several complex variables. Maurice Heins [11] showed the existence of a Blaschke product universal in the unit ball of $H^\infty(\mathbb{D})$, where $\mathbb{D}$ is the unit disc. Pak-Soong Chee [5] showed the existence of a function universal in the unit ball of $H^\infty(\mathbb{B}^n)$, where $\mathbb{B}^n$ is the unit ball in $\mathbb{C}^n$. XIAO Jie and I [7] showed the existence of such a universal function in $H^\infty(\mathbb{B}^n)$ which is inner.

It turns out (see [9]) that hypercyclicity is a generic phenomenon. For example, most entire functions are hypercyclic (universal). But no explicit example is known! The only known universal function in the sense of Birkhoff is the Riemann zeta-function $\zeta(s)$ (and some closely related zeta-functions). Of course, $\zeta(s)$ is not entire, but it is as close to being entire as possible. It has only one simple pole and that pole has residue 1. Let S be the strip $1/2 < \Re s < 1$, $\mathcal{O}(S)$ the set of functions holomorphic in S and $\mathcal{O}_o(S)$ the set of zero-free functions in $\mathcal{O}(S)$. The remarkable Universality Theorem of Sergei Mikhailovich Voronin (as extended by Steven M. Gonek and Bagchi) states that for any real number a, different from zero, there exists a sequence $\{t_k\}$ of real numbers such that the sequence of translates $\zeta(s + it_k a)$ comes arbitrarily close to each function in $\mathcal{O}_o(S)$. In fact, one can choose the sequence $\{t_k\}$ to be natural numbers $\{n_k\}$. Thus, the Riemann zeta-function $\zeta(s)$ is hypercyclic for the composition operator C_{ia} on the space $\mathcal{O}_o(S)$. That is, for each zero-free function g holomorphic in the strip S, there is a sequence $\{n_k\}$ of natural numbers, such that $\zeta(s + in_k a) \to g(s)$.

Frédéric Bayart and Sophie Grivaux [3] introduced the notion of *frequent* hypercyclicity [3]. For a subset E of $\mathbb{N}$, we denote by $\overline{d}_{\mathbb{N}}(E)$ and $\underline{d}_{\mathbb{N}}(E)$ respectively the upper and lower densities of E in $\mathbb{N}$, defined as follows.

$$\overline{d}_{\mathbb{N}}(E) := \limsup_{N\to+\infty} \frac{\#(E\cap\{1,\ldots,N\})}{N}$$

$$\underline{d}_{\mathbb{N}}(E) := \liminf_{N\to+\infty} \frac{\#(E\cap\{1,\ldots,N\})}{N}.$$

A vector x is frequently hypercyclic for an operator T on a space X if, for each open set U in X, the set of $n \in \mathbb{N}$ for which $T^n x \in U$ has positive lower density in $\mathbb{N}$. If such a frequently hypercyclic vector exists for T, then the operator T is said to be frequently hypercyclic. Bayart and Grivaux [3] gave a criterion for frequent hypercyclicity. In contrast to hypercyclicity, frequent hypercyclicity is not a generic phenomenon.

The following lemma is stated in [3].

Lemma 1 *If there is a frequent hypercyclic vector in X for T, then the set of frequent hypercyclic vectors for T is dense in X.*

Proof Let x be a frequently hypercyclic vector for T and let V be a fixed open set in X. Choose a positive integer p such that $T^p x \in V$. By Theorem 6.30 in [4], x is also a frequently hypercyclic vector for T^p. Consequently, for each open set $U \subset X$,

$$\underline{d}\{m : T^m(T^p x) \in U\} \geq \underline{d}\{n : (T^p)^n x \in U\} > 0.$$

This shows that $T^p x$ is frequently hypercyclic for T. Thus, for an arbitrary open set $V \subset X$, we have found a frequently hypercyclic vector for T in V. □

Birkhoff's theorem extends to frequent hypercyclicity by Example 2.5 in [3]. That is, for each $a \neq 0$, the composition operator C_a is frequently hypercyclic on the space of entire functions.

The above-mentioned Universality Theorem of Voronin was further refined by Reich, who showed that, for each real number a, not equal to 0, for each compact set $K \subset S$ with connected complement, for each holomorphic zero-free function on K and for each $\epsilon > 0$, the set of real t, such that $\max_K |f(s) - \zeta(s+ita)| < \epsilon$, is of positive lower density. He also showed that the t may be chosen from an arbitrary arithmetic progression $\{m\Delta\}$, $m \in \mathbb{N}$, $\Delta > 0$ with lower density taken with respect to $\mathbb{N}$. Thus, for each $\Delta > 0$, the Riemann zeta-function is frequently hypercyclic in $\mathcal{O}_o(S)$, for the composition operator $C_{i\Delta}$.

Markus Nieß in [12, 13], and [14] investigated approximation in a strip, by functions having universality properties *outside* the strip. The following theorem yields functions which approximate everywhere and have universality properties *inside* a strip.

Theorem 11 *Fix $-\infty \leq \alpha < \beta \leq +\infty$. Then, for each $f \in \mathcal{M}_{\mathbb{R}}(\mathbb{C})$ having no poles in $S(\alpha,\beta)$), for each compact set $L \subset \mathbb{C}$, for each $\delta > 0$, and for each $\Delta > 0$, there* *is a function $\varphi \in \mathcal{M}_{\mathbb{R}}(\mathbb{C})$ having the same poles and principal parts as f, such that φ is frequently hypercyclic in $\mathcal{O}(S(\alpha,\beta))$ for the vertical translation operator $C_{i\Delta}$ and moreover $|\varphi - f| < \delta$ on L.*

Proof We may assume that $\mathbb{C} \setminus L$ is connected and $\overline{L} = L$. Let $\ell = \max\{y : z = x + iy \in L\}$. For each $p \in \mathbb{N}^+$, choose a real number $N_p > (\ell + 2p)/\Delta$ and take pairwise disjoint subsets $\mathbb{N}_p$ of $\mathbb{N}$ as in Lemma 6.19 in [4], each of which is of positive lower density. For each $p \in \mathbb{N}^+$ and $n \in \mathbb{N}_p$, we form the closed intervals $I_n^+ = [n\Delta - p, n\Delta + p]$ and $I_n^- = \{y : -y \in I_n^+\}$. The sets $I_n^\pm, n \in \mathbb{N}_p, p \in \mathbb{N}^+$, form a locally finite family of disjoint closed intervals. Moreover, these intervals are all disjoint from the closed interval $[-\ell, \ell]$.

Let $\alpha_k \searrow \alpha$ and $\beta_k \nearrow \beta$. We arrange the $n \in \cup \mathbb{N}_p$ in an increasing sequence $\{n(k)\}$. Now fix p. For $n(k) \in \mathbb{N}_p$, let D_k^+, D_k^- and D_k be the closed rectangles

$$D_k^+ = \{z = x + iy : \alpha_k \le x \le \beta_k, y \in I_{n(k)}^+\}$$

$$D_k^- = \{z = x + iy : \alpha_k \le x \le \beta_k, y \in I_{n(k)}^-\}$$

$$D_k = \{z = x + iy : \alpha_k \le x \le \beta_k, |y| \le p\}.$$

Then,

$$D_k^+ - in(k)\Delta = D_k = D_k^- + in(k)\Delta.$$

Set

$$E_p = \bigcup_{n(k)\in\mathbb{N}_p} (D_k^+ \cup D_k^-).$$

We may do the same for each p and denoting the pole set of f by $f^{-1}(\infty)$, put

$$E = L \cup f^{-1}(\infty) \cup \bigcup_p E_p.$$

Then, $\overline{E} = E$ and, by Theorem 7, the set E is a set of uniform approximation.

Let $\mathcal{P}$ be the family of all polynomials with coefficients in $\mathbb{Q} + i\mathbb{Q}$. We arrange these polynomials in a sequence $\{P_p\}$ in such a way that each polynomial is repeated infinitely often. We define a function $h \in M_{\mathbb{R}}(E)$ as follows. For $z \in L \cup f^{-1}(\infty)$, we put $h = f$. If $z \in E_p$, then $z \in (D_k^+ \cup D_k^-)$, for some $n(k) \in \mathbb{N}_p$. We set $h(z) = P_p(z - in(k)\Delta)$, for $z \in D_k^+$ and $h(z) = \overline{P_p}(z + in(k)\Delta)$, for $z \in D_k^-$. By the Symmetric Approximation Theorem, there is a function $\varphi \in M_{\mathbb{R}}(\mathbb{C})$, whose poles are precisely those of h and with the same principle parts, such that

$$|\varphi(z) - h(z)| < \delta \exp(-|z|^{1/4}), \quad \text{for all } z \in E.$$

Thus, φ has the same poles as f and with the same principal parts and $|\varphi - f| < \delta$ on L.

We claim that the function φ is frequently hypercyclic for the vertical translation operator $C_{i\Delta}$, on the space $\mathcal{O}(S(\alpha,\beta))$. Indeed, suppose $g \in \mathcal{O}(S(\alpha,\beta))$,

$K \subset S(\alpha, \beta)$ is compact and $\epsilon > 0$. For all but finitely many values of p and for all but finitely many $n(k) \in \mathbb{N}_p$,

$$K \subset \{z = x + iy : \alpha_k \leq x \leq \beta_k, |y| \leq p\}.$$

We have already noted that this latter set is D_k, whenever $n(k) \in \mathbb{N}_p$. We may choose such a p for which

$$\max_K \{|g(z) - P_p(z)|\} < \epsilon/2.$$

Thus, if $n \in \mathbb{N}_p$ and $z \in K$, then $z + in\Delta \in E_p \subset E$, so

$$\begin{aligned} |\varphi(z + in\Delta) - g(z)| &\leq |\varphi(z + in\Delta) - h(z + in\Delta)| + |h(z + in\Delta) - g(z)| \\ &\leq \delta \exp(-|z + in\Delta|^{1/4}) + |P_p(z) - g(z)|. \end{aligned}$$

Hence, this is less than ϵ for all but finitely many $n \in \mathbb{N}_p$. Therefore,

$$\underline{d}\{n : |\varphi(z + in\Delta) - g(z)|_K < \epsilon\} \geq \underline{d}(\mathbb{N}_p) > 0. \qquad \square$$

4 Proof of Theorem 3

Lemma 2 *Fix* $-\infty \leq \alpha < \beta \leq +\infty$ *and* $\Delta > 0$. *If* $\phi \in \mathcal{O}(S(\alpha, \beta))$ *is frequently hypercyclic for the vertical translation operator* $C_{i\Delta}$, *then* ϕ *is strongly recurrent.*

Proof Let K be a compact subset of the strip $S(\alpha, \beta)$ and $\epsilon > 0$. Set $S = S(\alpha, \beta)$ and

$$U = \left\{ f \in S : \max_{z \in K} |f(z) - \phi(z)| < \epsilon \right\}.$$

Since U is an open subset of $\mathcal{O}(S)$ and ϕ is frequently hypercyclic for the operator $C_{i\Delta}$,

$$\begin{aligned} &\overline{d}_{\mathbb{Z}} \left\{ k \in \mathbb{Z} : \max_{z \in K} |\phi(z + k(i\Delta)) - \phi(z)| < \epsilon \right\} \\ &\geq \underline{d}_{\mathbb{Z}} \left\{ n \in \mathbb{N} : \max_{z \in K} |\phi(z + n(i\Delta)) - \phi(z)| < \epsilon \right\} \\ &= \frac{1}{2} \underline{d}_N \{ n \in \mathbb{N} : C_{i\Delta}^n \phi \in U \} > 0. \end{aligned}$$

Thus, ϕ is strongly recurrent modulo Δ and by the Inheritance Theorem 2, it is also recurrent. $\square$

Now, to prove Theorem 3, we may assume that $\Delta > 0$. For $n = 1, 2, \ldots,$ let $L_n = \{z : |z| \leq n\}$ and choose $\{\epsilon_n\}$ decreasing to zero. For each $n = 1, 2, \ldots,$ we invoke Theorem 11 for the strip $S = (1/2 < \Re z < 1)$, $f = \zeta$, $L = L_n$, and $\delta = \epsilon_n$ to obtain a function ϕ_n, frequently hypercyclic in the strip S for the translation operator $C_{i\Delta}$. By Lemma 2, the functions ϕ_n are also strongly recurrent. This concludes the proof of Theorem 3.

5 Proof of Theorem 4

It is not in general true that the product of strongly recurrent functions is strongly recurrent. It is not even true that if fg and hg are strongly recurrent, then $fghg$ is strongly recurrent, but the following lemma gives us a step in this direction.

Lemma 3 *Let S be the fundamental strip $0 < \Re z < 1$ and $g \in \mathcal{O}(S)$ be zero free. Let $\{D_k : k = 1, 2, \ldots\}$ be a regular exhaustion of S by closed (filled) rectangles centered at $1/2$. Let $\{n_k\}$ be an increasing sequence in $\mathbb{N}$ such that the rectangles $D_k^+ = D_k + in_k$ are disjoint and, for each k, the rectangles D_k and D_k^+ are also disjoint. Let $\{\epsilon_k\}$ be a sequence of positive numbers. Then, there exists $f \in \mathcal{O}(S)$, $f \not\equiv 0$, such that, for $k = 1, 2, \ldots$,*

$$\max_{z \in D_k} \left| f(z)f(1-z)g^2(z) - f(z+in_k)f(1-z-in_k)g^2(z+in_k) \right| < \epsilon_k. \tag{1}$$

Proof Set $D_k^- = D_k - in_k$ and let B_k be an exhaustion of S by closed (filled) rectangles, centered at $1/2$, containing D_j, $j \le k$ and $D_j^\pm$, $j < k$ but disjoint from $D_j^\pm$, $j \ge k$. We may assume that $\epsilon_1 < \max_{z \in B_2} |g(z)|$ and $\sum_{j>k} \epsilon_j < \epsilon_k$.

To prove (1) it is sufficient to obtain a function $f \in \mathcal{O}(S)$ satisfying

$$\max_{z \in D_k^+} \left| f(z) - \frac{f(z-in_k)f(1-z+in_k)g^2(z-in_k)}{f(1-z)g^2(z)} \right| < \frac{\epsilon_k}{\max_{z \in D_k^+} |f(1-z)g^2(z)|}. \tag{2}$$

We shall construct inductively a sequence $f_k, k = 0, 1, 2, \ldots$ of zero-free entire functions such that, for $k = 1, 2, \ldots$,

$$\max_{z \in B_k} \left| f_k(z) - f_{k-1}(z) \right| < \epsilon; \tag{3}$$

$$\max_{z \in D_k^+} \left| f_k(z) - \frac{f_k(z-in_k)f_k(1-z+in_k)g^2(z-in_k)}{f_k(1-z)g^2(z)} \right| < \frac{\epsilon_k}{\max_{z \in D_k^+} |f_k(1-z)g^2(z)|}. \tag{4}$$

First, we define an auxiliary function φ_1 on $D_1^- \cup B_1 \cup D_1^+$. We set $\varphi_1 = 1$ on $B_1 \cup D_1^-$. On D_1^+ we set φ_1 equal to a polynomial zero-free on D_1^+ and performing the approximation

$$\max_{z\in D_1^+}\left|\varphi_1(z)-\frac{\varphi_1(z-in_1)\varphi_1(1-z+in_1)g^2(z-in_1)}{\varphi_1(1-z)g^2(z)}\right|$$
$$<\frac{\epsilon_1}{\max_{z\in D_1^+}|\varphi_1(1-z)g^2(z)|}.$$

Now, let $f_0=1$ and let f_1 be a zero-free entire function which approximates φ_1 so well on $D_1^-\cup B_1\cup D_1^+$, that f_1 satisfies (3) and (4).

Now, suppose we have functions $f_1,\ldots, f_{k-1}$ satisfying (3) and (4). We define an auxiliary function φ_k on $D_k^-\cup B_k\cup D_k^+$. First, we set $\varphi_k=f_{k-1}$ on B_k. Then, we set $\varphi_k(z)=\varphi_k(z+in_k)$ on D_k^-. Finally, on D_k^+ we set φ_k equal to a polynomial zero-free on D_k^+ and performing the approximation

$$\max_{z\in D_k^+}\left|\varphi_k(z)-\frac{\varphi_k(z-in_k)\varphi_k(1-z+in_k)g^2(z-in_k)}{\varphi_k(1-z)g^2(z)}\right|$$
$$<\frac{\epsilon_k}{\max_{z\in D_k^+}|\varphi_k(1-z)g^2(z)|}.$$

Let f_k be a zero-free entire function which approximates φ_k so well on $D_k^-\cup B_k\cup D_k^+$, that f_k satisfies (3) and (4). By induction, we now have our sequence $f_k, k=0,1,\ldots$.

To prove the lemma, we need a single function that has the behavior of the sequence $\{f_k\}$. For this we shall employ tangential approximation. Set $E=\bigcup_k D_k^+$. Then, E is a set of tangential approximation. Thus, by Theorem 9, for an arbitrary sequence $\delta_k>0$, there is an entire function f such that $\max_{z\in D_k^+}|f(z)-f_k(z)|<\delta_k$, for each $k=1,2,\ldots$. Considering formula (4), we may choose δ_k sufficiently small so that f satisfies (2). Consequently, f also satisfies (1). Since $f_0=1$, if our approximation is sufficiently good, the function f is not identically zero. This completes the proof of the lemma. □

Lemma 4 *Let S be the fundamental strip $0<\Re z<1$ and $g\in\mathcal{O}(S)$ be zero free. Then, there exists $f\in O(S)$, $f\not\equiv 0$, such that $f(z)f(1-z)g^2(z)$ is strongly recurrent.*

Proof For $p\in\mathbb{N}$ choose disjoint sets of natural numbers $\mathbb{N}_p$ of positive lower density, as in the proof of Theorem 11 and writing $n_k=n(k)$, let D_k and D_k^+ be the corresponding sets. Note that D_k and D_k^+ are disjoint, since $N_p>2p$. Thus, the sets D_k and D_k^+ satisfy the hypotheses of Lemma 3.

Fix a compact set $K\subset S$ and $\epsilon>0$. Choose p so that $\max_{z\in K}|\Im z|<p$. Then, for all but finitely many $n=n_k\in\mathbb{N}_p$,

$$K\subset D_k.$$

For all but finitely many k, we have $\epsilon_k < \epsilon$. Thus, from (1),

$$2\overline{d}_{\mathbb{Z}}\left\{n \in \mathbb{Z} : \max_{z \in K}\left| f(z)f(1-z)g^2(z) - f(z+in)f(1-z-in)g^2(z+in)\right| < \epsilon\right\}$$

$$> \underline{d}_{\mathbb{N}}\left\{n_k \in \mathbb{N}_p : \max_{z \in D_k}\left| f(z)f(1-z)g^2(z)\right.\right.$$

$$\left.\left. - f(z+in_k)f(1-z-in_k)g^2(z+in_k)\right| < \epsilon_k\right\}$$

$$= \underline{d}_{\mathbb{N}}(\mathbb{N}_p) > 0.$$

Thus, $f(z)f(1-z)g^2(z)$ is strongly recurrent modulo 1, and therefore strongly recurrent by Theorem 2. This concludes the proof of the lemma. □

Finally we present the proof of Theorem 4. Namely, we show the existence of a strongly recurrent function f in the fundamental strip, which satisfies the functional equation of the Riemann zeta-function.

Proof Since the zeros of $\zeta(z)$ are symmetric with respect to the point $1/2$, there is an entire function $\phi(z)$ such that $\phi(z) = \phi(1-z)$ and $\zeta(z)/\phi(z)$ has no zeros in the fundamental strip S. Let $g = \sqrt{\zeta/\phi}$ be a branch of $(\zeta/\phi)^{1/2}$ in S. By Lemma 4, there is a function $f \in \mathcal{O}(S)$, $f \not\equiv 0$, such that $h(z) = f(z)f(1-z)g^2(z)$ is strongly recurrent in S. Now, set $\mu(z) = f(z)f(1-z)/\phi(z)$. Then, $h = \mu\zeta$, with $\mu(z) = \mu(1-z)$ and therefore h satisfies the functional equation of the Riemann zeta function. That is,

$$h(z)\pi^{-z/2}\Gamma\left(\frac{z}{2}\right) = h(1-z)\pi^{-(1-z)/2}\Gamma\left(\frac{1-z}{2}\right).$$ □

Acknowledgements Supported in part by NSERC (Canada).

References

1. Bagchi, B.: A joint universality theorem for Dirichlet L-functions. Math. Z. **181**, 319–334 (1982)
2. Bagchi, B.: Recurrence in topological dynamics and the Riemann hypothesis. Acta Math. Hung. **50**, 227–240 (1987)
3. Bayart, F., Grivaux, S.: Frequently hypercyclic operators. Trans. Am. Math. Soc. **358**(11), 5083–5117 (2006)
4. Bayart, F., Matheron, É.: Dynamics of Linear Operators. Cambridge Tracts in Mathematics, vol. 179. Cambridge University Press, Cambridge (2009)
5. Chee, P.S.: Universal functions in several complex variables. J. Aust. Math. Soc., Ser. A **28**, 189–196 (1979)
6. Gaier, D.: Lectures on Complex Approximation, vol. XV. Birkhäuser, Boston (1987). Transl. from the German by Renate McLaughlin
7. Gauthier, P.M., Xiao, J.: The existence of universal inner functions on the unit ball of $\mathbb{C}^n$. Can. Math. Bull. **48**(3), 409–413 (2005)

8. Gottschalk, W.H., Hedlund, G.A.: Topological Dynamics. Colloquium Publications of the American Mathematical Society (AMS), vol. 36. American Mathematical Society (AMS), Providence (1955). VIII
9. Grosse-Erdmann, K.-G.: Universal families and hypercyclic operators. Bull. Am. Math. Soc., New Ser. **36**(3), 345–381 (1999)
10. Hamburger, H.: Über einige Beziehungen, die mit der Funktionalgleichung der Riemannschen ζ-Funktion äquivalent sind. Math. Ann. **85**, 129–140 (1922)
11. Heins, M.: A universal Blaschke product. Arch. Math. **6**, 41–44 (1954)
12. Nieß, M.: Universal approximants of the Riemann zeta-function. Comput. Methods Funct. Theory **9**(1), 145–159 (2009)
13. Nieß, M.: Close universal approximants of the Riemann zeta-function. In: Steuding, R., et al. (eds.) Proceedings of the Conference New Directions in Value-Distribution Theory of Zeta and L-Functions, Würzburg, Germany, October 6–10, 2008, pp. 295–303. Shaker Verlag, Aachen (2009)
14. Nieß, M.: On universal relatives of the Riemann zeta-function. J. Contemp. Math. Anal., Armen. Acad. Sci. **44**(5), 335–339 (2009); translation from Izv. Nats. Akad. Nauk Armen., Mat. (5), 83–88 (2009)
15. Seidel, W., Walsh, J.L.: On approximation by euclidean and non-euclidean translations of an analytic function. Bull. Am. Math. Soc. **47**, 916–920 (1941)

A Survey on Blaschke-Oscillatory Differential Equations, with Updates

Janne Heittokangas

Abstract In the celebrated 1949 paper due to Nehari, necessary and sufficient conditions are given for a locally univalent meromorphic function to be univalent in the unit disc $\mathbb{D}$. The proof involves a second order differential equation of the form

$$f'' + A(z)f = 0, \tag{$\dagger$}$$

where $A(z)$ is analytic in $\mathbb{D}$. As an immediate consequence of the proof, it follows that if $|A(z)| \leq 1/(1-|z|^2)^2$ for every $z \in \mathbb{D}$, then any non-trivial solution of ($\dagger$) has at most one zero in $\mathbb{D}$.

Since 1949 a number of papers provide with different types of growth conditions on the coefficient $A(z)$ such that the solutions of ($\dagger$) have at most finitely many zeros in $\mathbb{D}$. If there exists at least one solution with infinitely many zeros in $\mathbb{D}$, then ($\dagger$) is oscillatory. If the zeros still satisfy the classical Blaschke condition, then ($\dagger$) is called Blaschke-oscillatory. This concept was introduced by the author in 2005, but the topic was considered by Hartman and Wintner already in 1955 (Trans. Am. Math. Soc. 78:492–500). This semi-survey paper provides with a collection of results and tools dealing with Blaschke-oscillatory equations.

As for results, necessary and sufficient conditions are given, and notable effort has been put in dealing with prescribed zero sequences satisfying the Blaschke condition. The concept of Blaschke-oscillation also extends to differential equations of arbitrary order. Many of the results given in this paper have been published earlier in a weaker form. All questions regarding the zeros of solutions can be rephrased for the critical points of solutions. This gives rise to a new concept called Blaschke-critical equations. To intrigue the reader, several open problems are pointed out in the text.

Some classical tools and closely related topics that are often related to the finite oscillation case include the Schwarzian derivative, properties of univalent functions, Green's identity, conformal mappings, and a certain Hardy-Littlewood inequality. The Blaschke-oscillatory case also makes use of interpolation theory, various growth

J. Heittokangas (✉)
Department of Physics and Mathematics, University of Eastern Finland, P.O. Box 111, 80101 Joensuu, Finland
e-mail: janne.heittokangas@uef.fi

J. Mashreghi, E. Fricain (eds.), *Blaschke Products and Their Applications*, Fields Institute Communications 65, DOI 10.1007/978-1-4614-5341-3_3,

estimates for logarithmic derivatives of Blaschke products, Bank-Laine functions and recently updated Wiman-Valiron theory.

Keywords Blaschke-critical · Blaschke-oscillatory · Blaschke product · Differential equation · Logarithmic derivative · Oscillation theory · Prescribed zeros · Zero sequence

Mathematics Subject Classification Primary 34M10 · Secondary 30J10 · 30H15

1 Prologue

Two basic topics in the theory of complex linear differential equations in the case of the unit disc $\mathbb{D}$ are the growth and oscillation of solutions. Indeed, if these two properties are known for a given solution, then by a classical factorization of an analytic function into a product of inner and outer functions, we have a pretty good understanding on how the solution looks like.

In the case of the complex plane $\mathbb{C}$, the importance of zeros of solutions is often justified by the fact that the value zero is the only possible deficient value for the solutions [4]. This makes the zero distribution of solutions all the more interesting. Indeed, in the case of $\mathbb{D}$, it is not uncommon that a solution is a bounded analytic function, and hence has uncountably many Picard values. This is just one aspect that makes oscillation theory in the cases of $\mathbb{C}$ and $\mathbb{D}$ different from one another. When comparing the existing literature in the cases of $\mathbb{C}$ and $\mathbb{D}$, see [55], the oscillation results in the case of $\mathbb{D}$ seem to call for further attention.

The early results on oscillation theory in the case of $\mathbb{D}$ go back to the work of Nehari and his students Beesack and Schwarz in the 1940's and 1950's. In the 1960's and 1970's results on non-oscillation were obtained by Hadass, Kim, Lavie and London, to name a few. After a quiet period, the unit disc oscillation theory begins to flourish again, starting from the late 1990's. In particular, a sequence of papers due to Chuaqui, Duren, Osgood and Stowe continue the classical considerations in oscillation theory, while Belaidi, Cao and Yi are inspired by the complex plane case, and consider oscillation of solutions in terms of the exponent of convergence. Oscillation results in terms of Blaschke sequences can be seen to form a midway between these two extremes.

Blaschke sequences and Blaschke products go hand in hand with many classical function spaces. Meanwhile, function spaces are associated to solutions of differential equations in [39, 67, 69].

2 Introduction

It is well-known that if $A(z)$ belongs to $\mathcal{H}(\mathbb{D})$, the space of all analytic functions in $\mathbb{D}$, then all solutions of the differential equation

$$f'' + A(z)f = 0 \tag{1}$$

belong to $\mathcal{H}(\mathbb{D})$ as well. In 1982 Pommerenke showed [67] that if in addition

$$\int_{\mathbb{D}} |A(z)|^{\frac{1}{2}}\, dm(z) < \infty, \tag{2}$$

where $dm(z)$ is the usual Euclidean area measure, then all solutions f of (1) belong to the Nevanlinna class N, that is, $T(r, f) = O(1)$ [29]. By the classical factorization theorem in N, it follows that the zero sequence $\{z_n\}$ of f satisfies the Blaschke condition

$$\sum_n (1 - |z_n|) < \infty. \tag{3}$$

If this is true in general, then (1) is called Blaschke-oscillatory [34, 35]. This makes (2) a sufficient condition for (1) to be Blaschke-oscillatory. Conversely, if $A \in \mathcal{H}(\mathbb{D})$ is such that (1) is Blaschke-oscillatory, then a reasoning based on Nevanlinna's second fundamental theorem and on certain logarithmic derivative estimates for meromorphic functions yield

$$\int_{\mathbb{D}} |A(z)|^{\alpha}\, dm(z) < \infty \tag{4}$$

for every $\alpha \in (0, 1/2)$, see [35].

Example 1 ([40, Example 5.3]) Let $A(z) = C/(1-z)^4$ for some $C \in \mathbb{C} \setminus \{0\}$, and denote $D(0, r) = \{z \in \mathbb{D} : |z| < r\}$. Then the reasoning in [40] shows that

$$\int_{D(0,r)} |A(z)|^{\frac{1}{2}}\, dm(z) = |C|^{\frac{1}{2}} \pi \log \frac{1}{1-r^2},$$

and that (1) is Blaschke-oscillatory if and only if $\arg(C) = \pi$. This illustrates that neither (2) nor a condition of the form

$$\int_{D(0,r)} |A(z)|^{\frac{1}{2}}\, dm(z) = O\left(\log \frac{e}{1-r}\right)$$

is necessary for (1) to be Blaschke-oscillatory.

In Sect. 4 we will improve (4) to

$$\int_{D(0,r)} |A(z)|^{\frac{1}{2}}\, dm(z) = O\left(\log^2 \frac{e}{1-r}\right). \tag{5}$$

The reasoning relies on recent estimates for integrated logarithmic derivatives of meromorphic functions and on a special treatment of exceptional sets. These findings were not available at the time when [35] was written. Keeping Example 1 in mind, the estimate (5) does not seem to be too far from the best possible one. We will also prove the following somewhat surprising statement: There exists Blaschke-oscillatory equations of the form (1) with all solutions being of unbounded Nevanlinna characteristic.

In addition to (3) there are plenty of other requirements for $\{z_n\}$ needed in oscillation theory. Typical cases are when $\{z_n\}$ is separated, uniformly separated, exponential, or belongs to a Stolz angle, to name a few. These requirements, plus a recently developed concept of q-separation, will be reviewed in Sect. 3.

The equation (1) is called non-oscillatory if all non-trivial solutions of (1) have at most finitely many zeros in $\mathbb{D}$. If (1) possesses a non-trivial solution with infinitely many zeros in $\mathbb{D}$, then (1) is called oscillatory. The early results due to Nehari [59] and Hille [46] indicate that

$$|A(z)| = O\left((1-|z|)^{-2}\right) \tag{6}$$

is an extremal growth rate between non-oscillatory and oscillatory. In Sect. 5 we will state a number of results and examples on equation (1) in the case when the coefficient $A(z)$ satisfies (6). Some of these findings are new.

The extremal growth rate (6) led Cima and Pfaltzgraff [13] to raise the question whether (1) can be oscillatory if $A(z)$ belongs to the Schlicht class S of normalized univalent functions. This discussion, together with some updates on geometric distribution of zeros, will be outlined in Sect. 6.

Various estimates for logarithmic derivatives of meromorphic functions have proven to be valuable tools in the field of complex differential equations both in the plane and in the unit disc. Such estimates for Blaschke products are indispensable in the field of Blaschke-oscillatory equations. Section 7 contains a variety of pointwise and integrated estimates for Blaschke products, and culminates in proving a new result of the form

$$\int_{\mathbb{D}} \left| \frac{B^{(k)}(z)}{B(z)} \right|^{\frac{1}{k}} dm(z) < \infty, \quad k \in \mathbb{N}, \tag{7}$$

valid for certain interpolating Blaschke products $B(z)$.

Given a Blaschke sequence $\{z_n\}$, it is natural to ask whether a function $A \in \mathcal{H}(\mathbb{D})$ can be found such that (1) possesses a solution f having zeros precisely at the points z_n. In such a case, $\{z_n\}$ is called a prescribed zero sequence [34, 42]. Writing (1) in the form $A(z) = -f''/f$, we see that each z_n must be simple, for otherwise $A(z)$ fails to be analytic at the points z_n.

For the sake of an argument, let $B(z)$ be a Blaschke product with $\{z_n\}$ as its zero sequence, and write $f = Be^g$ for a candidate solution of (1), where $g \in \mathcal{H}(\mathbb{D})$ is yet to be found. A simple substitution of f into (1) gives us

$$B'' + 2g'B' + \left((g')^2 + g'' + A\right)B = 0. \tag{8}$$

We conclude that $f = Be^g$ is a solution of (1) if and only if (8) holds. However, it is not clear at the outset that $A \in \mathcal{H}(\mathbb{D})$. Considering (8) at the points z_n, we find that

$$g'(z_n) = -B''(z_n)/2B'(z_n) =: \sigma_n, \quad n \in \mathbb{N}. \tag{9}$$

Writing (8) in the form

$$A = \frac{-B'' - 2g'B'}{B} - \left(g'\right)^2 - g'', \tag{10}$$

we conclude that if $g \in \mathcal{H}(\mathbb{D})$ can be found such that its derivative g' satisfies the interpolation problem (9), then $A(z)$ in (10) belongs to $\mathcal{H}(\mathbb{D})$, and $f = Be^g$ is a solution of (1). For example, if $\{z_n\}$ is uniformly separated, then standard interpolating results from the theory of H^p spaces become available. It is desirable to aim for the target growth (2) since that forces (1) to be Blaschke-oscillatory. This is where the estimate (7) kicks in. In general, we see that the distribution of the zeros z_n will affect the distribution of the points σ_n in (9), and eventually contribute to the growth of $A(z)$. More details are carried out in Sect. 8.

Recall that any solution base of (1) consists of two linearly independent solutions $f_1, f_2 \in \mathcal{H}(\mathbb{D})$. This gives raise to the following problem involving two prescribed zero sequences: Given two Blaschke sequences $\{a_n\}$ and $\{b_n\}$, find $A \in \mathcal{H}(\mathbb{D})$ such that (1) possesses two linearly independent solutions f_1, f_2 having zeros precisely at the points a_n and b_n, respectively. Again the points a_n and b_n must be simple for otherwise $A(z)$ fails to be analytic in $\mathbb{D}$. Moreover, the sequences $\{a_n\}$ and $\{b_n\}$ must be pairwise disjoint for otherwise the solutions f_1, f_2 fail to be linearly independent. Indeed, the Wronskian determinant of f_1, f_2 vanishes at any possible common zero of f_1, f_2.

A solution to the problem of two prescribed zero sequences exists [35] the reasoning being based on Bank-Laine functions. So far the best known growth condition for the solution $A(z)$ of this problem seems to be $A \in N$. In Sect. 9 we will show that the solution $A(z)$ has in fact a growth rate not too far from (6), which is extremal between non-oscillatory and oscillatory cases. In addition to Bank-Laine functions, this new approach relies on recent developments of Wiman-Valiron theory [19] as well as on BMOA-interpolation [63].

We note that prescribed zeros for solutions of (1) in the case when $A(z)$ is entire have been studied much earlier. The early results go back to Borůvka and Šeda in the 1950's, and the research was continued independently by Bank in the 1980's and by Sauer in the 1990's. See [42] for a historical review to these studies.

Section 10 contains a short and elementary discussion about finite prescribed zero sequences. In Sect. 11 we give necessary and sufficient conditions for Blaschke-oscillatory equations of arbitrary order, and give a condition under which a solution f and a few of its first derivatives belong to N. Recall that the zeros of f' are called the critical points of f. If $f, f' \in N$, then the zeros and the critical points of f both satisfy the Blaschke condition. This gives raise to the study of critical points of solutions, to be carried out in Sect. 12. If the sequence of critical points of any non-trivial solution of (1) satisfies the Blaschke condition, then (1) is called Blaschke-critical. Section 13 contains concluding remarks about replacing Blaschke products with Horowitz products in the oscillation theory.

The topic of Blaschke-oscillatory equations is relatively new and still evolving. The author apologizes if some of the open problems stated in the text are trivial, as this may indeed be the case.

3 Typical Requirements for Blaschke Sequences

A sequence $\{z_n\}$ of non-zero points in $\mathbb{D}$ satisfying (3) is called a Blaschke sequence. This requirement on $\{z_n\}$ ensures that the Blaschke product

$$B(z) = \prod_n \frac{|z_n|}{z_n} \frac{z_n - z}{1 - \overline{z_n} z}$$

converges uniformly on compact subsets of $\mathbb{D}$, and hence represents an analytic function in $\mathbb{D}$ having zeros precisely at the points z_n.

3.1 *Measurements for the Quantity of Points*

In the literature the condition (3) is typically strengthened by requiring that

$$S := \sum_n \big(1 - |z_n|\big)^\alpha < \infty \tag{11}$$

holds for some $\alpha \in (0, 1]$. This is in particular the case when the membership of B' in the Bergman spaces or in the Hardy spaces is considered [23, 50, 68]. For $p > 0$ the Bergman space A^p [17, 32] consists of functions $g \in \mathcal{H}(\mathbb{D})$ satisfying

$$\int_D \big|g(z)\big|^p \, dm(z) < \infty,$$

while the Hardy space H^p [15] consists of functions $g \in \mathcal{H}(\mathbb{D})$ such that

$$\sup_{0 \le r < 1} M_p(r, g) < \infty, \quad p < \infty,$$

$$\sup_{z \in \mathbb{D}} \big|g(z)\big| < \infty, \quad p = \infty,$$

where

$$M_p(r, g) = \left(\frac{1}{2\pi} \int_0^{2\pi} \big|g\big(re^{i\theta}\big)\big|^p \, d\theta \right)^{\frac{1}{p}}.$$

Integral means of derivatives of Blaschke products have been widely investigated recently. To name a few, Gotoh, Kutbi, Protas, and the complex analysis research group in Spain (Girela, Peláez, Vukotić, et al.) have been among the key authors in this field, with publications too numerous to be listed here.

Let h be a continuous and positive function on $(0, 1)$ satisfying $h(t) \to 0$ as $t \to 0^+$. In 2008 Fricain and Mashreghi introduced [21] a general convergence condition for a Blaschke sequence $\{z_n\}$ by requiring that

$$\sum_n h\big(1 - |z_n|\big) < \infty. \tag{12}$$

For example, by choosing $h(t) = t^{\alpha}$, we obtain (11).

It is well-known that Blaschke products have radial limits or even angular limits almost everywhere on $\partial\mathbb{D}$, and that $|B| = 1$ almost everywhere on $\partial\mathbb{D}$. Angular limits are taken within Stolz angles. For $\xi \in \partial\mathbb{D}$ and $\sigma \in (1, \infty)$, the set

$$\Omega_{\sigma}(\xi) = \left\{z \in \mathbb{D} : |1 - \bar{\xi}z| \leq \sigma\left(1 - |z|\right)\right\} \tag{13}$$

is called a Stolz angle with vertex at ξ. If the constant σ is close to 1, then the angle $\Omega_{\sigma}(\xi)$ is acute. We note that there are other ways to define a Stolz angle. The exact shape, however, is irrelevant. The important fact is that all points of the angle have bounded non-euclidean distance from the radius $[0, \xi]$.

A point on $\partial\mathbb{D}$ at which a radial limit of a given Blaschke product ceases to exist must be an accumulation point of the zeros due to the well-known Frostman condition. This makes it justified to consider Blaschke products with zeros z_n accumulating to a single boundary point $\xi \in \partial\mathbb{D}$. Typically the sequence $\{z_n\}$ converges to ξ either tangentially or non-tangentially. In the former case the zeros z_n may belong to some circle internally tangent to $\partial\mathbb{D}$ at ξ, while in the latter case the zeros z_n always belong to some Stolz angle $\Omega_{\sigma}(\xi)$. The special case when the zeros z_n lie on the interval $[0, \xi]$ has also drawn abundant attention.

3.2 Measurements for the Density of Points

Alongside the various convergence conditions presented above, the density of points in a given Blaschke sequence turns out to be very important also. We say that an arbitrary sequence $\{z_n\}$ of non-zero points in $\mathbb{D}$ is separated if

$$\inf_{n \neq k} \rho(z_n, z_k) > 0, \tag{14}$$

and uniformly separated if

$$\delta := \inf_{k} \prod_{n \neq k} \rho(z_n, z_k) > 0. \tag{15}$$

Here $\rho(z, \zeta) = |\zeta - z|/|1 - \bar{\zeta}z|$ is the pseudo-hyperbolic metric between the points $z, \zeta \in \mathbb{D}$, see [17, Sect. 2.5]. It turns out that a uniformly separated sequence is always a separated Blaschke sequence, while a separated sequence need not be a Blaschke sequence. In this connection we recall the following result, which will be useful later on in the paper.

Theorem 1 ([24, Theorem 1.3]) *Let $\mathbb{D}_1 \subset \mathbb{D}$ be a disc internally tangent to $\partial\mathbb{D}$ at $z = 1$. Let $\{z_n\}$ be any separated sequence in $\mathbb{D}_1$. Then*

$$n(r) = O\left(\left(\frac{1}{1-r}\right)^{\frac{1}{2}}\right), \quad r \to 1^{-}, \tag{16}$$

where $n(r)$ denotes the number of elements of $\{z_n\}$, counted according to multiplicity, in $D(0,r)\cap\mathbb{D}_1$.

Supposing that (16) holds, a simple reasoning based on Riemann-Stieltjes integration and integration by parts reveals that

$$\sum_{|z_n|\leq r}(1-|z_n|)=\int_0^r(1-t)\,dn(t)=O(1)+\int_0^r n(t)\,dt=O(1).$$

Hence the sequence $\{z_n\}$ in Theorem 1 is a Blaschke sequence.

A classical result due to Carleson [17, p. 157] states that $\{z_n\}$ is an interpolating sequence for H^∞ if and only if it satisfies (15). This means that, given any sequence $\{w_n\}\in\ell^\infty$, there exists a function $f\in H^\infty$ such that $f(z_n)=w_n$ for all n. In addition, Tse has proved [74] that if $\{z_n\}$ lies in a Stolz angle, then the requirement (15) in Carleson's theorem can be weakened to (14). Due to these connections to interpolation theory, a Blaschke product with a uniformly separated zero sequence is known as an interpolating Blaschke product in the literature. A search from the MathSciNet database produces over 110 matches on interpolating Blaschke products. In particular, an interested reader is encouraged to get acquainted with papers due to Gorkin, Mortini and Nicolau.

A sequence $\{z_n\}$ of points in $\mathbb{D}$ is called exponential if there is a constant $q\in(0,1)$ such that

$$1-|z_{n+1}|\leq q\big(1-|z_n|\big),\quad n\in\mathbb{N}. \tag{17}$$

An exponential sequence is uniformly separated by [15, Theorem 9.2]. The converse is also true if the points z_n lie in a Stolz angle. In fact, a separated sequence in a Stolz angle is a finite union of exponential sequences [24, Proposition 3.1].

Let $\phi:[1,\infty)\to[1,\infty)$ be a continuous and increasing function such that

$$\int_1^\infty\frac{dx}{\phi(x)}<\infty. \tag{18}$$

It is observed in [37] that an exponential sequence $\{z_n\}$ then satisfies (12) for $h(t)=\phi(\log(\frac{1}{t}))^{-1}$, and that the function ϕ cannot in general be chosen to be the identity mapping. Due to this reason, a sequence $\{z_n\}$ is called strongly exponential [37] if it is exponential and satisfies (12) for $h(t)=(\log(\frac{1}{t}))^{-1}$.

New density conditions for a point sequence, generalizing the concepts of separation and uniform separation, are introduced in [27] as follows. A sequence $\{z_n\}$ in $\mathbb{D}$ is called q-separated if there exist constants $q\geq 0$ and $\delta\in(0,1)$ such that

$$\rho(z_n,z_k)\geq\delta\big(1-|z_k|\big)^q \tag{19}$$

for all pairs of indices k,n with $k\neq n$, and uniformly q-separated if

$$\inf_{n\in\mathbb{N}}\left\{\left(\frac{1}{1-|z_k|}\right)^q\prod_{k\neq n}\rho(z_n,z_k)\right\}>0. \tag{20}$$

The case $q = 0$ reduces to the classical separation concepts. Note that the case $q < 0$ is impossible since every term of the form $\rho(z_n, z_k)$ as well as the Blaschke product in (20) are bounded above by the constant 1. The convergence of the Blaschke product implies, in particular, that a uniformly q-separated sequence is a Blaschke sequence. Moreover, every uniformly q-separated sequence is also q-separated. A concrete example is given in [27].

For $q \geq 0$ the weighted Hardy space H_q^∞ consists of functions $g \in \mathcal{H}(\mathbb{D})$ such that

$$\sup_{z\in\mathbb{D}} \left(1 - |z|^2\right)^q \left|g(z)\right| < \infty.$$

Recall that the Korenblum space $\mathcal{A}^{-\infty}$ is defined as $\mathcal{A}^{-\infty} = \bigcup_{q\geq 0} H_q^\infty$, see [52]. The following result deals with interpolation in the weighted space H_q^∞, and reduces to Carleson's result on H^∞ interpolation in the case when $q = 0$.

Lemma 1 ([27, Lemma 6.1]) *Let $s \geq 0$, and suppose that $\{\sigma_n\}$ is any sequence of points in $\mathbb{C}$ (not necessarily distinct) satisfying*

$$\sup_{n\in\mathbb{N}} \left(1 - |z_n|^2\right)^s |\sigma_n| < \infty. \tag{21}$$

(a) *If $\{z_n\}$ is a uniformly q-separated Blaschke sequence in $\mathbb{D}$ satisfying* (11) *for some $\alpha \in (0, 1]$, then there exists a function $G \in H^\infty_{\alpha+q+s}$ such that $G(z_n) = \sigma_n$ for all $n \in \mathbb{N}$.*
(b) *If $\{z_n\}$ is a uniformly q-separated sequence, which consist of a finite union of separated sequences, then there exists a function $G \in H^\infty_{q+s}$ such that $G(z_n) = \sigma_n$ for all $n \in \mathbb{N}$.*

4 The Converse Direction

Assuming that (1) is Blaschke-oscillatory, we investigate the properties of the coefficient function $A(z)$ and of the solutions of (1).

4.1 *Improvement of (4)*

We state and prove an improvement of (4) under the assumption that (1) is Blaschke-oscillatory.

Theorem 2 *Let $A \in \mathcal{H}(\mathbb{D})$ be such that* (1) *is Blaschke-oscillatory. Then*

$$\int_{D(0,r)} \left|A(z)\right|^{\frac{1}{2}} dm(z) = O\left(\log^2 \frac{e}{1-r}\right). \tag{22}$$

The proof of Theorem 2 requires integrated estimates for logarithmic derivatives of meromorphic functions. An exceptional set of arbitrarily small upper density is involved. For a measurable set $E \subset [0,1)$, the upper and lower densities are given respectively by

$$\overline{d}(E) = \limsup_{r\to 1^-} \frac{m(E\cap[r,1))}{1-r} \quad \text{and} \quad \underline{d}(E) = \liminf_{r\to 1^-} \frac{m(E\cap[r,1))}{1-r},$$

where $m(F)$ is the Lebesgue measure of the set F. It is clear that

$$0 \le \underline{d}(E) \le \overline{d}(E) \le 1$$

for any measurable set $E \subset [0,1)$. If $\underline{d}(E) = \overline{d}(E)$, then $d(E) = \underline{d}(E) = \overline{d}(E)$ is called the density of E.

Theorem 3 ([11, Theorem 5]) *Let k and j be integers satisfying $k > j \ge 0$, and let $b, \delta \in (0,1)$. Let f be a meromorphic function in $\mathbb{D}$ such that $f^{(j)}$ does not vanish identically. Then there exists a measurable set $E \subset [0,1)$ with $\overline{d}(E) \le \delta$, and a constant $C = C(b,\delta,k,j) > 0$, such that*

$$\int_0^{2\pi} \left| \frac{f^{(k)}(re^{i\theta})}{f^{(j)}(re^{i\theta})} \right|^{\frac{1}{k-j}} d\theta \le C \frac{T(1-b(1-r),f) - \log(1-r)}{1-r}, \quad r \notin E. \tag{23}$$

Moreover, if $k=1$ and $j=0$, then the logarithm in (23) can be dropped out.

The next auxiliary result allows us to avoid exceptional sets E with $\overline{d}(E) < 1$. This is an important property in the applications involving logarithmic derivatives.

Lemma 2 *Let $g(r)$ and $h(r)$ be nondecreasing real-valued functions on $[0,1)$ such that $g(r) \le h(r)$ for every $r \in [0,1)\setminus E$, where $\overline{d}(E) < 1$. Then there exist constants $b \in (0,1)$ and $r_0 \in [0,1)$ such that $g(r) \le h(1-b(1-r))$ for all $r \in [r_0, 1)$.*

The proof of Lemma 2 is a rather simple modification of that of [3, Lemma C]. To begin with, let $F = [0,1)\setminus E$ and $b = (1-\overline{d}(E))/2$. We claim that there exists a constant $r_0 \in [0,1)$ such that, for every $r \in [r_0,1)$, the interval $[r, 1-b(1-r)]$ meets the set F. Suppose on the contrary to this claim that there exists a sequence $\{r_n\}$ on $[0,1)$ such that $r_n \to 1^-$ and $[r_n, 1-b(1-r_n)] \subset E$ for every $n \in \mathbb{N}$. Define

$$I = \bigcup_{n=1}^{\infty} \big[r_n, 1-b(1-r_n)\big].$$

Then $I \subset E$, but

$$\overline{d}(I) \ge \lim_{n\to\infty} \frac{m(I\cap[r_n,1))}{1-r_n} \ge \lim_{n\to\infty} \frac{m([r_n,1-b(1-r_n)])}{1-r_n} = 1-b = \frac{1+\overline{d}(E)}{2},$$

and hence $\overline{d}(I) > \overline{d}(E)$, which leads to a contradiction. Finally, let $r \in [r_0, 1)$, and take $t \in [r, 1 - b(1-r)] \cap F$. Then

$$g(r) \le g(t) \le h(t) \le h\big(1 - b(1-r)\big)$$

by the monotonicity of $g(r)$ and $h(r)$. This completes the proof of Lemma 2.

The class $\mathcal{F}$ of non-admissible functions consists of meromorphic functions g in $\mathbb{D}$ such that

$$\limsup_{r \to 1^-} \frac{T(r,g)}{-\log(1-r)} < \infty.$$

Note that $N \subset \mathcal{F}$, and that this inclusion is clearly proper even for analytic functions. Some basic properties of $\mathcal{F}$ are obtained in [73].

To prove Theorem 2, let $\{f_1, f_2\}$ be a fundamental solution base of (1), and define $F = f_2/f_1$. Since (1) is Blaschke-oscillatory, a reasoning based on Nevanlinna's second fundamental theorem[1] reveals that $F \in \mathcal{F}$, see the proof of [35, Theorem 1.1]. Moreover, it follows by [10, Lemma 5.3] that all derivatives $F^{(n)}$ are non-admissible also. Note that $2A(z) = S_F(z)$, where

$$S_F = \frac{F'''}{F'} - \frac{3}{2}\left(\frac{F''}{F'}\right)^2 \tag{24}$$

is the Schwarzian derivative of F. If $\delta \in (0,1)$, we conclude by Theorem 3 that there exists a measurable set $E \subset [0,1)$ with $\overline{d}(E) \le \delta$ such that

$$\begin{aligned}\int_0^{2\pi} \left|A\big(te^{i\theta}\big)\right|^{\frac{1}{2}} d\theta &= O\left(\int_0^{2\pi} \left|\frac{F'''(te^{i\theta})}{F'(te^{i\theta})}\right|^{\frac{1}{2}} d\theta + \int_0^{2\pi} \left|\frac{F''(te^{i\theta})}{F'(te^{i\theta})}\right| d\theta\right)\\ &= O\left(\frac{1}{1-t}\log\frac{e}{1-t}\right), \quad t \notin E.\end{aligned} \tag{25}$$

The left-hand side of this is a nondecreasing function of t by [15, Theorem 1.5]. Hence Lemma 2 can be used to avoid the exceptional set E. We then multiply both sides by t and integrate with respect to t from 0 to r. This gives (22).

It remains to show that (22) actually improves (4). Indeed, we prove that (22) implies (4) for every $\alpha \in (0, 1/2)$. Suppose that $A \in \mathcal{H}(\mathbb{D})$ satisfies (22), and let $\alpha \in (0, 1/2)$. Then

$$\begin{aligned}\pi r(1-r) M_{\frac{1}{2}}(r,A)^{\frac{1}{2}} &= 2\pi r M_{\frac{1}{2}}(r,A)^{\frac{1}{2}} \int_r^{\frac{1+r}{2}} dt \le 2\pi \int_r^{\frac{1+r}{2}} M_{\frac{1}{2}}(t,A)^{\frac{1}{2}} t\, dt\\ &\le \int_{D(0,\frac{1+r}{2})} |A(z)|^{\frac{1}{2}}\, dm(z) = O\left(\log^2 \frac{e}{1-r}\right).\end{aligned}$$

---

[1] This part of the proof can be done without appealing to the second fundamental theorem, see Sect. 4.2

Using Hölder's inequality with conjugate indices $p = \frac{1}{2\alpha}$ and $q = \frac{1}{1-2\alpha}$, we now obtain

$$\int_{D(0,r)} |A(z)|^{\alpha}\, dm(z) \leq (2\pi)^{\frac{1}{1-2\alpha}} \int_0^r \left(\int_0^{2\pi} \left|A\left(te^{i\theta}\right)\right|^{\frac{1}{2}} d\theta \right)^{2\alpha} dt = O(1),$$

which implies (4). This completes the proof of Theorem 2.

The condition (2) is, by means of Herold's comparison theorem, sufficient for (1) to be Blaschke-oscillatory [67]. Alternatively, by relying on Gronwall's lemma, the condition

$$\int_{\mathbb{D}} |A(z)|\left(1-|z|\right) dm(z) < \infty \tag{26}$$

is sufficient also [33]. It is proved in [40, Theorem 4.1] that neither of the conditions (2) or (26) implies the other. Assuming that (1) is Blaschke-oscillatory, the best converse condition to (2) known to the author is (22), while finding a reasonably sharp converse to (26) is an open problem.

Remark 1 Prior to [34], the paper [28] due to Hartman and Wintner seems to be the one and only research associating infinite zero sequences of solutions of (1) to the Blaschke condition. Note that the concept of Blaschke-oscillation was not used in [28]. We recall the following result from [28]: Let $A \in \mathcal{H}(\mathbb{D})$, and let $\lambda(r)$ be a positive, continuously differentiable and nondecreasing function on $[0, 1)$ satisfying

$$\int_0^{2\pi} \left|A\left(re^{i\theta}\right)\right| d\theta \leq \lambda(r) \tag{27}$$

and $d\lambda/dr = O(\lambda(r)^2)$. If in addition

$$\int_0^1 \lambda(r)^2(1-r)\, dr < \infty, \tag{28}$$

then the zeros of any solution $f \not\equiv 0$ of (1) satisfies the Blaschke condition.

The integral $M_1(r, A)$ on the left-hand side of (27) is a nondecreasing function of r by [15, Theorem 1.5]. Keeping (28) in mind, we might as well assume that $\lambda(r)$ and $M_1(r, A)$ are both unbounded functions. Hence, if the term $\lambda(r)^2$ in (28) could be replaced by $\lambda(r)$, the result would become stronger. This possible improvement was presumed to be true in [28], but the case was left undecided. An elementary proof can be given by means of (26).

4.2 Basic Properties of Solutions

By making use of Nevanlinna's second fundamental theorem, we proved above that $F = f_2/f_1 \in \mathcal{F}$, where $\{f_1, f_2\}$ is an arbitrary solution base of (1). The next result shows that even more can be obtained.

Lemma 3 *Suppose that* $A \in \mathcal{H}(\mathbb{D})$, *and let* $\{f_1, f_2\}$ *be any solution base of* (1). *Then* $F = f_2/f_1 \in N$ *if and only if* (1) *is Blaschke-oscillatory.*

Suppose that $F = f_2/f_1 \in N$, and let $c \in \widehat{\mathbb{C}}$. Then the sequence of c-points of F satisfies the Blaschke condition by [75, Theorem V.7]. In particular, the zeros and the poles of F are Blaschke sequences. Let f be an arbitrary solution of (1). It is well-known that $f = C_1 f_2 + C_2 f_2$ for some constants $C_1, C_2 \in \mathbb{C}$. Then the zeros of f correspond to the c-points of F for $c = -C_1/C_2$, which we just noted to be a Blaschke sequence. This shows that (1) is Blaschke-oscillatory.

Conversely, suppose that (1) is Blaschke-oscillatory. Let $K \subset \mathbb{C}$ be a compact set, and let $c \in K$. If c is not a Picard value of $F = f_2/f_1$, denote the sequence of c-points of F by $\{z_n(c)\}$. Then $\{z_n(c)\}$ must be a Blaschke sequence, for otherwise the solution $f = f_2 - cf_1$ of (1) has a non-Blaschke sequence of zeros. Moreover, there must be a uniform upper bound $C > 0$ such that

$$\sum_n \left(1 - |z_n(c)|\right) < C$$

for every $c \in K$. If this were not true, there would be a sequence $\{c_k\}$ of c_k-points of F such that

$$k \leq \sum_n \left(1 - |z_n(c_k)|\right) < \infty, \quad k \in \mathbb{N}.$$

Since K is compact, the sequence $\{c_k\}$ converges to a point $c \in K$, while the sequence of c-points of F would not satisfy the Blaschke condition. This contradiction shows that the integrated Nevanlinna counting function $N(r, f, c)$ of the c-points of f in the disc $D(0, r)$ must be a uniformly bounded function of r for any $c \in K$. By choosing $K = \partial\mathbb{D}$ and by appealing to the classical Cartan identity [29, p. 8]

$$T(r, f) = \frac{1}{2\pi} \int_0^{2\pi} N\left(r, f, e^{i\theta}\right) d\theta + \log^+ |f(0)|,$$

we conclude that $F \in N$. This completes the proof of Lemma 3.

For curiosity, we give an alternative proof of the converse part. Suppose on the contrary to the assertion that $F = f_2/f_1 \notin N$. We conclude by [61, p. 276] that

$$\lim_{r \to 1^-} \frac{N(r, F, c)}{T(r, F)} = 1 \tag{29}$$

for all $c \in \mathbb{C}$ outside a set of zero capacity. Let $c \in \mathbb{C} \setminus \{0\}$ be a point satisfying (29), and let $\{z_n(c)\}$ be the corresponding sequence of c-points of F. Note that $\{z_n(c)\}$ is the zero sequence of the solution $f = f_2 - cf_1$ of (1). Since $T(r, F)$ is unbounded, it follows by (29) that $\{z_n(c)\}$ does not satisfy the Blaschke condition, and hence (1) is not Blaschke-oscillatory.

There are two further observations regarding the case $F \in N$. First, Theorem 3 does not give anything better than (25) even though $F \in N$. Second, we will prove that $F \in N$ does not necessarily imply $f_1, f_2 \in N$.

Theorem 4 *Suppose that $A \in \mathcal{H}(\mathbb{D})$.*

(a) *If* (1) *is Blaschke-oscillatory, then either all solutions of* (1) *belong to N or all solutions of* (1) *belong to $\mathcal{F} \setminus N$.*
(b) *If* (1) *possesses a solution $f \in N$, then* (22) *holds. If in addition $f' \in N$, then*

$$\int_{D(0,r)} |A(z)|^{\frac{1}{2}}\, dm(z) = O\left(\log \frac{e}{1-r}\right). \tag{30}$$

It is somewhat surprising that there are Blaschke-oscillatory equations with solutions in $\mathcal{F} \setminus N$, see Sect. 4.3. Conversely, not every equation (1) with solutions in $\mathcal{F} \setminus N$ is Blaschke-oscillatory, see Example 1 and its original reference in [40, Example 5.3]. We make two further remarks on Theorem 4 after its proof.

Suppose that (1) is Blaschke-oscillatory. By Lemma 3, all solutions lie either in N or in $\mathcal{H}(\mathbb{D}) \setminus N$. Indeed, something may cancel out in such a way that $f_2/f_1 \in N$ even though $f_1, f_2 \notin N$. Let f be any solution of (1). To prove (a), it suffices to show that $f \in \mathcal{F}$. Let g be any other solution of (1), linearly independent of f, and define $F = g/f$. Since $F \in N$, we have $F' \in \mathcal{F}$ by [10, Lemma 5.3]. It is well-known that $W(f,g) = fg' - f'g$, the Wronskian of f and g, reduces to a nonzero constant.[2] Since $F' = W(f,g)/f^2$, we have $f \in \mathcal{F}$.

Suppose then that (1) has a solution $f \in N$. If $\delta \in (0,1)$, we conclude by Theorem 3 that there exists a measurable set $E \subset [0,1)$ with $\overline{d}(E) \le \delta$ such that

$$\begin{aligned} \int_0^{2\pi} |A(te^{i\theta})|^{\frac{1}{2}}\, d\theta &= \int_0^{2\pi} \left| \frac{f''(te^{i\theta})}{f(te^{i\theta})} \right|^{\frac{1}{2}} d\theta \\ &= O\left(\frac{1}{1-t} \log \frac{e}{1-t}\right), \quad t \notin E. \end{aligned} \tag{31}$$

Similarly as above, the claim (22) follows by using Lemma 2 and by integrating (31) with respect to t. If in addition $f' \in N$, then the proof of Theorem 3 in [11] shows that (31) can be replaced with

$$\int_0^{2\pi} |A(te^{i\theta})|^{\frac{1}{2}}\, d\theta = O\left(\frac{1}{1-t}\right), \quad t \notin E.$$

This completes the proof of Theorem 4.

Remark 2

(1) Clunie has given [14] an elementary proof for the following result, originally due to Valiron and Whittaker: If f and $g \not\equiv 0$ are entire functions of orders ρ_f and ρ_g with $\rho_f > \rho_g$, then $f'g - g'f$ is of order ρ_f. Šeda has refined [71] this result to the case where the orders are equal but the types are different, that

[2] Since $W'(f,g)(z) \equiv 0$ by (1), we have $W(f,g)(z) \equiv c$. If $c = 0$, then $f'/f = g'/g$, and hence f and g are linearly dependent.

is, $\rho_f = \rho_g$ and $\tau(f) > \tau(g)$ implies that $\tau(f'g - g'f) = \tau(f)$. Noting that $W(g, f) = f'g - g'f$, we see that there is a close connection to Theorem 4(a) and its proof. In particular, any two linearly independent solutions f and g of (1) (in the plane case or in the unit disc case) must be of the same growth, or otherwise their Wronskian fails to be equal to a constant.

(2) We emphasize that if $f \in N$, then in general neither f' nor any primitive function of f need to belong to N, see [30]. However, if the condition (2) is slightly strengthened to

$$\sup_{0 \le r < 1} \int_0^{2\pi} \left|A\left(re^{i\theta}\right)\right|^{\frac{1}{4}} d\theta < \infty,$$

then $f, f'f'' \in N$ for any solution f of (1). This result follows as a special case of Corollary 3 below.

4.3 Solutions in $\mathcal{F} \setminus N$ are Possible

Our goal is to show that there are Blaschke-oscillatory equations of the form (1) with solutions in $\mathcal{F} \setminus N$. This proves the sharpness of Theorem 4(a).

The Bloch space $\mathcal{B}$ consists of functions $g \in \mathcal{H}(\mathbb{D})$ with norm

$$\|g\|_{\mathcal{B}} = \left|g(0)\right| + \sup_{z \in \mathbb{D}} \left(1 - |z|^2\right)\left|g'(z)\right| < \infty.$$

The little Bloch space $\mathcal{B}_0$ consists of all $g \in \mathcal{B}$ such that

$$\left(1 - |z|^2\right)|g(z)| \to 0 \quad \text{as } |z| \to 1^-.$$

Lemma 4 *If $h \in \mathcal{B}$ is univalent, then $h \in \bigcap_{0<p<\infty} H^p$ and $e^h \in N$.*

If $h \in \mathcal{B}$, then it is well-known [17, p. 43] that

$$\left|h(z)\right| \le \|h\|_{\mathcal{B}} \left(1 + \frac{1}{2} \log\left(\frac{1+|z|}{1-|z|}\right)\right), \quad z \in \mathbb{D}. \tag{32}$$

If in addition h is univalent, then by making use of (32), a simple modification of the proof of [15, Theorem 3.16] shows that $h \in H^p$ for any $p > 0$. In particular, $h \in H^1$, and hence $e^h \in N$. This completes the proof of Lemma 4.

The assumption on the univalence of h in Lemma 4 cannot be dropped out since there exists a function $g \in \mathcal{B}$ having radial limits almost nowhere on $\partial\mathbb{D}$, see [17, p. 80]. In particular this means that $g \notin N$. Further, a reasoning based on Privalov's uniqueness theorem as in [67, p. 28] shows that $e^g \notin N$. An alternative proof for $e^g \notin N$ relying on Kolmogorov's theorem is given in [44, p. 344].

By relying on a more general reasoning given in [44, Sect. 5], we outline a method for constructing Blaschke-oscillatory equations (1) with zero-free solution bases belonging to $\mathcal{F} \setminus N$. Let $g, h \in \mathcal{H}(\mathbb{D})$ be any non-constant functions depending on one another by means of the differential equation $h'' + 2g'h' = 0$. This means that

$$h' = Ce^{-2g} \tag{33}$$

for some $C \in \mathbb{C} \setminus \{0\}$. We find that the functions

$$f_j(z) = \exp\bigl(g(z) + (-1)^j h(z)\bigr), \quad j = 1, 2, \tag{34}$$

are zero-free and linearly independent solutions of (1), where

$$A = -g'' - (g')^2 - (h')^2. \tag{35}$$

If $g \in \mathcal{B}_0$, then

$$|g(z)| \leq \int_0^{|z|} |g'(\zeta)|\,|d\zeta| + |g(0)| = o\left(\log \frac{e}{1-|z|}\right), \quad |z| \to 1^-,$$

and so $h \in \mathcal{B}_0$ by (33). In such a case,

$$|A(z)| = \frac{o(1)}{(1-|z|^2)^2}, \quad |z| \to 1^-. \tag{36}$$

Multiplying g by a suitable constant, if necessary, we may suppose that

$$\bigl(1-|z|^2\bigr)|zg'(z)| \leq 1/2, \quad z \in \mathbb{D}. \tag{37}$$

Then

$$\bigl(1-|z|^2\bigr)\left|\frac{zh''(z)}{h'(z)}\right| \leq 1, \quad z \in \mathbb{D},$$

and h is univalent by the Becker condition [66, p. 172]. Choose, for example,

$$g(z) = \frac{1}{2B_2} \sum_{n=1}^{\infty} n^{-1/2} z^{2^n},$$

where B_2 is the constant from [25, Theorem 4] satisfying $2e^{-1} \leq B_2 \leq 2.8913$. Then (37) is valid, while $g \in \mathcal{B}_0$ has angular limits almost nowhere on $\partial\mathbb{D}$, see [25] or [6]. It follows that $g \notin N$, and hence $e^g \notin N$ by [44, Lemma 3]. However, $e^h \in N$ by Lemma 4, and so $e^{-h} \in N$ by the first fundamental theorem. This shows that $f_1, f_2 \notin N$, and hence all solutions lie outside of N. The ratio $f_2/f_1 = e^{2h}$ is clearly in N, and hence the differential equation in question is Blaschke-oscillatory by Lemma 3.

Remark 3 It will be seen in Sect. 5 that the condition (36) in general forces (1) to be non-oscillatory. The construction above has been used earlier in special cases. In particular, by choosing

$$g(z) = \log(1 - z) \quad \text{and} \quad h(z) = \frac{\sqrt{C}}{2}\frac{1+z}{1-z}$$

for $C \neq 0$, we have Example 1.

5 Disconjugate, Non-oscillatory and Oscillatory

Thus far our emphasis has been in Blaschke sequences as potential zero sequences for solutions of (1). The early results in oscillation theory, however, were in fact results on non-oscillation. In this section we will demonstrate that there is a fine line between non-oscillatory and oscillatory cases.

5.1 Pointwise Growth Restrictions for $A(z)$

Nehari proved in [59] that if $A \in \mathcal{H}(\mathbb{D})$ satisfies

$$\left|A(z)\right| \leq \frac{1}{(1 - |z|^2)^2} \tag{38}$$

for all $z \in \mathbb{D}$, then each non-trivial solution of (1) has at most one zero in $\mathbb{D}$. Equation (1) is then called disconjugate [28]. Nehari's basic idea combines disconjugacy with univalence: If f_1, f_2 are linearly independent solutions of (1), then $F = f_2/f_1$ is a solution of the differential equation $S_F(z) = 2A(z)$, where S_F is the Schwarzian of F given in (24), see [60, pp. 203–204]. It follows that $F = c \neq 0, \infty$ if and only if $f_2 - cf_1 = 0$, while the zeros and poles of F are the zeros of f_2 and of f_1, respectively. Hence (1) is disconjugate if and only if F is meromorphic univalent in $\mathbb{D}$. In addition, $F \in N$ by Lemma 3.

It is clear that disconjugacy implies non-oscillatory, which in turn implies Blaschke-oscillatory. Due to this connection, we will review some results involving disconjugate and non-oscillatory equations. The presentation that follows is by no means complete due to the vast literature in this field, but still gives some flavor of the existing results.

In 1955 Schwarz showed [70] that if Nehari's condition is relaxed in the sense that there exists a constant $R \in (0, 1)$ such that (38) holds for all $z \in \mathbb{D}$ with $R \leq |z| < 1$, then (1) is non-oscillatory. In this case, there exist constants $\alpha, \beta > 0$ such that the number of zeros of any solution f can be estimated as

$$\#\{z \in \mathbb{D} : f(z) = 0\} \leq \frac{\alpha}{1-R} + \beta \int_0^R \frac{\sqrt{M(r, A)}}{1-r}\, dr, \tag{39}$$

see [9, Theorem 1]. As proved in Sect. 2, all zeros are simple, and hence the left-hand side of (39) is the number of all zeros of f in $\mathbb{D}$, counting multiplicities. The constant R on the right-hand side of (39) depends only on the annulus $R < |z| < 1$ in which the estimate (38) holds. Schwarz also pointed out that if

$$A(z) = \frac{1+4\gamma^2}{(1-z^2)^2}, \tag{40}$$

where $\gamma > 0$ is a fixed constant, then (1) is oscillatory. We note that the example (40) is a special case of an earlier example due to E. Hille [46].

Example 2 ([37, Example 11]) If $A(z)$ is given by (40), then the reasoning in [37] reveals that (1) possesses a zero-free solution base $\{f_1, f_2\}$, and that solutions of the form $C_1 f_1 + C_2 f_2$ have infinitely many zeros in $\mathbb{D}$, provided that the inequalities $e^{-\frac{\gamma\pi}{2}} < |C_1/C_2| < e^{\frac{\gamma\pi}{2}}$ are satisfied. This example is in fact a special case of the construction given in Sect. 4.3 by choosing g, h to be the Bloch functions

$$g(z) = \log\sqrt{1-z^2} \quad \text{and} \quad h(z) = \frac{\gamma i}{2}\log\frac{1+z}{1-z}.$$

It is proved in [37] that every zero sequence is a union of two infinite exponential sequences converging non-tangentially to the points ± 1 along an arc of a certain circle passing through the points ± 1 and having its center on the imaginary axis. All constant multiples of the particular solution $f_1 + f_2$ have all zeros on the interval $(-1, 1)$ and accumulate to its end points ± 1.

It seems plausible that the condition above due to Schwarz could still be weakened. In fact, it is conjectured in [9, p. 564] that the condition

$$|A(z)| \le \frac{1 + C(1-|z|)}{(1-|z|^2)^2}, \quad z \in \mathbb{D}, \tag{41}$$

for some $C > 0$ would imply finite oscillation. Note that if (41) is still weakened to

$$|A(z)| \le \frac{1 + \beta(|z|)}{(1-|z|^2)^2}, \quad z \in \mathbb{D}, \tag{42}$$

where β satisfies

$$\lim_{r\to 1^-} \frac{\beta(r)}{1-r} = \infty,$$

then (1) can be oscillatory by [9, Theorem 5].

The following necessary and sufficient conditions for zero sequences to be separated are also due to Schwarz.

Theorem 5 ([70, Theorems 3 and 4]) *Let $A \in \mathcal{H}(\mathbb{D})$, and let f be any solution of* (1) *having at least two distinct zeros z_1, z_2. If there exists a constant $\alpha > 1$ such that*

$$|A(z)| \le \frac{\alpha}{(1-|z|^2)^2}, \quad z \in \mathbb{D}, \tag{43}$$

then

$$\rho(z_1, z_2) > \alpha^{-\frac{1}{2}}. \tag{44}$$

Conversely, if there exists a constant $\alpha > 1$ *such that* (44) *holds for any zeros* z_1, z_2 *of any solution* f, *then* (43) *holds with* α *being replaced by* 3α.

Herold has obtained [45] an improvement of (44) under the assumption (43). Regarding Theorem 5, it is an open problem to find necessary and sufficient conditions for the zero sequences of solutions of (1) to be uniformly separated.

We will show that the growth condition (43) does not imply (2). Therefore it is not clear at the outset whether the condition (43) is sufficient for (1) to be Blaschke-oscillatory. Our counterexample is the lacunary series

$$A(z) = \sum_{k=0}^{\infty} 2^{2k} z^{2^k}$$

with Hadamard gaps. We have

$$\sup_{z\in\mathbb{D}} \left(1 - |z|^2\right)^3 |A'(z)| < \infty$$

by [76, Theorem 1]. Using the representation

$$A(z) = \int_0^z A'(\zeta)\, d\zeta + A(0),$$

we see that (43) holds for some $\alpha > 1$. However, (2) fails by [5, Proposition 2.1]. Provided that the constant α in (43) is small enough, we are able to show that (1) is Blaschke-oscillatory.

Theorem 6 *Suppose that* $A \in \mathcal{H}(\mathbb{D})$ *satisfies* (43) *for some* $0 < \alpha < \sqrt{210}/2 \approx 7.246$. *Then* (1) *is Blaschke-oscillatory.*

Our proof covers the case $0 < \alpha < \sqrt{210}/2$ although by relying on Nehari's theorem it would suffice to consider the case $1 < \alpha < \sqrt{210}/2$. Let β be the positive solution of $\beta(\beta+1)(\beta+2)(\beta+3) = 16\alpha^2$, that is,

$$\beta = \frac{1}{2}\sqrt{5 + 4\sqrt{1 + 16\alpha^2}} - \frac{3}{2}.$$

If f solves (1), then [67, Example 1] shows that

$$\int_0^{2\pi} \left|f\left(re^{i\theta}\right)\right|^2 d\theta = O\left(\left(\frac{1}{1-r}\right)^{\beta}\right), \quad r \to 1^-.$$

Thus a simple application of Hölder's inequality yields

$$\int_0^{2\pi} \left|f\left(re^{i\theta}\right)\right|^{\frac{1}{2}} d\theta = O\left(\left(\frac{1}{1-r}\right)^{\frac{\beta}{4}}\right), \quad r \to 1^-.$$

Since $0 < \alpha < \sqrt{210}/2$, we have $0 < \beta < 4$. Therefore (2) holds, and so (1) is Blaschke-oscillatory.

Every separated sequence lying in a disc internally tangent to $\partial\mathbb{D}$ is a Blaschke sequence by Theorem 1. This gives rise to the following conclusion: If (1) under the assumption (43) possesses a solution with a non-Blaschke zero sequence $\{z_n\}$, then at least one of the following two claims must hold.

(1) The sequence $\{z_n\}$ has infinitely many accumulation points on $\partial\mathbb{D}$.
(2) A subsequence of $\{z_n\}$ converges tangentially with high degree of tangency to a boundary point on $\partial\mathbb{D}$, see [24, p. 165].

Also note that the growth condition (43) forces all solutions of (1) in the Korenblum space $\mathcal{A}^{-\infty}$ by [67, Lemma 2], while the zeros of a function in $\mathcal{A}^{-\infty}$ do not in general satisfy the Blaschke condition [17, 32, 52].

If (1) under the assumption (43) possesses two linearly independent solutions such that their ratio is not in the Nevanlinna class N, then (1) is not Blaschke-oscillatory by Lemma 3. The construction in Sect. 4.3 makes it tempting to reason towards this conclusion. In the notation of Sect. 4.3, we aim for $e^{2h} \notin N$. To achieve this, it suffices to have $h \notin N$. Recall that h is obtained by integrating in (33). However, even if g is a badly behaving lacunary series, it is not guaranteed that any primitive of e^{-2g} would be of unbounded characteristic [30]. Note that $h \in \mathcal{H}(\mathbb{D})$ satisfying $e^{2h} \notin N$ is easy to find. Take $h(z) = i/(1-z)$, for example. However, the resulting $A(z)$ then does not satisfy (43). On the other hand, the constant $\sqrt{210}/2$ in Theorem 6 is probably not the best possible. All this being said, the chips may fall either way on this matter.

5.2 *Integrated Growth Restrictions for* $A(z)$

So far the growth restrictions for the coefficient function $A(z)$ have been pointwise. In 1962 London showed [57] that if $A \in \mathcal{H}(\mathbb{D})$ satisfies

$$\int_{\mathbb{D}} \left|A(z)\right| dm(z) \le \pi, \tag{45}$$

then (1) is disconjugate. The condition (45) is sharp in the following sense. Using computer software on [34, Example 3.1], it follows that there exists a function $A \in \mathcal{H}(\mathbb{D})$ satisfying

$$\int_{\mathbb{D}} \left|A(z)\right| dm(z) \le 2.502\pi \tag{46}$$

such that (1) possesses a solution with two zeros in $\mathbb{D}$. Note that the upper bound $5\pi/(2z_1^2)$ for (46) as claimed in [34, Example 3.1] is incorrect. However, it is tempting to presume that the best possible upper bound in (45) is $5\pi/2$.

Moving on to non-oscillatory equations, we recall the following result due to London [57]: If $A \in \mathcal{H}(\mathbb{D})$ satisfies

$$\int_{\mathbb{D}} \left|A(z)\right| dm(z) < \infty, \tag{47}$$

then (1) is non-oscillatory. An alternative and a rather simple proof of this is given in [13]. As for the sharpness of (47), we note that if $A(z)$ is given by (40), then (1) is oscillatory, and a calculation based on Poisson's kernel gives us

$$\int_{D(0,r)} \left|A(z)\right| dm(z) = 2\pi \int_0^r \frac{1+4\gamma^2}{1-t^4} t\, dt = \left(1+4\gamma^2\right)\frac{\pi}{2} \log \frac{1+r^2}{1-r^2}.$$

It seems reasonable to presume, however, that (47) can still be weakened, for example to

$$\int_{D(0,r)} \left|A(z)\right| dm(z) = O\left(\log\log \frac{e}{1-r}\right),$$

and (1) would remain non-oscillatory. We take this opportunity to give the following slight improvement of (47).

Theorem 7 *Let $A \in \mathcal{H}(\mathbb{D})$ be such that there exists a constant $b \in (0,1)$ for which*

$$\lim_{r\to 1^-} \int_{D(0,s(r))\setminus D(0,r)} \left|A(z)\right| dm(z) = 0, \tag{48}$$

where $s(r) = 1 - b(1-r)$. Then (1) is non-oscillatory.

Our reasoning is a modification of the proof of [16, Theorem 3], and is also reminiscent to the reasoning given at the end of Sect. 4.1. We have

$$\begin{aligned} 2\pi\rho(1-\rho)M_1(\rho, A) &= (1-b)^{-1} 2\pi\rho M_1(\rho, A) \int_\rho^{s(\rho)} dt \\ &\le (1-b)^{-1} 2\pi \int_\rho^{s(\rho)} M_1(t, A) t\, dt \\ &\le (1-b)^{-1} \int_{D(0,s(\rho))\setminus D(0,\rho)} \left|A(z)\right| dm(z) = o(1), \end{aligned}$$

as $\rho \to 1^-$. Let $z \in \mathbb{D}$, and denote $\rho = (1+|z|)/2$. Then Cauchy's formula gives

$$\left|A(z)\right| \le \frac{1}{2\pi} \int_{|\zeta|=\rho} \frac{|A(\zeta)|}{|\zeta - z|} |d\zeta| \le \frac{M_1(\rho, A)}{2\pi(1-\rho)}.$$

By combining these observations, we conclude that

$$\big|A(z)\big| = \frac{o(1)}{(1-\rho)^2} = \frac{o(1)}{(1-|z|^2)^2}, \qquad |z| \to 1^-.$$

Hence there exists a constant $R \in (0,1)$ such that (38) holds for all $z \in \mathbb{D}$ with $R \le |z| < 1$, so that (1) is non-oscillatory.

It remains to show that (48) actually improves (47). Supposing that (47) holds, it follows that the right-hand side of

$$0 \le \int_{D(0,s(r))\setminus D(0,r)} \big|A(z)\big|\, dm(z) \le \int_{\mathbb{D}\setminus D(0,r)} \big|A(z)\big|\, dm(z)$$

tends to zero as $r \to 1^-$. Hence (47) implies (48) for any $b \in (0,1)$. This completes the proof of Theorem 7.

Let $A \in \mathcal{H}(\mathbb{D})$ be any function satisfying the asymptotic relation

$$M_1(r,A) \sim \frac{O(1)}{(1-r)\log\frac{e}{1-r}}. \tag{49}$$

Then it is easily seen that (47) is not valid, while (48) holds for any $b \in (0,1)$. Functions $A \in \mathcal{H}(\mathbb{D})$ satisfying (49) do exist, as is seen next.

Example 3 Define

$$A(z) = \frac{1}{(1-z)^2 \log\frac{e}{1-z}},$$

which clearly belongs to $\mathcal{H}(\mathbb{D})$. Writing

$$M_1(r,A) = \frac{1}{2\pi}\int_{|\theta|\le 1-r} \big|A\big(re^{i\theta}\big)\big|\, d\theta + \frac{1}{2\pi}\int_{1-r\le|\theta|\le\pi} \big|A\big(re^{i\theta}\big)\big|\, d\theta = J_1(r) + J_2(r)$$

and observing that $\log\frac{e}{1-z} = \log\frac{e}{|1-z|} + i\arg(\frac{e}{1-z})$, where $\arg(\frac{e}{1-z}) \in (-\frac{\pi}{2}, \frac{\pi}{2})$, a simple computation as in [75, p. 226] yields (49). In particular, since $\cos(x) \ge 1 - x^2$ for all $x \in \mathbb{R}$, we have $\cos(1-r) \ge 2r - r^2$ for all $r \in [0,1)$, and hence

$$\begin{aligned}(1-r)^2 \le \big|1-re^{i\theta}\big|^2 &= 1 - 2r\cos\theta + r^2\\ &\le 1 - 2r\cos(1-r) + r^2\\ &= (1-r)^2 + 2r\big(1-\cos(1-r)\big)\\ &\le (1-r)^2(1+2r) \le 3(1-r)^2\end{aligned}$$

for $|\theta| \le 1-r$. It follows that $J_1(r)$ has precisely the asymptotic growth in (49), while $J_2(r) \ge 0$ has at most the asymptotic growth in (49).

6 Univalent Coefficient Function

Let S denote the Schlicht class consisting of univalent functions $g \in \mathcal{H}(\mathbb{D})$ satisfying the normalization $g(0) = 0 = g'(0) - 1$. In 1972 Cima and Pfaltzgraff considered [13] the oscillatory behavior of (1) under the assumption that $A \in S$. It is well-known [31, Chap. 1] that if $A \in S$, then

$$\left|A(z)\right| \leq \frac{|z|}{(1-|z|)^2}, \quad z \in \mathbb{D}, \tag{50}$$

and

$$\frac{1}{2\pi}\int_0^{2\pi} \left|A\left(re^{i\theta}\right)\right| d\theta \leq \frac{r}{1-r}, \quad 0 \leq r < 1. \tag{51}$$

By (38) the question of the oscillatory behavior of (1) for $A \in S$ is of interest only for those functions $A(z)$ that attain the maximum growth $O((1-|z|)^{-2})$ near the boundary $\partial\mathbb{D}$. The following result is a consequence of the discussion above.

Corollary 1 *If $A \in S$, then the zero sequence of any solution of* (1) *is a separated Blaschke sequence. In particular,* (1) *is Blaschke-oscillatory.*

Using Hölder's inequality on (51), we conclude that (2) holds, and therefore (1) is Blaschke-oscillatory. Since (50) implies (43) for $\alpha = 4$, it follows that every zero sequence is separated. Alternatively, (1) is Blaschke-oscillatory by Theorem 6 because $4 < \sqrt{210}/2$.

It may seem at this point that the condition $A \in S$ would make (1) non-oscillatory. This is not true in general. Indeed, the functions $h_\varepsilon \in S$ given by

$$h_\varepsilon(z) = \frac{z}{1-z} + \frac{\varepsilon z^2}{(1-z)^2}, \quad 0 \leq \varepsilon \leq 1, \tag{52}$$

are taken as the candidates for univalent coefficients of (1) in [13]. The special cases $h_0(z) = z/(1-z)$ (Loewner function) and $h_1(z) = z/(1-z)^2$ (Koebe function) are worth pointing out. If $A(z) = h_\varepsilon(z)$, then (1) is non-oscillatory for $0 \leq \varepsilon < 1/4$ and oscillatory for $1/4 < \varepsilon \leq 1$. Indeed, if $1/4 < \varepsilon \leq 1$, then the corresponding real differential equation $f'' + h_\varepsilon(x)f = 0$ is oscillatory on $[0,1)$ by [13, Lemma 1]. It follows then that there are non-trivial complex solutions of $f'' + h_\varepsilon(z)f = 0$ which have infinitely many zeros on $[0,1)$. If $\varepsilon = 1/4$, then the corresponding real differential equation $f'' + h_{1/4}(x)f = 0$ has at most finitely many zeros on $[0,1)$ by [8, Theorem 1]. However, a complete description of the case $\varepsilon = 1/4$ is still an open problem.

The following result gives a partial description of the geometric distribution of zeros of solutions of (1) in the case when $A(z) = h(z)/(1-z)^q$, where $h \in \mathcal{H}(\mathbb{D})$ has restricted growth. This covers the special cases $A(z) = h_\varepsilon(z)$.

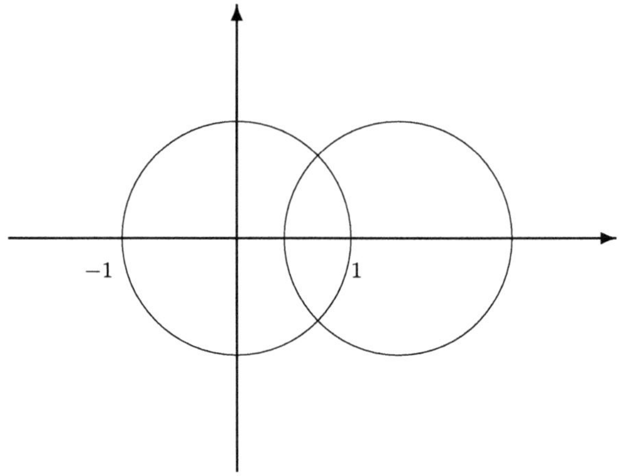

Fig. 1 The unit disc minus the disc $|z-\sqrt{2}|\le 1$

Theorem 8 *Let $A(z)=h(z)/(1-z)^q$, where $q\ge 0$ and $h\in\mathcal{H}(\mathbb{D})$ is such that*

$$\big|h(z)\big|\le\frac{o(1)}{(1-|z|^2)^2},\quad |z|\to 1^-. \tag{53}$$

Then $z=1$ is the only possible accumulation point of zeros of solutions of (1).

The following auxiliary result is needed.

Lemma 5 ([48, p. 394], [67, Lemma 1]) *Let T map $\mathbb{D}$ conformally into $\mathbb{D}$. If f is a solution of* (1), *then $g(z)=f(T(z))T'(z)^{-\frac{1}{2}}$ solves*

$$g''+B(z)g=0, \tag{54}$$

where

$$B(z)=A\big(T(z)\big)T'(z)^2+S_T(z)/2, \tag{55}$$

and S_T denotes the Schwarzian derivative of T. In particular, if (54) *is non-oscillatory in $\mathbb{D}$, then* (1) *is non-oscillatory in $T(\mathbb{D})$.*

We will be dealing with circular domains bounded by an arc of $\partial\mathbb{D}$ and by another circular arc lying in $\mathbb{D}$ and meeting $\partial\mathbb{D}$ at angles $\alpha\pi$ for some $\alpha\in(0,1)$. Such a domain is obtained for example by subtracting the closure of some disc that intersects with $\mathbb{D}$ from $\mathbb{D}$, see Fig. 1. A suitable linear fractional transformation will map this circular domain onto a sector of opening $\alpha\pi$. A rotation and the principal branch of the power function $z^{1/\alpha}$ will map this sector onto the right half-plane, which is then mapped conformally onto $\mathbb{D}$ by $(z-1)/(z+1)$. An inverse of this composition mapping is a conformal map from $\mathbb{D}$ onto the given circular domain. Such mappings turn out to be of the form

$$T(z)=\frac{a(1-z)^\alpha+b(1+z)^\alpha}{c(1-z)^\alpha+d(1+z)^\alpha}, \tag{56}$$

where $\alpha \in (0,1)$ and $a,b,c,d \in \mathbb{D}$ are such that $bc-ad \neq 0$. This requirement is necessary so that the derivative

$$T'(z) = \frac{2\alpha(bc-ad)}{(1-z^2)^{1-\alpha}(c(1-z)^\alpha + d(1+z)^\alpha)^2}$$

would not vanish identically. It is stated as an exercise in [60, pp. 208–209] that the Schwarzian of T reduces to just

$$S_T(z) = \frac{2(1-\alpha^2)}{(1-z^2)^2}. \tag{57}$$

This can be verified by using standard mathematical software. Note in particular that T is univalent for every $\alpha > 0$ by [59, Theorem I].

To prove Theorem 8, let $\phi \in (0,\pi/4)$, and let D_ϕ be the lens shaped domain bounded by $\partial\mathbb{D}$ and by the hyperbolic line L_ϕ intersecting $\partial\mathbb{D}$ orthogonally at $z = e^{\pm i\phi}$ and lying in $\mathbb{D}$. The limiting case $D_{\pi/4}$ is illustrated in Fig. 1. We will show that (1) is non-oscillatory in $\mathbb{D} \setminus \overline{D_\phi}$. As $\phi > 0$ can be chosen arbitrarily small, the assertion follows. The key idea is to construct a suitable conformal mapping of the form (56), and then use Lemma 5.

The linear fractional transformation $w_1 = (1+z)/(1-z)$ maps $\mathbb{D}$ conformally onto the right half-plane $H = \{z \in \mathbb{C} : \Re z > 0\}$. Choosing the principal branch for the square root, H will be mapped conformally onto the first quadrant $F = \{z \in \mathbb{C} : \arg(z) \in (0,\pi/2)\}$ by means of $w_2 = \sqrt{z}e^{i\frac{\pi}{4}}$. Then we use $w_3 = (z-1)/(z+1) = w_1^{-1}$ to map F conformally onto $\mathbb{D}_U = \{z \in \mathbb{D} : \Im z > 0\}$, which is the upper half of the unit disc. The image of $\mathbb{D}_U$ under the mapping $w_4 = -iz\sin\phi/(1-\cos\phi)$ is the half disc $\Delta = \{z \in \mathbb{C} : |z| < \sin\phi/(1-\cos\phi), \Re z > 0\}$ lying entirely in the right half-plane and intersecting the imaginary axis at the points $z = \pm i\sin\phi/(1-\cos\phi)$. Finally, we use w_3 again to map Δ onto $\mathbb{D} \setminus \overline{D_\phi}$. Note in particular that w_3 maps the points $z = \pm i\sin\phi/(1-\cos\phi)$ to $w = e^{\pm i\phi}$, respectively. The composition $T = w_3 \circ w_4 \circ w_3 \circ w_2 \circ w_1$ is therefore a conformal map from $\mathbb{D}$ onto $\mathbb{D} \setminus \overline{D_\phi}$. After simplification,

$$T(z) = \frac{a\sqrt{1-z} + b\sqrt{1+z}}{c\sqrt{1-z} + d\sqrt{1+z}}, \tag{58}$$

where $a = i(e^{i\phi}-1)$, $b = ie^{i\frac{\pi}{4}}(e^{-i\phi}-1)$, $c = -i(e^{-i\phi}-1)$, and $d = -ie^{i\frac{\pi}{4}} \times (e^{i\phi}-1)$.

Let $B(z)$ be the function given in (55). By symmetry, the equation for the circle that determines the hyperbolic line L_ϕ must be of the form $|z-a| = r$, where $a > 1$ and $r > 0$. Since L_ϕ meets $\partial\mathbb{D}$ orthogonally at $z = e^{i\phi}$, a simple implicit differentiation reveals that $a = 2\cos\phi$. This gives us $r = |e^{i\theta} - a| = 1$. The shortest distance from L_ϕ to the point $z = 1$ is $2 - 2\cos\phi$. Since $T(z) \notin D_\phi$, we have

$$\left|1 - T(z)\right|^{-q} \leq (2-2\cos\phi)^{-q} =: C(\phi,q), \quad z \in \mathbb{D}.$$

Note that $C(\phi,q)\to\infty$ as $\phi\to 0^+$. Then (53) and the Schwarz-Pick lemma yield

$$\left|A\big(T(z)\big)\right|\left|T'(z)\right|^2\le\frac{o(1)C(\phi,q)}{(1-|T(z)|^2)^2}\left|T'(z)\right|^2\le\frac{o(1)}{(1-|z|^2)^2},\quad |z|\to 1^-.$$

Next (57) for $\alpha=1/2$ gives us

$$\left|S_T(z)\right|\le\frac{3}{2(1-|z|^2)^2},\quad z\in\mathbb{D}. \tag{59}$$

Putting all together, we have

$$\left|B(z)\right|\le\frac{3+o(1)}{4(1-|z|^2)^2},\quad |z|\to 1^-. \tag{60}$$

This shows that $B(z)$ has the growth (38) for $R_\phi<|z|<1$, where $R_\phi\to 1^-$ as $\phi\to 0^+$. Hence (54) is non-oscillatory in $\mathbb{D}$. Lemma 5 then implies that (1) is non-oscillatory in $\mathbb{D}\setminus\overline{D_\phi}$. This completes the proof of Theorem 8.

Remark 4 We may replace the hyperbolic line L_ϕ by a circular arc of any circle of the form $|z-a|=r$, where $a>1$ and $r>0$. Suppose such an arc meets $\partial\mathbb{D}$ at angles $\alpha\pi$. Then the linear fractional transformation $(e^{-i\phi}-z)/(e^{i\phi}-z)$ will map the analogue of $\mathbb{D}\setminus\overline{D_\phi}$ onto a sector of opening $\alpha\pi$.

The discussion above on the univalent coefficient can be formulated in terms of Hayman's index. Recall from [31] that if $g\in S$, then the Hayman index

$$\alpha(g)=\limsup_{r\to 1^-}(1-r)^2M(r,g)$$

exists and satisfies $0\le\alpha(g)\le 1$. For example, $\alpha(h_\varepsilon)=\varepsilon$, where $h_\varepsilon(z)$ is given by (52). In general, we have $\alpha(g)=1$ if and only if g is a rotation of the Koebe function. If $\alpha(g)>0$, then g has a unique radius of maximal growth, i.e., there exists a unique $\theta\in[0,2\pi)$ such that $(1-r)^2|g(re^{i\theta})|\to\alpha(g)$ as $r\to 1^-$. In all other directions $\varphi\neq\theta$, we have $(1-r)^2|g(re^{i\varphi})|\to 0$ as $r\to 1^-$.

Theorem 9 *Let $A\in S$. If $\alpha(A)<1/4$, then* (1) *is non-oscillatory. If* (1) *is oscillatory and* $\arg(z)=\theta$ *is the unique radius of maximal growth of $A(z)$, then $z=e^{i\theta}$ is the only possible accumulation point of zeros of solutions of* (1).

Define $F(z)=f(ze^{-i\theta})$. Then F satisfies the equation $F''+C(z)F=0$, where $C(z)=e^{i2\theta}A(ze^{-i\theta})$ does not belong to S but is univalent and has $[0,1)$ as its radius of maximal growth. Since all growth properties of $A(z)$ can be applied to $C(z)$ without difficulty, we may, without loss of generality, assume that $\theta=0$.

Suppose that $\alpha(A)<1/4$. Then there exists an $r_0\in(0,1)$ such that

$$M(r,A)\le\frac{1}{4(1-r)^2}\le\frac{1}{(1-r^2)^2},\quad r_0<r<1.$$

Then (38) holds for $r_0 < |z| < 1$, and (1) is non-oscillatory by Schwarz' theorem.

Suppose next that (1) is oscillatory. Let $\phi \in (0, \pi/4)$, and let T be the conformal mapping given in (58). Let R be the unique inverse of T, and define $B(z)$ by (55). By the change of variable, we have

$$\int_{\mathbb{D}} |A(T(z))||T'(z)|^2\, dm(z) = \int_{\mathbb{D}\setminus\overline{D}_\phi} |A(w)||T'(R(w))|^2 |R'(w)|^2\, dm(w).$$

Using the Schwarz-Pick lemma twice, it follows that

$$\begin{aligned} |T'(R(w))|^2 |R'(w)|^2 &\leq \frac{|T'(R(w))|^2(1-|R(w)|^2)^2}{(1-|w|^2)^2} \\ &\leq \frac{(1-|T(R(w))|^2)^2}{(1-|w|^2)^2} = 1 \end{aligned}$$

for every $w \in \mathbb{D}$. Hence we conclude by [31, Theorem 2.7] that

$$\int_{\mathbb{D}} |A(T(z))||T'(z)|^2\, dm(z) \leq \int_{\mathbb{D}\setminus\overline{D}_\phi} |A(w)|\, dm(w) < \infty.$$

Finally, by making use of [17, Theorem 1, p. 80], we get

$$M\big(r, (A \circ T)(T')^2\big) \leq \frac{o(1)}{(1-r^2)^2}, \quad r \to 1^-.$$

Since (59) is again clearly valid, we conclude that (60) holds. Just as above, this shows that (54) is non-oscillatory in $\mathbb{D}$, and, a fortiori, (1) is non-oscillatory in $\mathbb{D} \setminus \overline{D}_\phi$. This completes the proof of Theorem 9.

We recall that (1) is oscillatory for $A(z) = h_\varepsilon(z)$ with $\varepsilon > 1/4$. The proof in [13] rests on the fact that $h_\varepsilon(z)$ is typically real [31, p. 13]. It seems quite likely that (1) is oscillatory for $A \in S$ with $\alpha(A) > 1/4$. In such a case, $A(z)$ is not necessarily typically real. This is left as an open problem.

Suppose that $A \in S$ with $[0, 1)$ being the radius of maximal growth. In light of Example 2 and of Theorem 9, it seems plausible that all zeros of a fixed solution of (1) lie in some Stolz angle with vertex at $z = 1$, with at most finitely many exceptions. Further it seems unlikely that any fixed Stolz angle at $z = 1$ would contain all zeros of all solutions of (1). However, a circle lying in $\mathbb{D}$ and being internally tangent to $\partial\mathbb{D}$ at $z = 1$ could be sufficient. The associated conformal mapping is given in the following example.

Example 4 Let $\mathbb{D}_1 \subset \mathbb{D}$ be any open disc internally tangent to $\partial\mathbb{D}$ at $z = 1$. We wish to construct a conformal mapping T from $\mathbb{D}$ onto $\mathbb{D} \setminus \overline{\mathbb{D}}_1$. First we map $\mathbb{D}$ onto the upper half-plane by means of $w_1 = (1+z)i/(1-z)$, which is then mapped onto the infinite strip $S = \{z \in \mathbb{C} : 0 < \Im(z) < \pi\}$ by $w_2 = \log z$. Second we use the rotation $w_3 = -iz$, a magnification $w_4 = \beta z$ for some suitable $\beta > 0$, and finally the linear fractional transformation $w_5 = (z-1)/(z+1)$. The composition $T = w_5 \circ w_4 \circ w_3 \circ w_2 \circ w_1$ is the conformal mapping we are looking for. After simplification,

$$T(z) = \frac{-1 - i\beta \log(\frac{1+z}{1-z} i)}{1 - i\beta \log(\frac{1+z}{1-z} i)}.$$

The construction already guarantees that $T \in \mathcal{H}(\mathbb{D})$. This can also be verified independently by observing that $1/i\beta = -i/\beta \notin \overline{S}$, and hence $\log(\frac{1+z}{1-z} i)$ cannot assume this value or even approach it asymptotically. By using standard mathematical software, we have

$$S_T(z) = \frac{2}{(1-z^2)^2}.$$

This is in line with (57) as $\partial\mathbb{D}_1$ meets $\partial\mathbb{D}$ at a zero angle. Note that T is univalent by [59, Theorem I].

Example 5 Let $\mathbb{D}_1$ and T be as in Example 4, and set $A(z) = g(z)/(1-z)^2$, where $g \in H^\infty$. This covers the special cases $A(z) = h_\varepsilon(z)$. Then

$$A\big(T(z)\big)T'(z)^2 = \frac{g(T(z))T'(z)^2}{(1-T(z))^2} = \frac{-4bg(T(z))}{(1-z^2)^2(1 - i\beta \log(\frac{1+z}{1-z} i))^2}.$$

It follows that this function is bounded in $\mathbb{D}$ with the exception of some neighborhoods of the points $z = \pm 1$. At the neighborhoods of the points $z = \pm 1$ (and hence in all of $\mathbb{D}$), we have

$$\big|A\big(T(z)\big)T'(z)^2\big| \le \frac{O(1)}{(1-|z|^2)^2 \log^2 \frac{e}{1-|z|}}, \quad |z| \to 1^-.$$

If now $B(z)$ is the mapping given by (55), we have shown that

$$\big|B(z)\big| \le \frac{1 + \beta(|z|)}{(1-|z|^2)^2}, \quad |z| \to 1^-,$$

where

$$\limsup_{r \to 1^-} \frac{\beta(r)}{\log^2 \frac{e}{1-r}} < \infty.$$

Unfortunately this doesn't seem to be enough for (1) to be non-oscillatory in $\mathbb{D} \setminus \overline{\mathbb{D}}_1$. The reader is invited to compare this situation to the Chuaqui-Stowe conjecture in (41) and to the growth in (42).

The conformal mappings considered in this section are not convex. For new results on the Schwarzian derivative of convex conformal mappings, see [7].

Finally we point out that Theorem 1 is closely related to all results and open problems listed in the present section.

7 Logarithmic Derivatives of Blaschke Products

Using estimates for logarithmic derivatives, both pointwise and integrated, is crucial in the theory of complex differential equations. Some estimates that can be found in the literature (e.g. Theorem 3) are valid for meromorphic functions. In this section we emphasize estimates for Blaschke products because they are directly related to Blaschke-oscillatory equations. Due to an extensive literature in this field, the selection of results is by no means complete.

Estimates for logarithmic derivatives are typically associated with exceptional sets of some kind. Two basic types for exceptional sets on the interval $[0, 1)$ are a set of finite logarithmic measure and a set of small upper density. We say that $E \subset [0, 1)$ is of finite logarithmic measure, provided that

$$\int_E d\log\frac{1}{1-r} = \int_E \frac{dr}{1-r} < \infty. \tag{61}$$

Exceptional sets of small upper density are treated in Sect. 4 above. Note that if E satisfies (61), then $m(E \cap [r, 1)) = o(1-r)$, and hence $\overline{d}(E) = 0$. This observation makes Lemma 2 applicable for sets of finite logarithmic measure.

7.1 Pointwise Estimates

We will begin with a pointwise estimate that follows by combining the methods in [20, 37]. The details are omitted.

Theorem 10 ([20, 37]) *Let $B(z)$ be a Blaschke product with zeros $z_n \neq 0$ such that* (12) *holds, and let $k \in \mathbb{N}$. Suppose that $\omega : [1, \infty) \to [1, \infty)$ is a continuous and increasing function such that*

$$\int_1^\infty \frac{x}{\omega(x)}\,dx < \infty. \tag{62}$$

Then there exists a set $E \subset [0, 1)$ which satisfies (61) *such that for all $z \in \mathbb{D}$ satisfying $|z| \notin E$, we have*

$$\left|\frac{B^{(k)}(z)}{B(z)}\right| = O\left(\left(\frac{1}{(1-|z|)h(1-|z|)}\right)^k \omega\left(\log\frac{e}{1-|z|}\right)^k\right). \tag{63}$$

If instead of (12) we assume that (11) holds for some $\alpha \in (0, 1]$, then (63) reduces to

$$\left|\frac{B^{(k)}(z)}{B(z)}\right| = O\left(\left(\frac{1}{1-|z|}\right)^{(1+\alpha)k} \omega\left(\log\frac{e}{1-|z|}\right)^k\right). \tag{64}$$

An estimate of the type (64) has been obtained in [36], but in a slightly weaker form. The sharpness of Theorem 10 is discussed in the next example.

Example 6 Let $0<\alpha<1$, and let $B(z)$ be the Blaschke product with zeros at the points $z_n=1-(\frac{1}{n})^{1/\alpha}$. If ω is as in Theorem 10, then the change of variable $x=1+\frac{1}{\alpha}\log t$ in (62) shows that

$$\int_1^\infty \frac{1+\frac{1}{\alpha}\log t}{t\cdot\omega(1+\frac{1}{\alpha}\log t)}\,dt<\infty.$$

Hence

$$\sum_{n=1}^\infty (1-|z_n|)^\alpha \frac{\log\frac{e}{1-|z_n|}}{\omega(\log\frac{e}{1-|z_n|})}=\sum_{n=1}^\infty \frac{1+\frac{1}{\alpha}\log n}{n\cdot\omega(1+\frac{1}{\alpha}\log n)}<\infty,$$

so that we may choose $h(t)=t^\alpha(\log\frac{e}{t})\omega(\log\frac{e}{t})^{-1}$ for (12) to hold. The estimate in (63) now gives

$$\left|\frac{B'(z)}{B(z)}\right|=O\left(\left(\frac{1}{1-|z|}\right)^{1+\alpha}\frac{\omega(\log\frac{e}{1-|z|})^2}{\log\frac{e}{1-|z|}}\right),\qquad |z|\notin E.$$

In comparison, by [36, Proposition 6.1] there exists a set $F\subset[0,1)$ satisfying $\int_F\frac{dr}{1-r}=\infty$, and a constant $C=C(\alpha)>0$, such that

$$\left|\frac{B'(x)}{B(x)}\right|\geq\frac{C}{(1-x)^{1+\alpha}}\log\frac{1}{1-x},\qquad x\in F\setminus E.$$

Since $\omega(x)$ can be chosen such that its asymptotic growth is not too far from that of x^2, we see that the estimate (63) is somewhat sharp.

For interpolating Blaschke products the function ω can be dropped out.

Theorem 11 ([20, 37]) *Let $B(z)$ be a Blaschke product with a uniformly separated sequence of zeros $z_n\neq 0$ such that* (12) *holds, and let $k\in\mathbb{N}$. Then there exists a set $E\subset[0,1)$ which satisfies* (61) *such that*

$$\left|\frac{B^{(k)}(z)}{B(z)}\right|=O\left(\left(\frac{1}{(1-|z|)h(1-|z|)}\right)^k\right),\qquad |z|\notin E. \tag{65}$$

Theorem 11 generalizes to uniformly q-separated sequences as follows.

Theorem 12 ([20, 27]) *Let $B(z)$ be a Blaschke product with a uniformly q-separated sequence of zeros $z_n\neq 0$ such that* (12) *holds, and let $k\in\mathbb{N}$. Then there exists a set $E\subset[0,1)$ which satisfies* (61) *such that*

$$\left|\frac{B^{(k)}(z)}{B(z)}\right|=O\left(\left(\frac{1}{(1-|z|)h(1-|z|)}\right)^k\left(1+q\log\frac{e}{1-|z|}\right)^k\right),\qquad |z|\notin E. \tag{66}$$

A simple geometric series argument shows that an exponential sequence $\{z_n\}$ satisfies (11) for any $\alpha > 0$. If $B(z)$ denotes the corresponding Blaschke product, it is natural to ask whether the function $h(t)$ in (65) could be dropped out. This was stated as an open problem in [20]. Unfortunately, Proposition 9 in [37] shows that this is not the case. The next result, however, shows that $h(t)$ in (65) can almost be ignored.

Theorem 13 ([37, Theorem 8 and Sect. 6]) *Let $B(z)$ be a Blaschke product whose zeros $z_n \neq 0$ form an exponential sequence, and let $k \in \mathbb{N}$. Suppose that $\phi : [1, \infty) \to [1, \infty)$ is a continuous and increasing function satisfying* (18). *Then there exists a set $E \subset [0, 1)$ which satisfies* (61) *such that*

$$\left|\frac{B^{(k)}(z)}{B(z)}\right| = O\left(\left(\frac{1}{1-|z|}\right)^k \phi\left(\log\frac{e}{1-|z|}\right)^k\right), \quad |z| \notin E. \tag{67}$$

If $\{z_n\}$ is a strongly exponential sequence, then the function ϕ in (67) *can be replaced with the identity mapping.*

7.2 Integrated Estimates

So far the estimates for logarithmic derivatives of Blaschke products that we have considered have been pointwise. The first integrated estimate known to the author is the following one due to Linden.

Theorem 14 ([56, Theorem 1]) *Let $k \in \mathbb{N}$, and let $B(z)$ be a Blaschke product with zeros $\{z_n\}$ such that* (11) *holds for some $\alpha \in (0, \frac{1}{k+1})$. Then, if $m = \frac{1-\alpha}{k}$, there is a constant $C = C(\alpha, k) > 0$ such that*

$$\int_0^{2\pi}\left|\frac{B^{(k)}(re^{i\theta})}{B(re^{i\theta})}\right|^m d\theta \leq CS, \quad \frac{1}{2} < r < 1.$$

In particular, $B^{(k)} \in H^p$ for each $p \in (0, m]$.

Further integrated estimates as well as areally integrated estimates can be found in [58]. We take this opportunity to introduce the following new result that will be applied to Blaschke-oscillatory equations (1) later on.

Theorem 15 *Let $k \in \mathbb{N}$, and let $B(z)$ be an interpolating Blaschke product with zeros $\{z_n\}$ satisfying* (11) *for $\alpha \in (0, 1)$. Then*

$$\int_{\mathbb{D}}\left|\frac{B^{(k)}(z)}{B(z)}\right|^{\frac{1}{k}} dm(z) < \infty.$$

Let $\delta > 0$ be the interpolation constant defined in (15). Let $K_n = K_n(z_n, \delta/2)$ for $n \in \mathbb{N}$ denote a pseudo-hyperbolic disc with center z_n and radius $\delta/2$, and define $K = \bigcup_n K_n$. Then [62, Lemma 3.4] implies

$$\frac{1}{|B(z)|} \leq \left(\frac{2}{\delta}\right)^2, \quad z \in \mathbb{D} \setminus K,$$

which gives us

$$\int_{\mathbb{D}\setminus K} \left|\frac{B^{(k)}(z)}{B(z)}\right|^{\frac{1}{k}} dm(z) \leq \left(\frac{2}{\delta}\right)^{\frac{2}{k}} \int_{\mathbb{D}} \left|B^{(k)}(z)\right|^{\frac{1}{k}} dm(z).$$

In the case $k = 1$ [26, Theorem 1.1] yields immediately

$$\int_0^{2\pi} \left|B'\left(re^{i\theta}\right)\right| d\theta = o\left(\left(\frac{1}{1-r}\right)^{\alpha}\right).$$

If $k \geq 2$, then we use Hölder's inequality and [26, Theorem 1.1] to obtain

$$\int_0^{2\pi} \left|B^{(k)}\left(re^{i\theta}\right)\right|^{\frac{1}{k}} d\theta \leq (2\pi)^{\frac{k-1}{k}} \left(\int_0^{2\pi} \left|B^{(k)}\left(re^{i\theta}\right)\right| d\theta\right)^{\frac{1}{k}} = o\left(\left(\frac{1}{1-r}\right)^{\frac{k+\alpha-1}{k}}\right),$$

where $\frac{k+\alpha-1}{k} < 1$ by the assumption $\alpha \in (0,1)$. Hence, for every $k \in \mathbb{N}$, we have

$$\int_{\mathbb{D}\setminus K} \left|\frac{B^{(k)}(z)}{B(z)}\right|^{\frac{1}{k}} dm(z) < \infty. \tag{68}$$

Integration over the discs K_n will be considered in two cases. We need the Green's function $g(z,a) = -\log \rho(z,a)$ with logarithmic singularity at $a \in \mathbb{D}$.

(1) Suppose $k = 1$. For $n \in \mathbb{N}$, we use Hölder's inequality to conclude that

$$\begin{aligned}
\int_{K_n} \left|\frac{B'(z)}{B(z)}\right| dm(z) &= \int_{K_n} \left|\frac{B'(z)}{B(z)}\right| \frac{g(z,z_n)}{g(z,z_n)}\, dm(z) \\
&\leq \left(\int_{K_n} \left|B'(z)\right|^2 g^2(z,z_n)\, dm(z)\right)^{\frac{1}{2}} \cdot \left(\int_{K_n} \frac{dm(z)}{|B(z)|^2 g^2(z,z_n)}\right)^{\frac{1}{2}} \\
&= I_1(1,n) \cdot I_2(1,n).
\end{aligned}$$

It is well-known that $B \in \mathcal{B}$, and hence, by [1, p. 43], there exists a constant $C(1) > 0$, depending only on $B(z)$, such that

$$I_1(1,n) \leq \sup_{a\in\mathbb{D}} \left(\int_{\mathbb{D}} \left|B'(z)\right|^2 g^2(z,a)\, dm(z)\right)^{\frac{1}{2}} \leq C(1).$$

A careful inspection of the proof of [62, Theorem A] shows that

$$I_2(1,n) \leq \frac{8(1-|z_n|^2)}{\delta(2-\delta)^2}\left(\int_0^{2\pi}\int_0^{\delta/2}\frac{dr\,d\theta}{r\log^2 r}\right)^{\frac{1}{2}} = \frac{8}{\delta(2-\delta)^2}\left(\frac{2\pi}{\log\frac{2}{\delta}}\right)^{\frac{1}{2}}(1-|z_n|^2).$$

(2) Suppose $k \geq 2$. For $n \in \mathbb{N}$, we use Hölder's inequality to conclude that

$$\begin{aligned}
\int_{K_n}\left|\frac{B^{(k)}(z)}{B(z)}\right|^{\frac{1}{k}} dm(z) &= \int_{K_n}\left|\frac{B^{(k)}(z)}{B(z)}\right|^{\frac{1}{k}}\left(\frac{(1-|z|)^{k-1}g(z,z_n)}{(1-|z|)^{k-1}g(z,z_n)}\right)^{\frac{1}{k}} dm(z)\\
&\leq \left(\int_{K_n}\left|B^{(k)}(z)\right|^2(1-|z|)^{2(k-1)}g^2(z,z_n)\,dm(z)\right)^{\frac{1}{2k}}\\
&\quad\cdot\left(\int_{K_n}\frac{dm(z)}{(|B(z)|(1-|z|)g(z,z_n))^{\frac{2}{2k-1}}}\right)^{\frac{2k-1}{2k}}\\
&= I_1(k,n)\cdot I_2(k,n).
\end{aligned}$$

Since $B \in \mathcal{B}$, we obtain by [1, Theorem 1] that there exists a constant $C(k) > 0$, depending only on $B(z)$ and k, such that

$$I_1(k,n) \leq \sup_{a\in\mathbb{D}}\left(\int_{\mathbb{D}}\left|B^{(k)}(z)\right|^2(1-|z|)^{2(k-1)}g^2(z,a)\,dm(z)\right)^{\frac{1}{2k}} \leq C(k).$$

By [17, Lemma 3, p. 41] there exists a constant $C(\delta) > 0$ such that

$$I_2(k,n) \leq C(\delta)^{\frac{1}{k}}\left(1-|z_n|^2\right)^{-\frac{1}{k}}\left(\int_{K_n}\frac{dm(z)}{(|B(z)|g(z,z_n))^{\frac{2}{2k-1}}}\right)^{\frac{2k-1}{2k}}.$$

Let $|K_n|$ denote the non-normalized area of the pseudo-hyperbolic disc K_n. Using Hölder's inequality, with conjugate indices $p = 2k-1$ and $q = \frac{2k-1}{2k-2}$, we find that

$$\begin{aligned}
I_2(k,n) &\leq C(\delta)^{\frac{1}{k}}\left(1-|z_n|^2\right)^{-\frac{1}{k}}|K_n|^{\frac{k-1}{k}}\left(\int_{K_n}\frac{dm(z)}{|B(z)|^2g^2(z,z_n)}\right)^{\frac{1}{2k}}\\
&= C(\delta)^{\frac{1}{k}}\left(1-|z_n|^2\right)^{-\frac{1}{k}}|K_n|^{\frac{k-1}{k}}I_2(1,n)^{\frac{1}{k}}\\
&\leq \left(\frac{8C(\delta)}{\delta(2-\delta)^2}\right)^{\frac{1}{k}}\left(\frac{2\pi}{\log\frac{2}{\delta}}\right)^{\frac{1}{2k}}|K_n|^{\frac{k-1}{k}}.
\end{aligned}$$

By [17, p. 40] it follows that

$$|K_n| = \frac{4\delta^2\pi}{(4-\delta^2|z_n|^2)^2}\left(1-|z_n|^2\right)^2 \leq \frac{4\delta^2\pi}{9}\left(1-|z_n|^2\right)^2,$$

and, a fortiori,

$$I_2(k,n) \leq \left(\frac{8C(\delta)}{\delta(2-\delta)^2}\right)^{\frac{1}{k}} \left(\frac{2\pi}{\log\frac{2}{\delta}}\right)^{\frac{1}{2k}} \left(\frac{4\delta^2\pi}{9}\right)^{\frac{k-1}{k}} \left(1-|z_n|^2\right).$$

In both cases (1) and (2) there exists a constant $C_0(\delta,k)>0$ independent on n such that

$$\int_{K_n} \left|\frac{B^{(k)}(z)}{B(z)}\right|^{\frac{1}{k}} dm(z) \leq C_0(\delta,k)\left(1-|z_n|^2\right),$$

and hence

$$\int_{K} \left|\frac{B^{(k)}(z)}{B(z)}\right|^{\frac{1}{k}} dm(z) \leq C_0(\delta,k)\sum_{n=1}^{\infty}\left(1-|z_n|^2\right) < \infty.$$

By combining this with (68), we complete the proof of Theorem 15.

By [64, Theorem 1] there exists an interpolating Blaschke product $B(z)$ (with zeros $\{z_n\}$ satisfying (11) for $\alpha=1$) such that $B' \notin B^1$. Since

$$\int_{\mathbb{D}} \left|\frac{B'(z)}{B(z)}\right| dm(z) \geq \int_{\mathbb{D}} \left|B'(z)\right| dm(z) = \infty,$$

the requirement $\alpha<1$ in Theorem 15 cannot be dropped in the case $k=1$. There is, however, some indication that the assumption (11) for $\alpha<1$ could be weakened to

$$\sum_{n=1}^{\infty}\left(1-|z_n|\right)\log^2\left(\frac{e}{1-|z_n|}\right) < \infty,$$

and the assertion in Theorem 15 would still hold. This requires updating the proof of [26, Theorem 1.1] accordingly. The details are left as an open problem.

8 One Infinite Prescribed Zero Sequence

As already observed in [34], for any given Blaschke sequence $\{z_n\}$ of distinct points in $\mathbb{D}$, the problem of one prescribed zero sequence has uncountably many solutions. We give the following alternative proof for this. Let $g \in \mathcal{H}(\mathbb{D})$ be a function satisfying the interpolation problem (9), and let $G \in \mathcal{H}(\mathbb{D})$ be arbitrary. Then the function $g_G(z) = g(z) + \int_0^z B(\zeta)G(\zeta)\,d\zeta$ satisfies

$$g_G'(z_n) = g'(z_n) = \sigma_n, \quad n \in \mathbb{N}, \tag{69}$$

and gives raise to a coefficient $A(z) = A_G(z)$ of (1) that depends on G, see (10). Moreover, given two distinct functions $G_1, G_2 \in \mathcal{H}(\mathbb{D})$, the associated coefficient functions $A_{G_1}(z)$ and $A_{G_2}(z)$ are also distinct. To see this, assume on the contrary to this claim that $A_{G_1} \equiv A_{G_2}$, and denote $f_j = Be^{g_{G_j}}$ for $j = 1, 2$. Then f_1, f_2 are solutions of (1) with the same zeros. Hence $f_1 \equiv f_2$, so that $g_{G_1} \equiv g_{G_2} + 2\pi ni$. Differentiating both sides gives us $G_1 \equiv G_2$, which is a contradiction.

We note that the growth of $A_G(z)$ is influenced by that of G. For now, we settle for finding just one coefficient $A(z)$ for (1) with the desired properties. Later on in Sect. 12 we are forced to make full use of $g_G(z)$ for $G \not\equiv 0$.

The problem of one prescribed zero sequence was solved in [37] with a pointwise growth estimate for $A(z)$. This result has recently been improved as follows.

Theorem 16 ([27, Corollary 2.3]) *If $\{z_n\}$ is a uniformly separated (or exponential) sequence of non-zero points in $\mathbb{D}$, then there exists a function $A \in H_2^\infty$ satisfying*

$$\limsup_{|z|\to 1^-} (1-|z|^2)^2 |A(z)| > 1 \tag{70}$$

such that (1) possesses a solution whose zero sequence is $\{z_n\}$.

We note that (70) follows directly by (38) since the prescribed zero sequence is infinite. Theorem 16 is sharp in the sense that the minimal growth rate in (70) has the same magnitude as the maximal growth rate allowed in the H_2^∞ space.

If the prescribed zero sequence consists of a union of two exponential sequences $\{a_n\}$ and $\{b_n\}$ that are pairwise close to one another in the sense that

$$0 < |a_n - b_n| < \exp(-2^n), \quad n \in \mathbb{N},$$

then the corresponding coefficient function $A(z)$ cannot belong to $\mathcal{A}^{-\infty}$ by [27, Theorem 2.1]. Recalling that all solutions of (1) are of finite order of growth, provided that $A \in \mathcal{A}^{-\infty}$ [67, Lemma 2], this observation justifies the need for the concept of q-separation stated earlier in Sect. 3. Note in particular that the union of $\{a_n\}$ and $\{b_n\}$ is not q-separated for any $q \geq 0$. We are now motivated to state a generalization of Theorem 16 as follows.

Theorem 17 ([27, Theorem 2.2]) *Let $\{z_n\}$ be a uniformly q-separated sequence of non-zero points in $\mathbb{D}$.*

(a) *Suppose that $\{z_n\}$ satisfies (11) for some $\alpha \in (0, 1]$. Then there exists a function $A \in H^\infty_{2(1+\alpha+2q)}$ satisfying (70) such that (1) possesses a solution whose zero sequence is $\{z_n\}$.*
(b) *Suppose that $\{z_n\}$ is a finite union of uniformly separated (or exponential) sequences in $\mathbb{D}$. Then there exists a function $A \in H^\infty_{2(1+2q)}$ satisfying (70) such that (1) possesses a solution whose zero sequence is $\{z_n\}$.*

The following result can be considered as a solution to a problem of a prescribed sequence of c-points, where $c \in \mathbb{C} \setminus \{0\}$.

Theorem 18 ([27, Theorem 2.4]) *Let $\{z_n\}$ be a Blaschke sequence of non-zero points in $\mathbb{D}$. Then there exists a function $A \in H_2^\infty$ such that for each $c \in \mathbb{C} \setminus \{0\}$, (1) possesses a solution $f_c \in H^\infty$ taking the value c precisely at the points z_n.*

The proof of Theorem 18 is so short and elementary that we repeat it here. Let $B(z)$ be the Blaschke product with zeros $\{z_n\}$, and define

$$A(z) = \frac{2B''(z) + B''(z)B(z) - 2(B'(z))^2}{(B(z)+2)^2}.$$

Using Cauchy's formula and the fact that $|B(z)+2| > 1$ for $z \in \mathbb{D}$, we have $A \in H_2^\infty$. For each fixed $c \in \mathbb{C} \setminus \{0\}$, the function $f_c(z) = 2c(B(z)+2)^{-1}$ is then the desired solution of (1).

So far the growth estimates for the coefficient function $A(z)$ have been pointwise. Some first attempts to solve the problem of one prescribed zero sequence such that the coefficient $A(z)$ belongs to the Bergman space B^α for some $\alpha \in (0, 1/2]$ were given in [34]. It is of particular interest to aim for $A \in B^{\frac{1}{2}}$ under weakest possible assumptions, as then (1) becomes Blaschke-oscillatory by (2). In [34, Theorem 4.2] this is achieved by assuming that the prescribed zero sequence is uniformly separated and satisfies (11) for $\alpha \in (0, 1/3)$. By using Theorem 15 we are able to improve this result for $\alpha \in (0, 1)$.

Theorem 19 *Let $\{z_n\}$ be an interpolating Blaschke sequence of nonzero points in $\mathbb{D}$ satisfying* (11) *for some $\alpha \in (0, 1)$. Then there exists a function $A \in B^{\frac{1}{2}}$ such that* (1) *possesses a solution f with zeros precisely at the points z_n.*

We rely heavily on the proof of [34, Theorem 4.2]. Let $B(z)$ be the Blaschke product with zeros at the points z_n. Then a function $g \in \mathcal{H}(\mathbb{D})$ can be found such that $f = Be^g$ solves (1), where

$$A(z) = -\frac{B''(z)}{B(z)} - 2g'(z)\frac{B'(z)}{B(z)} - g'(z)^2 - g''(z)$$

belongs to $\mathcal{H}(\mathbb{D})$. In particular, $g' \in B^1$ and $g'' \in B^{\frac{1}{2}}$ even if the assumption $\alpha \in (0, 1/3)$ in [34] is weakened to $\alpha \in (0, 1)$. Since

$$|A(z)|^{\frac{1}{2}} \le \left|\frac{B''(z)}{B(z)}\right|^{\frac{1}{2}} + \sqrt{2}|g'(z)|^{\frac{1}{2}}\left|\frac{B'(z)}{B(z)}\right|^{\frac{1}{2}} + |g'(z)| + |g''(z)|^{\frac{1}{2}},$$

we have by Theorem 15 and by Hölder's inequality that $A \in B^{\frac{1}{2}}$. This completes the proof of Theorem 19.

9 Two Infinite Prescribed Zero Sequences

The problem of two prescribed zero sequences is solved in [35] as follows.

Theorem 20 ([35, Theorem 1.2]) *Let $\{a_n\}$ and $\{b_n\}$ be two infinite Blaschke sequences in $\mathbb{D}$ such that the sequence $\{z_n\}$ defined by $z_{2n-1} = a_n$ and $z_{2n} = b_n$ is uniformly separated.*

(a) *If* $\{z_n\}$ *satisfies*

$$\sum_n (1-|z_n|)\log\frac{1}{1-|z_n|}<\infty,$$

then we may construct a function $A\in\mathcal{F}$ *such that* (1) *possesses linearly independent solutions* f_1, f_2 *with zeros precisely at the points* a_n, b_n*, respectively, such that* $E=f_1f_2\in N$.

(b) *If* $\{z_n\}$ *satisfies*

$$\sum_n (1-|z_n|)^{\frac{1}{3}}\log\frac{1}{1-|z_n|}<\infty, \tag{71}$$

then we may construct a function $A\in N$ *such that* (1) *possesses linearly independent solutions* f_1, f_2 *with zeros precisely at the points* a_n, b_n*, respectively, such that* $E=f_1f_2\in N$ *and* $E', E''\in N$.

We note that in [35] a stronger condition of the form (11) for $\alpha\in(0,1/3)$ is assumed instead of (71). However, the proof in [35] shows that (71) suffices. See, in particular, formula (4.8) in [35].

Just like in the case of one prescribed zero sequence, the problem of two prescribed zero sequences has uncountably many solutions. This will be proved at the end of Sect. 9.4. For now, we settle for finding just one coefficient $A(z)$ for (1) with the desired properties.

9.1 New Result Involving Two Sequences

So far the best growth condition for the solution $A(z)$ of the problem of two prescribed zeros seems to be $A\in N$. Note that $g(z)=\exp(\frac{1+z}{1-z})$ is an extremal function in N, and hence functions in N may be of exponential growth. It would be desirable to obtain a growth condition as close as possible to (2) since (2) forces (1) to be Blaschke-oscillatory. A growth condition on $|A(z)|$ as close as possible to the minimal growth in (70) would also be acceptable.

The proof of Theorem 20 in [35] culminates in showing that a Bank-Laine function E satisfies a certain interpolation property, which contributes to the growth of E, and, a fortiori, to the growth of $A(z)$. Recall that $g\in\mathcal{H}(\mathbb{D})$ is a Bank-Laine function if at every zero ζ of g we have either $g'(\zeta)=1$ or $g'(\zeta)=-1$.

We take this opportunity to get some new insight about the growth of $A(z)$ in the case of two prescribed zero sequences. First, instead of using interpolation in the H^p spaces as in [35], we apply interpolation in a certain subspace of the Bloch space $\mathcal{B}$ [63]. This approach forces E in the Korenblum space $\mathcal{A}^{-\infty}$. Second, we use an upgraded Wiman-Valiron theory introduced quite recently by Fenton and Rossi in [19]. This will show that the coefficient $A(z)$ has a growth rate not far from the minimal growth in (70). The new result is stated as follows.

Theorem 21 *Let $\{a_n\}$ and $\{b_n\}$ be two infinite Blaschke sequences in $\mathbb{D}$ such that the sequence $\{z_n\}$ defined by $z_{2n-1} = a_n$ and $z_{2n} = b_n$ is separated and satisfies*

$$\sup_{z\in\mathbb{D}} \sum_n \left(1-\rho(z,z_n)^2\right)^{\alpha} < \infty \tag{72}$$

for some $\alpha \in (0,1)$. Then we may construct a function $A \in G_2$ satisfying (70) *such that* (1) *possesses linearly independent solutions f_1, f_2 with zeros precisely at the points a_n, b_n, respectively, such that $E = f_1 f_2 \in \mathcal{A}^{-\infty} \cap N$.*

We recall [10] that the growth space G_p for $p > 0$ consists of functions $g \in \mathcal{H}(\mathbb{D})$ such that

$$\limsup_{r\to 1^-} \frac{\log^+ M(r,g)}{-\log(1-r)} = p.$$

The questions whether the function $A(z)$ in Theorem 21 satisfies (2) or belongs to H_2^∞ are left as open problems. We make three remarks on the assumptions.

(1) By taking $z = 0$ in (72), we have (11).
(2) By Lemma 6 below, we conclude that $\{z_n\}$ is uniformly separated.
(3) Let $B(z)$ be a Blaschke product with zeros $\{z_n\}$. Then $B' \in Q_\alpha$ if and only if $\{z_n\}$ satisfies (72), see [18] and the definition (73) below. Hence there might be a deeper connection between the condition (72) and Blaschke-oscillatory equations than what is described below.

Lemma 6 ([47, Lemmas 4–5]) *A sequence $\{z_n\}$ in $\mathbb{D}$ is uniformly separated if and only if the following two conditions hold*:

(a) *$\{z_n\}$ is separated,*
(b) $\sup_{z\in D} \sum_n (1-\rho(z,z_n)^2) < \infty$.

Moreover, $\{z_n\}$ is a finite union of uniformly separated sequences if and only if the condition (b) *holds.*

The proof of Theorem 21 is conducted in three steps. We build the necessary machinery in Sects. 9.2–9.3, while Sect. 9.4 contains the actual proof. Before entering the proof, we point out that a two sequences analogue of Theorem 18 for a single prescribed value $c \in \mathbb{C} \setminus \{0\}$ is still an open problem.

9.2 Interpolation in the BMOA Space

Recall that the hyperbolic distance between the points $z, w \in \mathbb{D}$ is

$$d(z,w) = \frac{1}{2}\log\frac{1+\rho(z,w)}{1-\rho(z,w)}.$$

For $p > 0$, the Q_p space [2] consists of functions $g \in \mathcal{H}(\mathbb{D})$ such that

$$\sup_{a\in\mathbb{D}} \int_{\mathbb{D}} |g'(z)|^2 \left(1 - \rho(z,a)^2\right)^p \, dm(z) < \infty. \tag{73}$$

Note in particular that $Q_1 =$ BMOA, the space of analytic functions of bounded mean oscillation, and that $Q_p = \mathcal{B}$ for every $p > 1$ [2]. This shows that BMOA $\subset \mathcal{B}$. In the early literature the space BMOA is defined as the intersection of the Hardy space H^1 and BMO($\partial\mathbb{D}$), so that BMOA $\subset H^1$.

Theorem 22 ([63, p. 2129]) *Let $\{z_n\}$ be a separated sequence in $\mathbb{D}$ satisfying* (72) *for some $\alpha \in (0, 1)$. Let $\{w_n\}$ be any sequence in $\mathbb{C}$ satisfying the Lipschitz condition*

$$|w_j - w_k| = O\big(d(z_j, z_k)\big). \tag{74}$$

Then there exists a function $f \in$ BMOA such that $f(z_n) = w_n$ for all $n \in \mathbb{N}$.

We make two remarks. First, for any $\alpha \in (0, 1)$, a concrete example of an interpolating sequence for $\mathcal{B}$ (in fact for BMOA) satisfying (72) is constructed in [22, Lemma 5.2]. Second, let $\{z_n\}$ be an exponential sequence. Since $\{z_n\}$ is uniformly separated, it satisfies (72) for $\alpha = 1$. In fact, it is shown in [77, Lemma 3] that $\{z_n\}$ satisfies (72) for any $\alpha \in (0, \infty)$. In particular, exponential sequences are interpolating for BMOA.

9.3 Wiman-Valiron Theory in $\mathbb{D}$

Let $f(z) = \sum_{n=1}^{\infty} a_n z^n$ be analytic in $\mathbb{D}$. The maximum term is defined as

$$\mu(r) = \mu(r, f) = \max_{n\geq 0} |a_n| r^n,$$

while the central index $\nu(r) = \nu(r, f)$ is the largest integer n for which the maximum is attained. Since a Taylor series converges uniformly in compact subsets of $\mathbb{D}$, the maximum term certainly exists. The M-order of f is given by

$$\sigma_M(f) = \limsup_{r\to 1^-} \frac{\log^+ \log^+ M(r, f)}{-\log(1-r)}.$$

One of the advantages of Wiman-Valiron theory is a representation of a logarithmic derivative of an analytic function in the form of an algebraic expression. The recent improvements due to Fenton and Rossi in [19] deal with the cases $\sigma_M(f) > 0$ and $\sigma_M(f) = 0$. We will make use of the case $\sigma_M(f) = 0$ only.

Theorem 23 ([19, Theorem 2]) *Let $f \in \mathcal{H}(\mathbb{D})$ with $\sigma_M(f) = 0$. Suppose that γ : $(0, 1) \to \mathbb{R}$ is positive such that $\gamma(t) \to 0$ as $t \to 1^-$. If $\zeta \in \mathbb{D}$ is such that $|f(\zeta)| \geq$*

$\nu(|\zeta|)^{-\gamma(|\zeta|)}M(|\zeta|, f)$, then, for every $k \in \mathbb{N}$ and $\varepsilon > 0$, we have

$$\frac{f^{(k)}(\zeta)}{f(\zeta)} = O\left(\left(\frac{1}{1-|\zeta|}\right)^{k+\varepsilon}\right)$$

as $|\zeta| \to 1^-$ outside of a set of zero density.

In proving Theorem 21 we will make use of Theorem 23 in the case $f \in \mathcal{A}^{-\infty}$. In addition, we will need the estimate

$$\nu(r) = O\left(\frac{1}{1-r}\log\frac{1}{1-r}\right), \tag{75}$$

which is valid for the central index $\nu(r)$ of a given function $f \in \mathcal{A}^{-\infty}$. To prove (75), we note that, for any $b > 2$, there exists a constant $r_0(b) \in (0,1)$ such that

$$(1-r)\nu(r) \le b \log \mu\big(r + (1-r)/b\big), \quad r_0(b) < r < 1,$$

see formula (1.5.6) in [49]. Combining this with the trivial estimate $\mu(r) \le M(r, f)$ and the fact that $f \in \mathcal{A}^{-\infty}$ proves the claim.

9.4 Proof of Theorem 21

Let $B(z)$ be the Blaschke product with zeros z_n. Since the points z_n are simple, we have $B'(z_n) \neq 0$. Define a sequence $\{w_n\}$ of complex points by setting

$$w_n = \log\left(\frac{(-1)^n}{B'(z_n)}\right), \quad n \in \mathbb{N},$$

where the principal branch for the logarithm has been chosen for every n. Now

$$w_j - w_k = \log\left|\frac{B'(z_k)}{B'(z_j)}\right| + i\big(\arg\big((-1)^k B'(z_k)\big) - \arg\big((-1)^j B'(z_j)\big)\big),$$

and hence

$$|w_j - w_k| \le \left|\log\left|\frac{B'(z_k)}{B'(z_j)}\right|\right| + 2\pi.$$

Suppose first that $|z_k| > |z_j|$. Since $\{z_n\}$ is uniformly separated, it follows by the proof of [35, Lemma 4.1] that

$$|w_j - w_k| \le \log\frac{1-|z_j|^2}{1-|z_k|^2} + 2\pi + \log\frac{1}{\delta}, \tag{76}$$

where $\delta > 0$ is the constant from (15). Since

$$\left(1-|z_j|^2\right)^2 \le 4\left(1-|z_j|\right)^2 \le 4\left(1-|z_k||z_j|\right)^2 \le 4|1-\overline{z_k}z_j|^2,$$

we have

$$\begin{aligned}\log\frac{1-|z_j|^2}{1-|z_k|^2} &\le \log\frac{4|1-\overline{z_k}z_j|^2}{(1-|z_k|^2)(1-|z_j|^2)}\\ &= \log\frac{1}{1-\rho(z_j,z_k)^2}+\log 4\\ &\le \log\frac{1+\rho(z_j,z_k)}{1-\rho(z_j,z_k)}+\log 4.\end{aligned}$$

Combining this with (76), we conclude that (74) holds in the case $|z_k| > |z_j|$. If $|z_k| < |z_j|$, then (76) is replaced with

$$|w_j - w_k| \le \log\frac{1-|z_k|^2}{1-|z_j|^2} + 2\pi + \log\frac{1}{\delta},$$

and a calculation similar to the one above shows that (74) again holds. Finally, if $|z_k| = |z_j|$, then $|w_j - w_k| \le 2\pi + \log\frac{1}{\delta} = O(d(z_j, z_k))$. It now follows by Theorem 22 that there exists a function $g \in \mathrm{BMOA}$ such that $g(z_n) = w_n$.

Define $E = Be^g$, so that $\{z_n\}$ is the zero sequence of E. Since $E' = B'e^g + Bg'e^g$, we have $E'(z_n) = (-1)^n$ by the interpolation property of g, and hence E is a Bank-Laine function. In particular, $E'(a_n) = -1$ and $E'(b_n) = 1$. Define $A(z)$ by the equation

$$-4A(z)E^2 = 1 - \left(E'\right)^2 + 2EE''.$$

Then it is known (see [35, Sect. 5] and the references therein) that $A \in \mathcal{H}(\mathbb{D})$, and that E is a product of two linearly independent solutions f_1, f_2 of (1) having zeros precisely at the points a_n, b_n, respectively.

We are down to estimate the growth of $A(z)$ and of E. Since $B \in H^\infty$ and BMOA $\subset \mathcal{B}$, it follows by (32) that $E \in \mathcal{A}^{-\infty}$. Since BMOA $\subset H^1$, we have $e^g \in N$, and hence $E \in N$. Finally, let $\varepsilon > 0$ and write $A(z)$ in the form

$$A(z) = -\frac{1}{4}\left(\frac{1}{E^2} - \left(\frac{E'}{E}\right)^2 + 2\frac{E''}{E}\right).$$

Let $\nu(r)$ be the central index of E. Using Theorem 23 on E, we conclude that

$$\left|A(z)\right| = \frac{\nu(|z|)^{2\gamma(|z|)}}{4M(|z|,E)^2} + O\left(\left(\frac{1}{1-|z|}\right)^{2+\varepsilon}\right),$$

as $|z| \to 1^-$ outside of a set of zero density. By (75) we have

$$M(r,A) = O\left(\left(\frac{1}{1-r}\right)^{2+\varepsilon}\right),$$

as $r \to 1^-$ outside of a set of zero density. The exceptional set can be dealt with Lemma 2. Since $\varepsilon > 0$ is arbitrary, and since $A(z)$ must satisfy (70), this completes the proof of Theorem 21.

Remark 5 The problem of two prescribed zero sequences has uncountably many solutions. This was observed in [42, Sect. 6] in the case of complex plane. We may cut corners in this reasoning. Let $g \in \mathrm{BMOA}$ and $B(z)$ be as in the proof above, and let $G \in \mathcal{H}(\mathbb{D})$ be arbitrary. Then the function $g_G = g + BG$ satisfies $g_G(z_n) = g(z_n) = w_n$ for all n. This gives raise to $E = E_G$, and finally to a coefficient $A(z) = A_G(z)$ of (1). Moreover, given two distinct functions $G_1, G_2 \in \mathcal{H}(\mathbb{D})$, the associated coefficient functions $A_{G_1}(z)$ and $A_{G_2}(z)$ are also distinct. Suppose on the contrary to this claim that $A_{G_1} \equiv A_{G_2}$. This gives raise to two fundamental systems of solutions for (1), one depending on G_1 and the other one on G_2. Since the solutions in these systems share the same zeros by the assumption $A_{G_1} \equiv A_{G_2}$, they must be pairwise equal, and hence $E_{G_1} \equiv E_{G_2}$. In particular, $g_{G_1} \equiv g_{G_2} + 2n\pi i$, so that $B(G_1 - G_2) \equiv 2n\pi i$. Since $B(z)$ has zeros, we must have $n = 0$, and hence $G_1 \equiv G_2$, which is a contradiction.

10 Finite Prescribed Zero Sequences

A zero-free function $f = e^g$ with $g \in \mathcal{H}(\mathbb{D})$ is a solution of (1) if and only if $A(z) = -g'' - (g')^2$. Methods for constructing equations (1) with zero-free solution bases $\{f_1, f_2\}$ can be found in [44], see also Sect. 4.3.

Let then $z_1, \ldots, z_N$ be distinct points in $\mathbb{D}$. It is easy to construct a function $A \in \mathcal{H}(\mathbb{D})$ such that (1) possesses a solution f with zeros precisely at the (finitely many) points z_n. To begin with, let $P(z) = (z - z_1) \cdots (z - z_N)$ be a polynomial with zeros at the points z_n, and let $f = Pe^g$ be a candidate solution of (1) for some $g \in \mathcal{H}(\mathbb{D})$. We require that the derivative of g satisfies the interpolation property

$$g'(z_n) = w_n = -\frac{P''(z_n)}{2P'(z_n)}, \quad n = 1, \ldots, N.$$

For example, we may take for g' the Lagrange interpolation polynomial

$$g'(z) = \sum_{n=1}^{N} \frac{P(z)}{(z - z_n)P'(z_n)} w_n.$$

As for the function g we may then take any primitive of g'. Then the function

$$A = \frac{-P'' - 2g'P'}{P} - \left(g'\right)^2 - g''$$

is analytic in $\mathbb{D}$ and in fact belongs to H^∞. Now $f = Pe^g \in H^\infty$ solves (1) and has the prescribed zeros $z_1, \dots, z_N$.

As for the case of two finite prescribed zero sequences, we may proceed as in Sect. 9 with Lagrange interpolation polynomial taking the role of BMOA interpolation. This results in the fact that the Bank-Laine function E in the construction belongs to H^∞. Regarding the growth of $A(z)$, this method does not seem to give anything better than $A \in G_2$.

11 Blaschke-Oscillatory Equations of Arbitrary Order

Let $A_0, \dots, A_{k-1} \in \mathcal{H}(\mathbb{D})$, where $k \in \mathbb{N}$. Then the solutions of

$$f^{(k)} + A_{k-1}(z)f^{(k-1)} + \cdots + A_1(z)f' + A_0(z)f = 0 \tag{77}$$

are analytic in $\mathbb{D}$. The (possible) zeros of solutions are of multiplicity at most $k-1$. For if f has a zero of multiplicity $p \geq k$ at z_0, then dividing (77) by f and noticing that (1) $f^{(k)}/f$ has a pole of multiplicity k at z_0, (2) all other terms of the form $A_j(z)f^{(j)}/f$ in (77) have a pole of multiplicity at most $j \leq k-1$ at z_0, we arrive at a contradiction.

Following the second order case above, we call (77) Blaschke-oscillatory if the zero sequence of any nontrivial solution of (77) satisfies the Blaschke condition. Problems related to prescribed zeros in the general case (77) seem quite difficult. There seems to be some hope to obtain such results in the case $f^{(k)} + A(z)f = 0$, but the author is unaware of any existing results.

We then proceed to finding necessary and sufficient conditions for (77) to be Blaschke-oscillatory. Based on the general growth estimates in [38], it is proved in [40] that if $A_0, \dots, A_{k-1} \in \mathcal{H}(\mathbb{D})$ are such that

$$\int_{\mathbb{D}} |A_j(z)|^{\frac{1}{k-j}}\, dm(z) < \infty, \quad j = 0, \dots, k-1, \tag{78}$$

then all solutions of (77) belong to N. Slightly more delicate conditions on the coefficients $A_0, \dots, A_{k-1} \in \mathcal{H}(\mathbb{D})$ given in [69, Theorem 1.7] imply that the solutions of (77) belong to the Hardy space H^p. In both cases, (77) is Blaschke-oscillatory. Conversely, if (77) is assumed to be Blaschke-oscillatory, then it would be desirable to get a growth condition for the coefficients of (77) as close as possible to (78). In particular, we aim to generalize Theorem 2, but first we make two remarks.

(1) When studying the oscillatory behavior of the solutions of (77), we may suppose that $A_{k-1}(z) \equiv 0$. For if ϕ denotes a primitive function of $A_{k-1}(z)$, then the standard substitution $g = fe^{-\frac{1}{k}\phi}$ has no effect on the zeros, and it transforms (77) to an equation where the coefficient function corresponding to the $(k-1)$th derivative vanishes.

(2) In the proof of Theorem 2, the converse part of Pommerenke's result, the crucial step is to represent the sole coefficient function $A(z)$ in terms of a ratio of two

linearly independent solutions of (1). An analogous maneuver can be done for the coefficients of (77), provided that $A_{k-1}(z) \equiv 0$. This was shown by Kim already in 1969, see [51].

Theorem 24 *Let $A_0, \dots, A_{k-2} \in \mathcal{H}(\mathbb{D})$ be such that*

$$f^{(k)} + A_{k-2}(z)f^{(k-2)} + \cdots + A_1(z)f' + A_0(z)f = 0 \tag{79}$$

is Blaschke-oscillatory. Then

$$\int_{D(0,r)} |A_j(z)|^{\frac{1}{k-j}}\, dm(z) = O\left(\log^2 \frac{e}{1-r}\right), \quad j = 0, \dots, k-2.$$

The basic idea of the proof is the same as that of Theorem 2, relying either on the second fundamental theorem or on the following analogue of Lemma 3.

Lemma 7 *Suppose that $A_0, \dots, A_{k-1} \in \mathcal{H}(\mathbb{D})$, and let $\{f_1, \dots, f_k\}$ be any solution base of (77). Then $f_n/f_m \in N$ for any pair $n, m \in \{1, \dots, k\}$ if and only if (77) is Blaschke-oscillatory.*

Suppose that $f_n/f_m \in N$ for any pair n, m. Since f_n and f_m have no common zeros for $n \neq m$, it follows by the assumption that the zeros of f_j satisfy the Blaschke condition for every $j \in \{1, \dots, k\}$. Let then f be an arbitrary solution of (77). It is clear that f can be written as $f = C_1 f_1 + \cdots + C_k f_k$ for some $C_1, \dots, C_k \in \mathbb{C}$. At least one of the coefficients C_j must be nonzero, for otherwise $f \equiv 0$. The case where precisely one coefficient C_j is nonzero has been treated above. Hence we may suppose that $C_{j_1} C_{j_2} \neq 0$ for some $j_1, j_2 \in \{1, \dots, k\}$. Then the zeros of f are precisely the $-C_{j_1}$-points of the nontrivial meromorphic function

$$F = \sum_{\substack{j=1 \\ j \neq j_1}}^{k} C_j \frac{f_j}{f_{j_1}}.$$

Since $T(r, F) = O(1)$ by the assumption (and by some arithmetic properties of the characteristic function), it follows by [75, Theorem V.7] that the $-C_{j_1}$-points of F is a Blaschke sequence. This shows that (77) is Blaschke-oscillatory.

Conversely, suppose that (77) is Blaschke-oscillatory. Then suppose on the contrary to the assertion that $f_n/f_m \notin N$ for some pair n, m. By the proof of Lemma 3 there exists a point $c \in \mathbb{C}$ such that the sequence of c-points of f_n/f_m do not satisfy the Blaschke condition. Hence the solution $f = f_n - cf_m$ of (77) has a non-Blaschke sequence of zeros, which is a contradiction. This completes the proof of Lemma 7.

Lemma 8 *Let $g_1, \dots, g_k$ be linearly independent and non-admissible meromorphic solutions of a linear differential equation*

$$g^{(k)} + B_{k-1}(z)g^{(k-1)} + \cdots + B_1(z)g' + B_0(z)g = 0$$

with coefficients $B_0(z), \ldots, B_{k-1}(z)$ *meromorphic in* $\mathbb{D}$. *Then*

$$\int_{D(0,r)} \left|B_j(z)\right|^{\frac{1}{k-j}} dm(z) = O\left(\log^2 \frac{e}{1-r}\right), \quad j = 0, \ldots, k-1.$$

We give an outline of the proof of Lemma 8. The proof of [43, Theorem E(b)] gives raise to the following conclusion: If g is meromorphic and non-admissible in $\mathbb{D}$, and if k, j are integers satisfying $k > j \geq 0$, then

$$\int_{D(0,r)} \left|\frac{g^{(k)}(z)}{g^{(j)}(z)}\right|^{\frac{1}{k-j}} dm(z) = O\left(\log^2 \frac{e}{1-r}\right). \tag{80}$$

Now the special case $B_j(z) \equiv 0$ for every $j \in \{1, \ldots, k-1\}$ in Lemma 8 is easy to prove. Indeed, it suffices to write $B_0(z) = -g^{(k)}(z)/g(z)$ for any non-admissible meromorphic solution g, and then make use of (80). In the general case one may follow the course of proof of [54, Lemma 7.7], keeping in mind that non-admissible meromorphic functions in $\mathbb{D}$ form a differential field [10, Lemma 5.3]. In particular, the function

$$h_{k+1} = \left(\frac{d}{dz}\frac{W(g_1, \ldots, g_{k+1})}{W(g_1, \ldots, g_k)}\right) \Big/ \frac{W(g_1, \ldots, g_{k+1})}{W(g_1, \ldots, g_k)}$$

on [54, p. 134] is a logarithmic derivative of a non-admissible meromorphic function, and hence satisfies

$$\int_{D(0,r)} \left|h_{k+1}(z)\right| dm(z) = O\left(\log^2 \frac{e}{1-r}\right)$$

by (80). The calculations culminate in (7.15) of [54], where (80) and Hölder's inequality are the key tools. We omit the details.

The next step in proving Theorem 24 is to represent the coefficients of (79) in terms of ratios of linearly independent solutions of (79). Such a representation is given in the case of complex plane in [51]. A unit disc analogue of this result is nothing but a trivial modification of the original result, see [43, Theorem D]. Note in particular that the coefficients of (79) have a representation in terms of functions of bounded characteristic by Lemma 7. The rest of the proof follows the method in [43, Sect. 4], where Lemma 8 takes the role of [43, Lemma 2.1(b)]. We omit the details.

The next result is a generalization of Theorem 4(a).

Theorem 25 *Let* $A_0, \ldots, A_{k-2} \in \mathcal{H}(\mathbb{D})$ *be such that* (79) *is Blaschke-oscillatory. Then either all solutions of* (79) *belong to* N *or all solutions of* (79) *belong to* $\mathcal{F} \setminus N$.

Suppose that (79) possesses two solutions, one in the class N and the other one in $\mathcal{H}(\mathbb{D}) \setminus N$. These two solutions are obviously linearly independent, while their ratio does not belong to N. Hence, by Lemma 7, (79) is not Blaschke-oscillatory, which is a contradiction. Therefore all solutions of (79) are either in N or they belong to $\mathcal{H}(\mathbb{D}) \setminus N$. Let f be an arbitrary non-trivial solution of (79). It suffices to show that $f \in \mathcal{F}$. We note that any fundamental base of (79) contains precisely k linearly independent solutions. Since f is one solution, there must be $k-1$ other solutions, say $f_2, \ldots, f_k$. Recall that the Wronskian determinant $W(f, f_2, \ldots, f_k)$ of $f, f_2 \ldots, f_k$ reduces to a non-zero constant c by [54, Proposition 1.4.8]. By appealing to [54, Proposition 1.4.3(e)], we may write

$$c = W(f, f_2, \ldots, f_k) = f^k W\left(\left(\frac{f_2}{f}\right)', \ldots, \left(\frac{f_k}{f}\right)'\right). \tag{81}$$

We have $f_j/f \in N$ for every $j \in \{2, \ldots, k\}$ by Lemma 7. Finally, using [10, Lemma 5.3] on (81), it follows that f^k is non-admissible, and hence $f \in \mathcal{F}$. This completes the proof of Theorem 25.

Assuming that the "middle" coefficients in (77) vanish identically, we are easily able write the following analogue of Theorem 4(b). The proof is nothing but an easy application of Theorem 3 and Lemma 2, and hence is omitted.

Corollary 2 *Suppose that $A \in \mathcal{H}(\mathbb{D})$. If the equation*

$$f^{(k)} + A(z)f = 0$$

possesses a solution $f \in N$, then

$$\int_{D(0,r)} |A(z)|^{\frac{1}{k}} \, dm(z) = O\left(\log^2 \frac{e}{1-r}\right).$$

If in addition $f', \ldots, f^{(k-1)} \in N$, then

$$\int_{D(0,r)} |A(z)|^{\frac{1}{k}} \, dm(z) = O\left(\log \frac{e}{1-r}\right).$$

We will now continue the discussion that was started at the end of Sect. 4.2. According to the isoperimetric inequality [17, p. 77], the inclusion $H^{\frac{p}{2}} \subset B^p$ holds for every $p \in (0, \infty)$. Hence, if $A_j \in H^{\frac{1}{2(k-j)}}$ for $j \in \{0, \ldots, k-1\}$, then (78) holds, and, a fortiori, all solutions of (77) belong to N. Even more can be said, as is seen next.

Corollary 3 *Suppose that $A_j \in H^{\frac{1}{2(k-j)}}$ for $j \in \{0, \ldots, k-1\}$. Then every solution f of (77) satisfies $f, f', \ldots, f^{(k)} \in N$.*

The fact that $f, f', \ldots, f^{(k-1)} \in N$ follows by a general growth estimate for the solutions (and for their derivatives), see [41, Theorem 5]. Recalling that $H^p \subset N$ for every $p \in (0, \infty)$, we then conclude that $f^{(k)} \in N$ by (77).

12 Critical Points and Blaschke-Critical Equations

We point out that if $A \in H^{\frac{1}{4}}$, then all solutions f of (1) satisfy $f, f' \in N$ by Corollary 3. In particular, the sequence of critical points of f (the zeros of f') satisfies the Blaschke condition (3). This observation leads to a bunch of new questions: Is it possible to find $A \in \mathcal{H}(\mathbb{D})$ such that (1) possesses a solution with a prescribed Blaschke sequence of critical points? If this can be done for one sequence of fixed points, then how about for two sequences? Note that two linearly independent solutions of (1) have distinct critical points, for otherwise their Wronskian vanishes identically. How is the geometric distribution of critical points like when compared to the geometric distribution of zeros? Note in particular that if $A(z)$ has no zeros, like in the case of (40), then f and f'' have precisely the same zeros by (1). What conditions are necessary for the critical points to satisfy the Blaschke condition?

We propose a new terminology: If the sequence of critical points of any nontrivial solution of (1) satisfies the Blaschke condition, then (1) is called Blaschke-critical. Obviously this concept generalizes to equations of the form (77). Partial answers to the questions above on Blaschke-critical equations of the form (1) are given below.

12.1 Zeros Versus Critical Points

In general, the zeros and the critical points of a given solution of (1) may be of different "category". To support this claim we give an example of a non-oscillatory equation (1) with a solution $f \in H^\infty$ such that the sequence of critical points of f does not satisfy the Blaschke condition. In particular, Blaschke-oscillatory equations (1) are not always Blaschke-critical.

Example 7 Define

$$g(z) = \sum_{n=1}^{\infty} \frac{1}{n^{1/2} 2^n} z^{2^n}.$$

Then $g \in H^\infty$, while $g' \in \mathcal{B}_0$ has radial limits almost nowhere on $\partial\mathbb{D}$, see [6]. Clearly $e^{-2g} \in H^\infty$ and $g' \notin N$. In particular, $g'e^{2g} \notin N$. By [61, p. 276] there are uncountably many choices for $C \neq 0$ such that the zeros of $g' - Ce^{-2g}$ do not satisfy the Blaschke condition. Let C_0 be such a constant, and let $h \in \mathcal{H}(\mathbb{D})$ be any function such that $h' = C_0 e^{-2g}$. Since $h' \in H^\infty$, it is well-known that h belongs to the disc algebra. Next we find that the functions f_1, f_2 in (34) are linearly independent solutions of (1), where $A(z)$ is given by (35). Moreover, $f_1, f_2 \in H^\infty$, and since $g' \in \mathcal{B}_0$, we have

$$|g''(z)| = o\left(\frac{1}{1-|z|}\right), \quad |z| \to 1^-.$$

Using the triangle inequality on (35) then gives us

$$\left|A(z)\right| = o\left(\frac{1}{1-|z|}\right), \quad |z| \to 1^-.$$

This is a stronger condition than (38) near the boundary $\partial\mathbb{D}$, and hence (1) is non-oscillatory. Meanwhile, the zeros of

$$f_1' = (g' - h')e^{g-h} = (g' - C_0 e^{-2g})e^{g-h}$$

do not satisfy the Blaschke condition.

The question whether a Blaschke-critical equation (1) is always Blaschke-oscillatory is still open. In this task the following two observations may be useful. First, recall that $f' \in N$ does not in general imply that $f \in N$, see [30]. Second, let $A \in \mathcal{H}(\mathbb{D})$, and let $\{f, g\}$ be a fundamental solution base of (1). If (1) is both Blaschke-oscillatory and Blaschke-critical, then Lemma 3 and Lemma 9 below show that $(f'/f)/(g'/g) \in N$.

12.2 Necessary and Sufficient Conditions

The proof of Lemma 3 yields the following necessary and sufficient condition for Blaschke-critical equations.

Lemma 9 *Suppose that $A \in \mathcal{H}(\mathbb{D})$, and let $\{f_1, f_2\}$ be any solution base of* (1). *Then $F_* = f_2'/f_1' \in N$ if and only if* (1) *is Blaschke-critical.*

As noted above, the condition $A \in H^{\frac{1}{4}}$ is sufficient for (1) to be Blaschke-critical as well as Blaschke-oscillatory. This condition is sharp for both equation types in the sense of Example 1. The following result contains necessary conditions for Blaschke-critical equations, resembling to Theorems 2 and 4 about Blaschke-oscillatory equations.

Theorem 26 *Let $A \in \mathcal{H}(\mathbb{D})$ be such that* (1) *is Blaschke-critical. Then every solution f of* (1) *satisfies either $f' \in N$ or $f' \in \mathcal{F} \setminus N$.*

(a) *If* (1) *has a solution f with $f' \in N$, then $f \in \mathcal{F}$ and $A(z)$ satisfies* (22).
(b) *If* (1) *has a solution f with $f' \in \mathcal{F} \setminus N$, then*

$$T(r, f) = O\left(\log^2 \frac{e}{1-r}\right) \tag{82}$$

and

$$\int_{D(0,r)} \left|A(z)\right|^{\frac{1}{2}} dm(z) = O\left(\log^{\frac{5}{2}} \frac{e}{1-r}\right). \tag{83}$$

Suppose that (1) possesses solutions f and g such that $f' \in N$ and $g' \notin N$. Then f and g are obviously linearly independent, and $g'/f' \notin N$. According to Lemma 9 the equation (1) is not Blaschke-critical, which is a contradiction. Hence every solution f of (1) satisfies either $f' \in N$ or $f' \in \mathcal{H}(\mathbb{D}) \setminus N$.

Let f be any solution of (1). We then proceed to prove that $f' \in \mathcal{F}$. Let g be a solution of (1), linearly independent of f. Denote $F_* = g'/f'$. Then $F_* \in N$ by Lemma 9, while $F_*' \in \mathcal{F}$ by [10, Lemma 5.3]. Using (1), we have

$$F_*' = \frac{g''f' - g'f''}{(f')^2} = A \cdot \frac{g'f - gf'}{(f')^2} = A \cdot \frac{W(f,g)}{(f')^2} = -\frac{f''}{f} \cdot \frac{W(f,g)}{(f')^2}, \tag{84}$$

where $W(f,g)$ is the Wronskian of f and g. As noted in Sect. 4.2, $W(f,g)$ reduces to a non-zero constant. In general, the functions f and f' have the same T-order of growth [75, Theorem V.28], defined in terms of the Nevanlinna characteristic as follows:

$$\sigma_T(f) = \limsup_{r \to 1^-} \frac{\log^+ T(r,f)}{-\log(1-r)}.$$

Moreover, f and $A(z)$ do not have the same T-order, unless $\sigma_T(f) = 0 = \sigma_T(A)$. This claim is easily verified by means of the standard lemma on the logarithmic derivative: We have

$$T(r,A) = m(r,A) = m\left(r, \frac{f''}{f}\right) = O\left(\log^+ T(r,f) + \log\frac{e}{1-r}\right)$$

outside of a possible exceptional set of finite logarithmic measure. Hence the equation $F_*' = AW(f,g)/(f')^2$ in (84) shows that $\sigma_T(f') = 0$, for otherwise we have a contradiction. Applying the lemma on the logarithmic derivative to (84) in the finite order case, it follows that

$$\begin{aligned} 2T(r,f') &= m\left(r,(f')^2\right) \le m\left(r,\frac{f''}{f}\right) + m\left(r,\frac{W(f,g)}{F_*'}\right) \\ &= O\left(\log\frac{e}{1-r}\right), \quad r \notin E, \end{aligned}$$

where $E \subset [0,1)$ is of finite logarithmic measure. The exceptional set E can be avoided by means of [3, Lemma C] or by means of Lemma 2. Hence $f' \in \mathcal{F}$.

Next we turn to the proofs of (a) and (b). By [30] we conclude the following: If $f' \in N$, then $f \in \mathcal{F}$, while if $f' \in \mathcal{F} \setminus N$, then (82) holds. Suppose first that $f' \in N$, and write

$$A = -\frac{f''}{f} = -\frac{f''}{f'} \cdot \frac{f'}{f}.$$

By making use of Theorem 3 and Hölder's inequality, we have

$$\begin{aligned}\int_0^{2\pi} |A(te^{i\theta})|^{\frac{1}{2}}\,d\theta &\leq \left(\int_0^{2\pi}\left|\frac{f''(te^{i\theta})}{f'(te^{i\theta})}\right|d\theta\right)^{\frac{1}{2}}\cdot\left(\int_0^{2\pi}\left|\frac{f'(te^{i\theta})}{f(te^{i\theta})}\right|d\theta\right)^{\frac{1}{2}}\\ &= O\left(\frac{1}{1-t}\log\frac{e}{1-t}\right),\quad t\notin E,\end{aligned}$$

where $E \subset [0,1)$ is of arbitrarily small upper density. The exceptional set E can be avoided by means of Lemma 2. A simple integration then yields (22). A similar reasoning applies to the case $f' \in \mathcal{F} \setminus N$. This completes the proof of Theorem 26.

12.3 Prescribed Critical Points

We construct a coefficient $A \in \mathcal{H}(\mathbb{D})$ of (1) such that a solution of (1) has prescribed Blaschke sequences of zeros and critical points. This construction is in fact a rather simple modification of the analogous reasoning due to Šeda [72], who considered the case of entire solutions in terms of Riccati differential equations.

Let $\{z_n\}$ and $\{\zeta_n\}$ be two given Blaschke sequences of distinct points with no points in common. Let $B(z)$ and $B_c(z)$ be the corresponding Blaschke products, where the subindex c refers to "critical". We prove that a function $g \in \mathcal{H}(\mathbb{D})$ satisfying (9) exists. The Mittag-Leffler type of argument allows us to construct a meromorphic function F with principal parts $\sigma_n/B'(z_n)(z-z_n)$ at every z_n, and no other poles. The product $P = BF$ is analytic in $\mathbb{D}$ except perhaps at the points z_n. For a fixed z_n, the Taylor series expansion of $B(z)$ at z_n gives us the representation

$$B(z) = B'(z_n)(z-z_n)\big(1+B_n(z)\big),$$

where $B_n(z)$ is analytic at z_n and satisfies $B_n(z_n) = 0$. By the Laurent series expansion for the function F at z_n, we find that $F_n(z) = (z-z_n)F(z)$ is analytic at z_n and satisfies $F_n(z_n) = \sigma_n/B'(z_n)$. Now $P(z_n) = \sigma_n$, and hence $P \in \mathcal{H}(\mathbb{D})$. If g is any primitive of P, then g' satisfies (9).

Now $g_G = g + \int_0^z B(\zeta)G(\zeta)\,d\zeta$ satisfies (69), where $G \in \mathcal{H}(\mathbb{D})$ will be constructed later on. A simple computation reveals that the function $f = Be^{g_G}$ is a solution of (1), where

$$A = \frac{-B'' - 2g_G'B'}{B} - \left(g_G'\right)^2 - g_G''. \tag{85}$$

Moreover, f has prescribed zeros at the points z_n, and due to (69) the coefficient $A(z)$ in (85) belongs to $\mathcal{H}(\mathbb{D})$. This is true for any $G \in \mathcal{H}(\mathbb{D})$.

We proceed to define G in such a way that $f = Be^{g_G}$ has prescribed critical points ζ_n. Since $f' = (B' + Bg_G')e^{g_G}$, we must have

$$B' + Bg_G' = B_c e^h \tag{86}$$

for some suitable $h \in \mathcal{H}(\mathbb{D})$. Solving this for G yields

$$G = \frac{B_c e^h - B' - Bg'}{B^2} =: \frac{H}{B^2}. \tag{87}$$

For G to be analytic in $\mathbb{D}$, the function H needs to have at least a double zero at every point z_n. Hence $H(z_n) = 0$ and $H'(z_n) = 0$ must hold for every n.

(1) The requirement $H(z_n) = 0$ leads to the interpolation property

$$h(z_n) = \log\left(\frac{B'(z_n)}{B_c(z_n)}\right) =: s_n, \quad n \in \mathbb{N}, \tag{88}$$

because $B(z_n) = 0$.

(2) Since $H' = (B_c' + B_c h')e^h - B'' - B'g' - Bg''$, we obtain the interpolation property

$$h'(z_n) = \frac{B''(z_n)}{2B'(z_n)} - \frac{B_c'(z_n)}{B_c(z_n)} =: t_n, \quad n \in \mathbb{N}, \tag{89}$$

by using (9) and (88) together with the fact that $B(z_n) = 0$.

Define $h = h_1 + Bh_2$, where $h_1, h_2 \in \mathcal{H}(\mathbb{D})$ are constructed such that $h_1(z_n) = s_n$ and $h_2(z_n) = (t_n - h_1'(z_n))/B'(z_n)$ for all n. Obviously h_1 needs to be constructed first. Then it is clear that h satisfies (88) and (89). Hence the function G in (87) belongs to $\mathcal{H}(\mathbb{D})$, and (86) holds in $\mathbb{D}$. This means that f' has zeros precisely at the points ζ_n, that is, f has prescribed critical points ζ_n.

Remark 6 The distribution of the points z_n and ζ_n will affect to the growth of the functions h_1 and h_2, and eventually to the growth of the coefficient $A(z)$ in (85). Target growth $A \in \mathcal{A}^{-\infty}$ seems more realistic, yet difficult, than either (2) or $A \in G_2$. This would follow from $G \in \mathcal{A}^{-\infty}$, which in turn would be a result of $h \in \mathcal{B}$. Assuming that $\{z_n\}$ and $\{\zeta_n\}$ are both uniformly separated such that the union $\{z_n\} \cup \{\zeta_n\}$ is separated seems to yield

$$|s_n| \leq \log \frac{1}{1 - |z_n|} + O(1), \quad n \in \mathbb{N},$$

see [35, Lemma 4.1] and [12, Theorem 1]. This may enable BMOA-interpolation for h_1, as described in Sect. 9.2. The function h_2 is another story. Finding suitable conditions (even harsh ones) for the sequences $\{z_n\}$ and $\{\zeta_n\}$ such that $h \in \mathcal{B}$ satisfying (88) and (89) can be found is left as an open problem.

13 Concluding Remarks

The previous sections contain several open problems relating Blaschke products and Blaschke sequences to linear differential equations. However, neither the zeros nor the critical points of functions in the Bergman spaces B^p or in the Korenblum space $\mathcal{A}^{-\infty}$ need to satisfy the Blaschke condition [17, 32, 52]. Meanwhile, functions of this character have been associated to solutions of linear differential equations (77) in [39, 69]. This indicates that the next step in the unit disc oscillation theory could be to replace Blaschke products with Horowitz products

$$H(z)=\prod_{n=1}^{\infty}\frac{|z_n|}{z_n}\frac{z_n-z}{1-\overline{z_n}z}\left(2-\frac{|z_n|}{z_n}\frac{z_n-z}{1-\overline{z_n}z}\right)$$

which act as universal divisors in the Bergman spaces [17, Chap. 4]. It is known [17, p. 102] that if

$$\sum_n (1-|z_n|)^2<\infty, \tag{90}$$

then the associated Horowitz product $H(z)$ converges uniformly in compact subsets of $\mathbb{D}$, and hence represents an analytic function in $\mathbb{D}$ with zeros precisely at the points z_n. Replacing (3) with (90), the definitions for Horowitz-oscillatory and Horowitz-critical differential equations are obvious. For example, [53, Theorem 5] can now be restated as follows: If $A_0,\ldots,A_{k-1}\in\mathcal{H}(\mathbb{D})$ are such that

$$\int_{\mathbb{D}}|A_j(z)|^{\frac{1}{k-j}}(1-|z|^2)\,dm(z)<\infty,\qquad j=0,\ldots,k-1,$$

then (77) is Horowitz-oscillatory. As noted above, when oscillation of solutions of (77) is considered, we might as well restrict to considering (79). Now [43, Theorem 1.3] yields the following converse claim: Let $A_0,\ldots,A_{k-2}\in\mathcal{H}(\mathbb{D})$ be such that (79) is Horowitz-oscillatory. Then, for every $\varepsilon>0$, we have

$$\int_{\mathbb{D}}|A_j(z)|^{\frac{1}{k-j}}(1-|z|^2)^{1+\varepsilon}\,dm(z)<\infty,\qquad j=0,\ldots,k-2. \tag{91}$$

A more complicated approach reveals that the constant ε in (91) can be dropped out, see the discussion in [65] following the proof of [65, Theorem 7.9].

The discussion in Sect. 12 shows that Blaschke-oscillatory and Blaschke-critical equations are not the same in general. In particular, see Example 7. However, these two concepts are tied together by means of Horowitz products.

Corollary 4 *Let $A\in\mathcal{H}(\mathbb{D})$. Concerning* (1), *Blaschke-oscillatory implies Horowitz-critical, while Blaschke-critical implies Horowitz-oscillatory.*

Suppose first that (1) is Blaschke-oscillatory. Then all solutions f of (1) belong to $\mathcal{F}$ by Theorem 4. Let $n(r)$ denote the counting function of the zeros of f'. Since $f'\in\mathcal{F}$, [73, Theorem 1] yields

$$n(r)=O\left(\frac{1}{1-r}\log\frac{e}{1-r}\right).$$

A simple Riemann-Stieltjes integration now gives us

$$\sum_{|z_n|<r} \left(1-|z_n|\right)^2 = \int_0^r (1-t)^2\, dn(t) = O(1),$$

and hence (1) is Horowitz-critical. Suppose then that (1) is Blaschke-critical. Now every solution f of (1) satisfies (82). A simple modification of the reasoning above reveals that (1) is Horowitz-oscillatory.

As for existing tools, we point out that the interpolation theory has been connected to the spaces B^p and $\mathcal{A}^{-\infty}$, and hence to Horowitz products, see [17, Chap. 6] and [32, Chap. 5]. The author is unaware of any existing estimates for logarithmic derivatives of Horowitz products.

References

1. Aulaskari, R., Nowak, M., Zhao, R.: The nth derivative characterisation of Möbius invariant Dirichlet space. Bull. Aust. Math. Soc. **58**(1), 43–56 (1998)
2. Aulaskari, R., Xiao, J., Zhao, R.: On subspaces and subsets of BMOA and UBC. Analysis **15**(2), 101–121 (1995)
3. Bank, S.: A general theorem concerning the growth of solutions of first-order algebraic differential equations. Compos. Math. **25**, 61–70 (1972)
4. Bank, S., Laine, I.: On the oscillation theory of $f'' + Af = 0$ where A is entire. Trans. Am. Math. Soc. **273**(1), 351–363 (1982)
5. Buckley, S., Koskela, P., Vukotic, D.: Fractional integration, differentiation, and weighted Bergman spaces. Math. Proc. Camb. Philos. Soc. **126**(2), 369–385 (1999)
6. Carmona, J., Cufi, J., Pommerenke, Ch.: On the angular limits of Bloch functions. Publ. Mat. **32**(2), 191–198 (1988)
7. Chuaqui, M., Duren, P., Osgood, B.: Schwarzian derivatives of convex mappings. Ann. Acad. Sci. Fenn. Math. **36**, 449–460 (2011)
8. Chuaqui, M., Duren, P., Osgood, B., Stowe, D.: Oscillation of solutions of linear differential equations. Bull. Aust. Math. Soc. **79**(1), 161–169 (2009)
9. Chuaqui, M., Stowe, D.: Valence and oscillation of functions in the unit disk. Ann. Acad. Sci. Fenn. Math. **33**(2), 561–584 (2008)
10. Chyzhykov, I., Gundersen, G.G., Heittokangas, J.: Linear differential equations and logarithmic derivative estimates. Proc. Lond. Math. Soc. **86**(3), 735–754 (2003)
11. Chyzhykov, I., Heittokangas, J., Rättyä, J.: Finiteness of φ-order of solutions of linear differential equations in the unit disc. J. Anal. Math. **109**(1), 163–198 (2009)
12. Cima, J., Colwell, P.: Blaschke quotients and normality. Proc. Am. Math. Soc. **19**, 796–798 (1968)
13. Cima, J., Pfaltzgraff, J.: Oscillatory behavior of $u''(z) + h(z)u(z) = 0$ for univalent $h(z)$. J. Anal. Math. **25**, 311–322 (1972)
14. Clunie, J.: The derivative of a meromorphic function. Proc. Am. Math. Soc. **7**, 227–229 (1956)
15. Duren, P.: Theory of H^p Spaces. Academic Press, New York (1970)
16. Duren, P., Romberg, B., Shields, A.: Linear functionals on H^p spaces with $0 < p < 1$. J. Reine Angew. Math. **238**, 32–60 (1969)
17. Duren, P., Schuster, A.: Bergman Spaces. Mathematical Surveys and Monographs, vol. 100. Am. Math. Soc., Providence (2004)

18. Essén, M., Xiao, J.: Some results on Q_p spaces, $0 < p < 1$. J. Reine Angew. Math. **485**, 173–195 (1997)
19. Fenton, P., Rossi, J.: ODEs and Wiman-Valiron theory in the unit disc. J. Math. Anal. Appl. **367**(1), 137–145 (2010)
20. Fricain, E., Mashreghi, J.: Exceptional sets for the derivatives of Blaschke products. In: Proceedings of the St. Petersburg Mathematical Society. Vol. XIII. Amer. Math. Soc. Transl. Ser. 2, vol. 222, pp. 163–170. Am. Math. Soc., Providence (2008)
21. Fricain, E., Mashreghi, J.: Integral means of the derivatives of Blaschke products. Glasg. Math. J. **50**(2), 233–249 (2008)
22. Girela, D., Peláez, J., Pérez-González, F., Rättyä, J.: Carleson measures for the Bloch space. Integral Equ. Oper. Theory **61**(4), 511–547 (2008)
23. Girela, D., Peláez, J., Vukotic', D.: Integrability of the derivative of a Blaschke product. Proc. Edinb. Math. Soc. (2) **50**(3), 673–687 (2007)
24. Girela, D., Peláez, J., Vukotic', D.: Uniformly discrete sequences in regions with tangential approach to the unit circle. Complex Var. Elliptic Equ. **52**(2–3), 161–173 (2007)
25. Gnuschke-Hauschild, D., Pommerenke, Ch.: On Bloch functions and gap series. J. Reine Angew. Math. **367**, 172–186 (1986)
26. Gotoh, Y.: On integral means of the derivatives of Blaschke products. Kodai Math. J. **30**(1), 147–155 (2007)
27. Gröhn, J., Heittokangas, J.: New findings on Bank-Sauer approach in oscillation theory. Constr. Approx. **35**, 345–361 (2012)
28. Hartman, P., Wintner, A.: On linear second order differential equations in the unit circle. Trans. Am. Math. Soc. **78**, 492–500 (1955)
29. Hayman, W.K.: Meromorphic Functions. Oxford Mathematical Monographs. Clarendon Press, Oxford (1964)
30. Hayman, W.K.: On the characteristic of functions meromorphic in the unit disk and of their integrals. Acta Math. **112**, 181–214 (1964)
31. Hayman, W.K.: Multivalent Functions, 2nd edn. Cambridge Tracts in Mathematics, vol. 110. Cambridge University Press, Cambridge (1994)
32. Hedenmalm, H., Korenblum, B., Zhu, K.: Theory of Bergman Spaces. Graduate Texts in Mathematics, vol. 199. Springer, New York (2000)
33. Heittokangas, J.: On complex differential equations in the unit disc. Ann. Acad. Sci. Fenn. Math. Diss. **122**, 1–54 (2000)
34. Heittokangas, J.: Solutions of $f'' + A(z)f = 0$ in the unit disc having Blaschke sequences as the zeros. Comput. Methods Funct. Theory **5**(1), 49–63 (2005)
35. Heittokangas, J.: Blaschke-oscillatory equations of the form $f'' + A(z)f = 0$. J. Math. Anal. Appl. **318**(1), 120–133 (2006)
36. Heittokangas, J.: Growth estimates for logarithmic derivatives of Blaschke products and of functions in the Nevanlinna class. Kodai Math. J. **30**, 263–279 (2007)
37. Heittokangas, J.: On interpolating Blaschke products and Blaschke-oscillatory equations. Constr. Approx. **34**(1), 1–21 (2011)
38. Heittokangas, J., Korhonen, R., Rättyä, J.: Growth estimates for solutions of linear complex differential equations. Ann. Acad. Sci. Fenn. **29**(1), 233–246 (2004)
39. Heittokangas, J., Korhonen, R., Rättyä, J.: Linear differential equations with solutions in the Dirichlet type subspace of the Hardy space. Nagoya Math. J. **187**, 91–113 (2007)
40. Heittokangas, J., Korhonen, R., Rättyä, J.: Linear differential equations with coefficients in weighted Bergman and Hardy spaces. Trans. Am. Math. Soc. **360**(2), 1035–1055 (2008)
41. Heittokangas, J., Korhonen, R., Rättyä, J.: Growth estimates for solutions of nonhomogeneous linear differential equations. Ann. Acad. Sci. Fenn. **34**(1), 145–156 (2009)
42. Heittokangas, J., Laine, I.: Solutions of $f'' + A(z)f = 0$ with prescribed sequences of zeros. Acta Math. Univ. Comen. **124**(2), 287–307 (2005)
43. Heittokangas, J., Rättyä, J.: Zero distribution of solutions of complex linear differential equations determines growth of coefficients. Math. Nachr. **284**(4), 412–420 (2011)

44. Heittokangas, J., Tohge, K.: A unit disc analogue of the Bank-Laine conjecture does not hold. Ann. Acad. Sci. Fenn. **36**(1), 341–351 (2011)
45. Herold, H.: Nichteuklidischer Nullstellenabstand der Lösungen von $w'' + p(z)w = 0$. Math. Ann. **287**, 637–642 (1990)
46. Hille, E.: Remarks on a paper by Zeev Nehari. Bull. Am. Math. Soc. **55**, 552–553 (1949)
47. Horowitz, C.: Factorization theorems for functions in the Bergman spaces. Duke Math. J. **44**(1), 201–213 (1977)
48. Ince, E.: Ordinary Differential Equations. Dover, New York (1956)
49. Juneja, O., Kapoor, G.: Analytic Functions—Growth Aspects. Research Notes in Mathematics, vol. 104. Pitman Adv. Publ. Prog., Boston (1985)
50. Kim, H.: Derivatives of Blaschke products. Pac. J. Math. **114**(1), 175–190 (1984)
51. Kim, W.: The Schwarzian derivative and multivalence. Pac. J. Math. **31**, 717–724 (1969)
52. Korenblum, B.: An extension of the Nevanlinna theory. Acta Math. **135**(3–4), 187–219 (1975)
53. Korhonen, R., Rättyä, J.: Finite order solutions of linear differential equations in the unit disc. J. Math. Anal. Appl. **349**(1), 43–54 (2009)
54. Laine, I.: Nevanlinna Theory and Complex Differential Equations. de Gruyter, Berlin (1993)
55. Laine, I.: Complex differential equations. In: Handbook of Differential Equations: Ordinary Differential Equations. Vol. IV. Handb. Differ. Equ., pp. 269–363. Elsevier/North-Holland, Amsterdam (2008) (English summary)
56. Linden, C.: H^p-derivatives of Blaschke products. Mich. Math. J. **23**, 43–51 (1976)
57. London, D.: On the zeros of solutions of $w''(z) + p(z)w(z) = 0$. Pac. J. Math. **12**, 979–991 (1962)
58. Mashreghi, J., Shabankhah, M.: Integral means of the logarithmic derivative of Blaschke products. Comput. Methods Funct. Theory **9**(2), 421–433 (2009)
59. Nehari, Z.: The Schwarzian derivative and Schlicht functions. Bull. Am. Math. Soc. **55**, 545–551 (1949)
60. Nehari, Z.: Conformal Mapping. Dover, New York (1975). Reprinting of the 1952 edn.
61. Nevanlinna, R.: Analytic Functions. Die Grundlehren der mathematischen Wissenschaften, vol. 162. Springer, New York (1970). Translated from the second German edn. by Phillip Emig
62. Nolder, C.: An L^p definition of interpolating Blaschke products. Proc. Am. Math. Soc. **128**(6), 1799–1806 (2000)
63. Pascuas, D.: A note on interpolation by Bloch functions. Proc. Am. Math. Soc. **135**(7), 2127–2130 (2007)
64. Peláez, J.: Sharp results on the integrability of the derivative of an interpolating Blaschke product. Forum Math. **20**(6), 1039–1054 (2008)
65. Peláez, J., Rättyä, J.: Weighted Bergman spaces induced by rapidly increasing weights. Mem. Am. Math. Soc. (to appear)
66. Pommerenke, Ch.: Univalent Functions. Studia Mathematica/Mathematische Lehrbücher, vol. XXV. Vandenhoeck & Ruprecht, Göttingen (1975). With a chapter on quadratic differentials by Gerd Jensen
67. Pommerenke, Ch.: On the mean growth of the solutions of complex linear differential equations in the disk. Complex Var. Theory Appl. **1**(1), 23–38 (1982)
68. Protas, D.: Blaschke products with derivative in H^p and B^p. Mich. Math. J. **20**, 393–396 (1973)
69. Rättyä, J.: Linear differential equations with solutions in Hardy spaces. Complex Var. Elliptic Equ. **52**(9), 785–795 (2007)
70. Schwarz, B.: Complex nonoscillation theorems and criteria of univalence. Trans. Am. Math. Soc. **80**, 159–186 (1955)
71. Šeda, V.: A note to a paper by Clunie. Acta Fac. Nat. Univ. Comen. **4**, 255–260 (1959)
72. Šeda, V.: On some properties of solutions of the differential equation $y'' = Q(z)y$, where $Q(z) \neq 0$ is an entire function. Acta Fac. Nat. Univ. Comen. Math. **4**, 223–253 (1959) (Slovak)
73. Shea, D., Sons, L.: Value distribution theory for meromorphic functions of slow growth in the disk. Houst. J. Math. **12**(2), 249–266 (1986)

74. Tse, K.-F.: Nontangential interpolating sequences and interpolation by normal functions. Proc. Am. Math. Soc. **29**, 351–354 (1971)
75. Tsuji, M.: Potential Theory in Modern Function Theory. Chelsea, New York (1975). Reprinting of the 1959 edn.
76. Yamashita, S.: Gap series and α-Bloch functions. Yokohama Math. J. **28**, 31–36 (1980)
77. Zabulionis, A.: Separation of points of the unit disc. Lith. Math. J. **23**(3), 271–274 (1983)

Bi-orthogonal Expansions in the Space $L^2(0, \infty)$

André Boivin and Changzhong Zhu

Abstract In this paper we deduce bi-orthogonal expansions in the space $L^2(0, \infty)$ with respect to two special systems of functions from the corresponding expansions in the Hardy space H_+^2 for the upper half-plane.

Keywords $L^2(0, \infty)$ · Bi-orthogonal expansion · System of functions

Mathematics Subject Classification Primary 30E10 · Secondary 30B60

1 Introduction

Assume that the sequence of complex numbers

$$\{\lambda_k\} \quad (k = 1, 2, \ldots) \tag{1}$$

satisfies the conditions: $\text{Re}(\lambda_k) > 0$, $\lambda_k \neq \lambda_j$ for $k \neq j$, and

$$\sum_{k=1}^{\infty} \frac{\text{Re}(\lambda_k)}{1 + |\lambda_k|^2} < +\infty. \tag{2}$$

It is known that under this (Blaschke) condition, the exponential system

$$\left\{e^{-\lambda_k x}\right\} \quad (k = 1, 2, \ldots) \tag{3}$$

is incomplete in $L^2(0, \infty)$ (see [5] for example). Hence the closed linear span of the system (3), namely $\mathbf{E}$, is a proper subspace of $L^2(0, \infty)$. It is also well known that,

A. Boivin (✉) · C. Zhu
Department of Mathematics, University of Western Ontario, London, Ontario, Canada N6A 5B7
e-mail: boivin@uwo.ca

C. Zhu
e-mail: czhu28@uwo.ca

J. Mashreghi, E. Fricain (eds.), *Blaschke Products and Their Applications*,
Fields Institute Communications 65, DOI 10.1007/978-1-4614-5341-3_4,

under this condition, the Blaschke product

$$W(\xi)=\prod_{k=1}^{\infty}\left[\frac{\xi-\lambda_k}{\xi+\overline{\lambda}_k}\cdot\frac{|1-\lambda_k^2|}{1-\lambda_k^2}\right] \tag{4}$$

converges to an analytic function $W(\xi)$ in the right half-plane $\mathrm{Re}(\xi)>0$.

Let us consider the following system of functions introduced in [5]:

$$\psi_k(x)=-\frac{1}{W'(\lambda_k)}\cdot\frac{1}{2\pi}\int_{-\infty}^{+\infty}\frac{e^{-i\tau x}}{W(i\tau)(i\tau+\overline{\lambda}_k)}d\tau \quad (k=1,2,\ldots), \tag{5}$$

i.e.,

$$\psi_k(x)=\frac{1}{2\pi}\int_{-\infty}^{\infty}\frac{e^{-i\tau x}}{W(i\tau)}\overline{\left[\frac{1}{2\pi i}\int_{c_k}\frac{d\xi}{W(\xi)(i\tau-\xi)}\right]}d\tau \quad (k=1,2,\ldots), \tag{6}$$

where c_k is a circle with centre at λ_k, lying entirely in the right half-plane $\mathrm{Re}(\xi)>0$ and containing no other λ_j $(j\neq k)$, and traced in the counterclockwise direction.

It is not difficult (see [5]) to show that:

(i) the systems (3) and (5) are bi-orthogonal in $L^2(0,+\infty)$, i.e.

$$\int_0^{+\infty}e^{-\lambda_k x}\cdot\overline{\psi_j(x)}dx=\int_0^{+\infty}\overline{e^{-\lambda_k x}}\cdot\psi_j(x)dx=\delta_{kj}=\begin{cases}1 & \text{if } k=j;\\ 0 & \text{if } k\neq j.\end{cases} \tag{7}$$

(ii) all the elements of the system (5) belong to the closed linear span $\mathbf{E}$ of the system (3); hence the system (5) is also incomplete in $L^2(0,+\infty)$, and the closed linear span, namely $\mathbf{P}$, of the system (5), is also a proper subspace of $L^2(0,+\infty)$.

The representation of the projection $(P_{\mathbf{E}})f(x)$ of any function $f(x)\in L^2(0,+\infty)$ onto $\mathbf{E}$ was also obtained in [5]. The same representation was deduced in [6] using the Fourier transform. The purpose of this paper is to give the bi-orthogonal expansions of $f(x)$ in $L^2(0,+\infty)$ with respect to the systems (3) and (5), respectively. To do so, we will also use the Fourier transform as a "bridge", and borrow the corresponding results in the Hardy space H_+^2 of the upper half-plane with respect to some corresponding systems of functions.

2 Two Systems of Analytic Functions in H_+^2

Recall that (see, for example, [3] and [1, Chap. 11]) the Hardy space H_+^2 is the set of functions $F(z)$ analytic in the upper half-plane $\mathrm{Im}(z)>0$ such that

$$\|F(z)\|_{H_+^2}=\sup_{0<y<+\infty}\left[\int_{-\infty}^{+\infty}|F(x+iy)|^2dx\right]^{1/2}<+\infty. \tag{8}$$

Moreover, the function $F(z) \in H_+^2$ has non-tangential boundary values $F(x)$ for almost every $x \in (-\infty,\infty)$ with $F(x) \in L^2(-\infty,\infty)$.

With the inner product

$$(F_1, F_2) = \int_{-\infty}^{+\infty} F_1(x)\overline{F_2(x)}dx$$

for $F_1(z), F_2(z) \in H_+^2$ and the norm $\|F(z)\|_{H_+^2} = (F,F)^{1/2}$, H_+^2 is a Hilbert space, and

$$\|F(z)\|_{H_+^2} = \|F(x)\|_{L^2(-\infty,+\infty)} = \left[\int_{-\infty}^{+\infty} |F(x)|^2 dx\right]^{1/2}. \tag{9}$$

We will use the following known properties (see, for example, [3, Theorems A and C], [1]):

Lemma 1 *If $G(z) \in H_+^2$, then*

$$\frac{1}{2\pi i}\int_{-\infty}^{+\infty} \frac{G(x)}{x-z}dx = \begin{cases} G(z), & \mathrm{Im}(z) > 0, \\ 0, & \mathrm{Im}(z) < 0. \end{cases} \tag{10}$$

If $g(x) \in L^2(-\infty,+\infty)$, then

$$G(z) := \frac{1}{2\pi i}\int_{-\infty}^{+\infty} \frac{g(x)}{x-z}dx \tag{11}$$

belongs to H_+^2 and

$$\|G(x)\|_{L^2(-\infty,+\infty)} \le C\|g(x)\|_{L^2(-\infty,+\infty)}, \tag{12}$$

where C is a constant independent of g.

Lemma 2 (Paley-Wiener) *If $F(z) \in H_+^2$, then its Fourier Transform (FT) satisfies*

$$\frac{1}{\sqrt{2\pi}}\int_{-\infty}^{+\infty} F(t)e^{-itx}dt = \begin{cases} f(x), & x \in (0,+\infty); \\ 0, & x \in (-\infty, 0) \end{cases} \tag{13}$$

with $f(x) \in L^2(0,+\infty)$. And conversely, if $f(x) \in L^2(0,+\infty)$, then its Inverse Fourier Transform (FT^{-1})

$$F(z) := \frac{1}{\sqrt{2\pi}}\int_0^{+\infty} f(x)e^{ixz}dx \tag{14}$$

belongs to H_+^2 and

$$\int_0^{\infty} |f(x)|^2 dx = \int_{-\infty}^{\infty} |F(t)|^2 dt.$$

The class of functions in H_+^2 coincides with the set of functions representable in the form (14) *where* $f(x) \in L^2(0,+\infty)$.

Let $z = i\overline{\xi}$ which transforms the right half-plane $\mathrm{Re}(\xi) > 0$ onto the upper half-plane $\mathrm{Im}(z) > 0$, let $\{\lambda_k\}$ be the sequence considered in (1) and let $a_k = i\overline{\lambda}_k$ $(k = 1,2,\ldots)$, then all elements of the sequence

$$\{a_k\} \quad (k = 1,2,\ldots) \tag{15}$$

are located in the upper half-plane and are pairwise distinct. And, since $|\lambda_k| = |a_k|$ and $\mathrm{Re}(\lambda_k) = \mathrm{Im}(a_k)$ $(k = 1,2,\ldots)$, condition (2) becomes

$$\sum_{k=1}^{\infty} \frac{\mathrm{Im}(a_k)}{1+|a_k|^2} < +\infty, \tag{16}$$

and the Blaschke product

$$B(z) = \prod_{k=1}^{\infty} \left[\frac{z-a_k}{z-\overline{a}_k} \cdot \frac{|1+a_k^2|}{1+a_k^2} \right] = \overline{W(i\overline{z})} \tag{17}$$

converges to an analytic function $B(z)$ in the upper half-plane $\mathrm{Im}(z) > 0$. It is well known that $B(z) \in H_+^2$, $|B(z)| \leq 1$ for $\mathrm{Im}(z) > 0$; and $|B(x)| = 1$ for almost every $x \in (-\infty,+\infty)$. Let

$$W_n(\xi) := \prod_{k=1}^{n} \left[\frac{\xi-\lambda_k}{\xi+\overline{\lambda}_k} \cdot \frac{|1-\lambda_k^2|}{1-\lambda_k^2} \right] \tag{18}$$

and

$$B_n(z) := \prod_{k=1}^{n} \left[\frac{z-a_k}{z-\overline{a}_k} \cdot \frac{|1+a_k^2|}{1+a_k^2} \right] = \overline{W_n(i\overline{z})}. \tag{19}$$

It is easy to verify that, for $1 \leq k \leq n$,

$$\overline{\left(\frac{B(a_k)}{B_n(a_k)} \right)} = \frac{W(\lambda_k)}{W_n(\lambda_k)}. \tag{20}$$

We also have (see [4]):

Lemma 3 *For any $g(x) \in L^2(-\infty,+\infty)$, as $n \to \infty$,*

$$\left\| g(x)\left[B_n(x) - B(x)\right] \right\|_{L^2(-\infty,+\infty)} \longrightarrow 0. \tag{21}$$

For $F(z) \in H_+^2$, denote

$$H_F(z) := \frac{B(z)}{2\pi i} \int_{-\infty}^{+\infty} \frac{F(x)}{B(x)(x-z)} dx, \quad \mathrm{Im}(z) > 0. \tag{22}$$

It is known (see, for example, [4]) that $H_F(z) \in H_+^2$.

For the above sequence $a_k = i\overline{\lambda_k}$ $(k = 1, 2, \ldots)$, consider the following two systems of functions:

$$e_k(z) = \frac{1}{2\pi i} \cdot \frac{1}{z - \overline{a}_k} \quad (k = 1, 2, \ldots), \tag{23}$$

$$\phi_k(z) = -\frac{B(z)}{(z - a_k)B'(a_k)} \quad (k = 1, 2, \ldots). \tag{24}$$

By the Residue Theorem, $\phi_k(z)$ has the following integral expression:

$$\phi_k(z) = \frac{B(z)}{2\pi i} \int_{c_k'} \frac{d\xi}{B(\xi)(\xi - z)}, \quad k = 1, 2, \ldots, \tag{25}$$

where c_k' is a small circle in the upper half-plane $\mathrm{Im}(z) > 0$ and centred at a_k containing no points a_j different from a_k, and the closed disc bounded by c_k' does not contain z. Indeed, $-c_k'$ is the image of c_k (which appears in (6)) under the map $z = i\overline{\xi}$ and traced in the clockwise direction.

It is known that (see, for example, [3, 4])

(i) under the condition (16), both the systems (23) and (24) are incomplete in H_+^2. Hence the closed linear span of the system (23) in H_+^2, namely E, is a proper subspace of H_+^2, and so is the closed linear span in H_+^2 of the system (24), namely Φ.

(ii)

$$E = \Phi. \tag{26}$$

(iii) the systems (23) and (24) are bi-orthogonal on $(-\infty, +\infty)$:

$$\int_{-\infty}^{+\infty} e_k(x)\overline{\phi_j(x)}dx = \int_{-\infty}^{+\infty} \overline{e_k(x)}\phi_j(x)dx = \delta_{kj}. \tag{27}$$

3 Bi-orthogonal Expansions in H_+^2

In [4],[1] we find the bi-orthogonal expansion with respect to the system (23):

Lemma 4 *For any $F(z) \in H_+^2$, the bi-orthogonal expansion of $F(z)$ with respect to the system* (23) *is*

$$F(z) = \lim_{n\to\infty} \sum_{k=1}^{n} \alpha(F, \phi_k) \overline{\left(\frac{B(a_k)}{B_n(a_k)}\right)} e_k(z) + H_F(z), \quad \mathrm{Im}(z) > 0, \tag{28}$$

[1] See (15) in [4] with $m_k = 1$ for $k = 1, 2, \ldots$.

where the limit is in the sense of $L^2(-\infty, +\infty)$,

$$\alpha(F, \phi_k) = \int_{-\infty}^{+\infty} F(x)\overline{\phi_k(x)}dx, \quad k = 1, 2, \ldots, \tag{29}$$

and the function $H_F(z)$ *is given by* (22).

In the expression (28),

$$(P_E F)(z) = \lim_{n\to\infty} \sum_{k=1}^{n} \alpha(F, \phi_k)\overline{\left(\frac{B(a_k)}{B_n(a_k)}\right)} e_k(z), \quad \mathrm{Im}(z) > 0 \tag{30}$$

is the orthogonal projection of $F(z)$ onto E, since

$$\begin{aligned}
&\int_{-\infty}^{+\infty} \big(F(x) - (P_E F)(x)\big)\overline{e_k(x)}dx \\
&= \int_{-\infty}^{+\infty} H_F(x)\overline{e_k(x)}dx \\
&= \frac{1}{2\pi i} \int_{-\infty}^{+\infty} \frac{H_F(x)}{x - a_k} dx = H_F(a_k) = 0 \quad (k = 1, 2, \ldots)
\end{aligned} \tag{31}$$

by Lemma 1.

Now we give a parallel result with respect to the system (24):[2]

Lemma 5 *For any* $F(z) \in H_+^2$, *the bi-orthogonal expansion of* $F(z)$ *with respect to the system* (24) *is*

$$F(z) = \lim_{n\to\infty} \sum_{k=1}^{n} \alpha(F, e_k) \frac{B(a_k)}{B_n(a_k)} \phi_k(z) + H_F(z), \quad \mathrm{Im}(z) > 0, \tag{32}$$

where the limit is in the sense of $L^2(-\infty, +\infty)$,

$$\alpha(F, e_k) = \int_{-\infty}^{+\infty} F(x)\overline{e_k(x)}dx = -F(a_k), \quad k = 1, 2, \ldots, \tag{33}$$

and $H_F(z)$ *is given by* (22).

In the expression (32),

$$(P_\Phi F)(z) = \lim_{n\to\infty} \sum_{k=1}^{n} \alpha(F, e_k) \frac{B(a_k)}{B_n(a_k)} \phi_k(z), \quad \mathrm{Im}(z) > 0 \tag{34}$$

is the orthogonal projection of $F(z)$ *onto* Φ.

---

[2] A similar result can be found in [2, Theorem 2.3.1], but here we use a different method to prove the result.

Proof First, we point out that if $z = a_j$ $(j = 1, 2, \ldots)$, then (32) must hold. In fact, since $\phi_k(a_j) = -\delta_{kj}$, $\alpha(F, e_j) = -F(a_j)$, $H_F(a_j) = 0$, we have

$$\lim_{n\to\infty} \sum_{k=1}^{n} \alpha(F, e_k) \cdot \frac{B(a_k)}{B_n(a_k)} \cdot \phi_k(a_j) + H_F(a_j) = \lim_{n\to\infty} F(a_j) \cdot \frac{B(a_j)}{B_n(a_j)} = F(a_j).$$

Now assume that $\mathrm{Im}(z) > 0$ and $z \neq a_k$ $(k = 1, 2, \ldots)$. Choose sufficiently small circle c_k' as in (25). We have

$$\begin{aligned}
(P_{\Phi_{(n)}} F)(z) &= \sum_{k=1}^{n} \alpha(F, e_k) \cdot \frac{B(a_k)}{B_n(a_k)} \cdot \phi_k(z) \\
&= \sum_{k=1}^{n} [-F(a_k)] \cdot \frac{B(a_k)}{B_n(a_k)} \cdot \left[-\frac{B(z)}{B'(a_k)(z - a_k)} \right] \\
&= \sum_{k=1}^{n} B(z) \cdot \frac{F(a_k) B(a_k)}{B_n(a_k) B'(a_k)(z - a_k)} \\
&= \sum_{k=1}^{n} \frac{-B(z)}{2\pi i} \int_{c_k'} \frac{F(\xi) B(\xi)}{B_n(\xi) B(\xi)(\xi - z)} d\xi \quad \text{(Residue Theorem)} \\
&= \sum_{k=1}^{n} \frac{-B(z)}{2\pi i} \int_{c_k'} \frac{F(\xi)}{B_n(\xi)(\xi - z)} d\xi \\
&= \sum_{k=1}^{n} \frac{-B(z)}{2\pi i} \int_{c_k'} \left[\frac{1}{2\pi i} \int_{-\infty}^{+\infty} \frac{F(x) dt}{x - \xi} \right] \frac{d\xi}{B_n(\xi)(\xi - z)} \quad \text{(Lemma 1)} \\
&= \int_{-\infty}^{+\infty} F(x) \left[\sum_{k=1}^{n} \frac{-B(z)}{(2\pi i)^2} \int_{c_k'} \frac{d\xi}{B_n(\xi)(\xi - z)(x - \xi)} \right] dx \\
&= \int_{-\infty}^{+\infty} F(x) G_n(z, x) dx,
\end{aligned}$$

where for $\mathrm{Im}(z) > 0$, $-\infty < x < +\infty$,

$$G_n(z, x) = \sum_{k=1}^{n} \frac{-B(z)}{(2\pi i)^2} \int_{c_k} \frac{d\xi}{B_n(\xi)(\xi - z)(x - \xi)}.$$

Note that the function

$$g(\xi) = \frac{1}{B_n(\xi)(\xi - z)(\xi - x)}$$

is analytic in the ξ plane except at a_k $(k = 1, 2, \ldots, n)$, z and x which are all poles of $g(\xi)$ with order 1. Thus, by the Residue Theorem,

$$\sum_{k=1}^{n} \frac{1}{2\pi i} \int_{c_k} g(\xi) d\xi = -Res(g, z) - Res(g, x) - Res(g, \infty)$$
$$= \frac{1}{B_n(z)(x-z)} - \frac{1}{B_n(x)(x-z)} - 0,$$

where $Res(g, \infty) = 0$ since, as $\xi \to \infty$,

$$|g(\xi)| = O\left(\frac{1}{|\xi|^2}\right).$$

Thus, we have

$$G_n(z, x) = \frac{B(z)}{2\pi i}\left[\frac{1}{B_n(z)(x-z)} - \frac{1}{B_n(x)(x-z)}\right],$$

and by Lemma 1,

$$(P_{\Phi_{(n)}} F)(z) = \frac{B(z)}{B_n(z)} F(z) - \frac{B(z)}{2\pi i} \int_{-\infty}^{+\infty} \frac{F(x)dx}{B_n(x)(x-z)}.$$

It is seen that as $n \to \infty$, $(P_{\Phi_{(n)}} F)(z)$ converges to $(P_\Phi F)(z) = F(z) - H_F(z)$ in the sense of $L^2(-\infty, +\infty)$. Indeed, by Lemmas 1 and 3, and noting that $|B_n(x)| = 1$ on $(-\infty, +\infty)$, and $|B(x)| = 1$ almost everywhere on $(-\infty, +\infty)$, we have

$$\left\| \frac{B(z)}{2\pi i} \int_{-\infty}^{+\infty} \frac{F(x)dx}{B_n(x)(x-z)} - \frac{B(z)}{2\pi i} \int_{-\infty}^{+\infty} \frac{F(x)dx}{B(x)(x-z)} \right\|_{L^2(-\infty,+\infty)}$$
$$= \left\| \frac{B(z)}{2\pi i} \int_{-\infty}^{+\infty} \frac{[B(x) - B_n(x)]F(x)}{B_n(x)B(x)} \cdot \frac{dx}{x-z} \right\|_{L^2(-\infty,+\infty)}$$
$$\leq C \left\| F(x)[B(x) - B_n(x) \right\|_{L^2(-\infty,+\infty)} \longrightarrow 0 \quad \text{as } n \to \infty,$$

and

$$\left\| \frac{B(z)}{B_n(z)} F(z) - F(z) \right\|_{L^2(-\infty,+\infty)}$$
$$= \left\| F(x)\big[B(x) - B_n(x)\big] \right\|_{L^2(-\infty,+\infty)} \longrightarrow 0 \quad \text{as } n \to \infty.$$

Thus, (32) holds.

By (28) and (32),

$$(P_\Phi F)(z) = (P_E F)(z) = F(z) - H_F(z), \quad \text{Im}(z) > 0.$$

As mentioned in Lemma 4, $(P_\Phi F)(z)$ is the orthogonal projection of $F(z)$ onto E (i.e. Φ). The proof is complete. □

By [4], if $F(z) \in E$ (i.e. Φ), then $H_F(z) \equiv 0$ for $\mathrm{Im}(z) > 0$. Thus, by Lemma 4 or Lemma 5, we have

Lemma 6 *$F(z) \in E$ (i.e. Φ) if and only if $H_F(z) \equiv 0$ for* $\mathrm{Im}(z) > 0$.

4 Main Results and Their Proofs

Our main results are:

Theorem 1 *For any $f(x) \in L^2(0, +\infty)$, the bi-orthogonal expansion of $f(x)$ with respect to the system* (3) *is*

$$f(x) = \lim_{n\to\infty} \sum_{k=1}^{n} \beta(f, \psi_k) \frac{W(\lambda_k)}{W_n(\lambda_k)} e^{-\lambda_k x} + h_f(x), \quad 0 < x < +\infty, \tag{35}$$

where the limit is in the sense of $L^2(0, +\infty)$, $\beta(f, \psi_k)$ is defined by

$$\beta(f, \psi_k) = \int_0^{+\infty} f(x)\overline{\psi_k(x)}dx, \quad k = 1, 2, \ldots, \tag{36}$$

and $h_f(x)$ is determined by the following steps: (i) *Inverse Fourier Transform* (14) *of $f(x)$ to obtain $F(z) \in H_+^2$*, (ii) *using* (22) *to obtain $H_F(z) \in H_+^2$ from $F(z)$, and* (iii) *the Fourier Transform* (13) *of $H_F(z)$ to obtain $h_f(x) \in L^2(-\infty, +\infty)$*:

$$\frac{1}{\sqrt{2\pi}} \int_{-\infty}^{+\infty} H_F(t)e^{-itx}dt = \begin{cases} h_f(x), & x \in (0, +\infty); \\ 0, & x \in (-\infty, 0). \end{cases} \tag{37}$$

In the expression (35),

$$(P_{\mathbf{E}} f)(x) = \lim_{n\to\infty} \sum_{k=1}^{n} \beta(f, \psi_k) \frac{W(\lambda_k)}{W_n(\lambda_k)} e^{-\lambda_k x}, \quad 0 < x < +\infty \tag{38}$$

is the orthogonal projection of $f(x)$ onto the subspace $\mathbf{E}$.

Theorem 2 *For any $f(x) \in L^2(0, +\infty)$, the bi-orthogonal expansion of $f(x)$ with respect to the system* (5) *is*

$$f(x) = \lim_{n\to\infty} \sum_{k=1}^{n} \beta\left(f(x), e^{-\lambda_k x}\right) \overline{\left(\frac{W(\lambda_k)}{W_n(\lambda_k)}\right)} \psi_k(x) + h_f(x), \quad 0 < x < +\infty, \tag{39}$$

where the limit is in the sense of $L^2(0,+\infty)$,

$$\beta\big(f(x), e^{-\lambda_k x}\big) = \int_0^{+\infty} f(x) e^{-\overline{\lambda}_k x} dx, \quad k = 1, 2, \ldots, \tag{40}$$

and $h_f(x)$ *is the same as in Theorem* 1.

In the expression (39),

$$(P_{\mathbf{P}} f)(x) = \lim_{n\to\infty} \sum_{k=1}^{n} \beta\big(f(x), e^{-\lambda_k x}\big) \overline{\left(\frac{W(\lambda_k)}{W_n(\lambda_k)}\right)} \psi_k(x), \quad 0 < x < +\infty \tag{41}$$

is the orthogonal projection of $f(x)$ *onto the subspace* $\mathbf{P}$.

We can prove Theorems 1 and 2 using Lemmas 4 and 5, respectively. We will only prove Theorem 2, the proof of Theorem 1 being similar.

Before proving Theorem 2, first, we point out that the two pairs of bi-orthogonal systems (3) and (5), and (23) and (24), are connected via the Fourier Transform (14) and (13). Indeed we have (FT^{-1})

$$\phi_k(z) = \frac{1}{\sqrt{2\pi}} \int_0^{+\infty} [-\sqrt{2\pi}\,\psi_k(x)] e^{ixz} dx \in H_+^2, \tag{42}$$

and (FT)

$$\frac{1}{\sqrt{2\pi}} \int_{-\infty}^{+\infty} \phi_k(t) e^{-itx} dt = \begin{cases} -\sqrt{2\pi}\,\psi_k(x), & x \in (0,+\infty); \\ 0, & x \in (-\infty, 0). \end{cases} \tag{43}$$

We also have (FT^{-1})

$$e_k(z) = \frac{1}{\sqrt{2\pi}} \int_0^{+\infty} \frac{-1}{\sqrt{2\pi}} e^{-\lambda_k x} e^{ixz} dx \in H_+^2, \tag{44}$$

and (FT)

$$\frac{1}{\sqrt{2\pi}} \int_{-\infty}^{+\infty} e_k(t) e^{-itx} dt = \begin{cases} \frac{-1}{\sqrt{2\pi}} e^{-\lambda_k x}, & x \in (0,+\infty); \\ 0, & x \in (-\infty, 0). \end{cases} \tag{45}$$

Indeed, for example, we can verify (43) as follows: Noting that $W(i\tau) = \overline{B(\tau)}$ for $\tau \in (-\infty,\infty)$, and under $z = i\overline{\xi}$, $W(\xi) = \overline{B(z)}$, c_k is changed to $-c_k'$, by (6), we have for $x \in (0,+\infty)$,

$$\begin{aligned} \psi_k(x) &= \frac{1}{2\pi} \int_{-\infty}^{\infty} \frac{e^{-i\tau x}}{\overline{B(\tau)}} \cdot \overline{\left[\frac{1}{2\pi i} \int_{-c_k'} \frac{i d\overline{z}}{\overline{B(z)}(i\tau - i\overline{z})}\right]} d\tau \\ &= -\frac{1}{2\pi} \int_{-\infty}^{\infty} B(\tau) e^{-i\tau x} \left[\frac{1}{-2\pi i} \int_{c_k'} \frac{-i dz}{B(z)(-i\tau + iz)}\right] d\tau \end{aligned}$$

$$= \frac{1}{2\pi}\int_{-\infty}^{\infty}\left[-\frac{B(\tau)}{2\pi i}\int_{c_k'}\frac{dz}{B(z)(z-\tau)}\right]e^{-i\tau x}d\tau$$
$$= \frac{1}{2\pi}\int_{-\infty}^{\infty}\left[-\phi_k(\tau)\right]e^{-i\tau x}d\tau$$

and by Paley-Wiener's Theorem [3, Theorem C], since $\phi_k(z) \in H_+^2$, we have that for $x \in (-\infty, 0)$,

$$\frac{1}{\sqrt{2\pi}}\int_{-\infty}^{\infty}\phi_k(\tau)e^{-i\tau x}d\tau = 0.$$

Note that by (26), and the continuity of Fourier Transform, this implies:

Corollary 1

$$\mathbf{E} = \mathbf{P}. \tag{46}$$

(Note that this also follows immediately from [5, Theorems 1 and 2].)
Now we prove Theorem 2:

Proof Let $f(x) \in L^2(0, +\infty)$, and $F(z)$ be defined by (14), i.e. the Fourier inverse transform of $f(x)$, then $F(z) \in H_+^2$. By Lemma 5, we have

$$F(z) = \lim_{n\to\infty}\sum_{k=1}^{n}\alpha\big(F(x), e_k(x)\big)\frac{B(a_k)}{B_n(a_k)}\phi_k(z) + H_F(z), \quad \mathrm{Im}(z) > 0. \tag{47}$$

By the generalized Parseval equality, and (45) and (40),

$$\begin{aligned}\alpha\big(F(x), e_k(x)\big) &= \int_{-\infty}^{+\infty} F(x)\overline{e_k(x)}dx \\ &= \int_0^{+\infty} f(x)\overline{\left(\frac{-1}{\sqrt{2\pi}}e^{-\lambda_k x}\right)}dx \\ &= \frac{-1}{\sqrt{2\pi}}\beta\big(f(x), e^{-\lambda_k x}\big) \quad (k = 1, 2, \ldots).\end{aligned}$$

By (20), we have

$$\frac{B(a_k)}{B_n(a_k)} = \overline{\left(\frac{W(\lambda_k)}{W_n(\lambda_k)}\right)}.$$

Thus, noting (43), the Fourier transform of the summation

$$\sum_{k=1}^{n}\alpha\big(F(x), e_k(x)\big)\frac{B(a_k)}{B_n(a_k)}\phi_k(z)$$

is the summation

$$\sum_{k=1}^{n} \beta\big(f(x), e^{-\lambda_k x}\big)\overline{\left(\frac{W(\lambda_k)}{W_n(\lambda_k)}\right)}\psi_k(x).$$

And, by the continuity of the Fourier transform, the Fourier transform of the limit

$$\lim_{n\to\infty} \sum_{k=1}^{n} \alpha\big(F(x), e_k(x)\big)\frac{B(a_k)}{B_n(a_k)}\phi_k(z)$$

is the limit

$$\lim_{n\to\infty} \sum_{k=1}^{n} \beta\big(f(x), e^{-\lambda_k x}\big)\overline{\left(\frac{W(\lambda_k)}{W_n(\lambda_k)}\right)}\psi_k(x).$$

Noting that $h_f(x)$ is the Fourier transform of $H_F(z)$, and passing the Fourier transform for the two sides of (47), we obtain (39).

Using the same argument as above, we also can obtain (35). Thus, we have

$$f(x) - (P_{\mathbf{P}} f)(x) = f(x) - (P_{\mathbf{E}} f)(x) = h_f(x), \quad 0 < x < +\infty.$$

By the Parseval equality,

$$\int_0^{+\infty} h_f(x)\overline{e^{-\lambda_k(x)}}dx$$

$$= -\sqrt{2\pi}\int_{-\infty}^{+\infty} H_F(x)\overline{e_k(x)}dx = 0, \quad k = 1, 2, \ldots.$$

Hence $(P_{\mathbf{P}} f)(x) = (P_{\mathbf{E}} f)(x)$ is the orthogonal projection of $f(x)$ onto $\mathbf{E}$ (i.e. $\mathbf{P}$). The proof is complete. □

By Lemma 6 and Theorem 1 or Theorem 2, we have

Corollary 2 *$f(x) \in \mathbf{E}$ if and only if $h_f(x) \equiv 0$ for $x \in (0, +\infty)$.*

Remark 1 We can also give an integral expression for the function $h_f(x)$ by using of Laplace transform.[3] For $f(x) \in L^2(0, +\infty)$, the Laplace transform of $\overline{f(x)}$ is

$$\hat{f}(\xi) = \int_0^{\infty} \overline{f(t)}e^{-\xi t}dt, \quad \mathrm{Re}(\xi) > 0.$$

Denote

$$F(\lambda) = -\frac{1}{2\pi}\int_{-\infty}^{\infty} \frac{\overline{\hat{f}(i\tau)}d\tau}{\overline{W(i\tau)}(i\tau - \lambda)}, \quad \mathrm{Re}(\lambda) > 0.$$

---

[3]We are indebted to M. Martirosian for helpful discussions concerning this integral expression.

By [5], we have

$$\begin{aligned}
&\lim_{n\to\infty}\sum_{k=1}^{n}\beta(f,\psi_k)\frac{W(\lambda_k)}{W_n(\lambda_k)}e^{-\lambda_k x}\\
&=\lim_{n\to\infty}\left[-\frac{1}{2\pi}\int_{-\infty}^{\infty}\frac{F(i\tau)}{W_n(i\tau)}e^{-i\tau x}d\tau\right]\\
&=-\frac{1}{2\pi}\int_{-\infty}^{\infty}\frac{F(i\tau)}{W(i\tau)}e^{-i\tau x}d\tau
\end{aligned}$$

where the limits are in the sense of $L^2(0, +\infty)$ convergence, and the integrals are in the sense of principal value.

Noting that

$$\overline{f(x)}=\frac{1}{2\pi i}\int_{-i\infty}^{+i\infty}\hat{f}(\xi)e^{\xi x}d\xi=\frac{1}{2\pi}\int_{-\infty}^{+\infty}\hat{f}(i\tau)e^{i\tau x}d\tau,$$

hence

$$f(x)=\frac{1}{2\pi}\int_{-\infty}^{\infty}\overline{\hat{f}(i\tau)}e^{-i\tau x}d\tau,\quad 0<x<+\infty,$$

by (35), we get

$$h_f(x)=\frac{1}{2\pi}\int_{-\infty}^{\infty}\left[\overline{\hat{f}(i\tau)}+\frac{F(i\tau)}{W(i\tau)}\right]e^{-i\tau x}d\tau,\quad 0<x<+\infty. \tag{48}$$

The techniques in this paper can also be used to study when the above sequences are bases in the subspaces E (i.e. Φ) and $\mathbf{E}$ (i.e. $\mathbf{P}$).

Clearly, if $a_k = i\overline{\lambda}_k$ $(k = 1, 2, \ldots)$, then the sequence (15) satisfies the Carleson condition

$$\inf_k\prod_{j\neq k}\left|\frac{a_j-a_k}{a_j-\overline{a}_k}\right|\geq\delta>0 \tag{49}$$

if and only if the sequence (1) satisfies the Carleson condition

$$\inf_k\prod_{j\neq k}\left|\frac{\lambda_j-\lambda_k}{\lambda_j+\overline{\lambda}_k}\right|\geq\delta>0. \tag{50}$$

It is also well known (see for example [2, p. 1347]) that if (15) satisfies the Carleson condition then it satisfies (16); and if (1) satisfies the Carleson condition then it satisfies (2).

By [2, Theorems 4.1 and 5.2], we have

Lemma 7 *Both the systems* (23) *and* (24) *are bases of E if and only if the sequence* (15) *satisfies the Carleson condition.*

Thus, we have

Theorem 3 *Both the systems* (3) *and* (5) *are bases of* **E** *if and only if the sequence* (1) *satisfies the Carleson condition.*

It will be interesting to consider the above problems in $L^p(0, \infty)$.

Acknowledgements This research was supported in part by grants from NSERC of Canada.

References

1. Duren, P.: Theory of H^p Spaces. Academic Press, New York (1970). Dover, Mineola (2000)
2. Džrbašjan, M.M.: Basicity of some biorthogonal systems and the solution of a multiple interpolation problem in the H^p classes in the half plane. Izv. Akad. Nauk SSSR, Ser. Mat. **42**(6), 1322–1384 (1978), also see pp. 1439–1440 (Russian). Math. USSR Izv. **13**(3), 589–646 (1979) (English translation)
3. Džrbašjan, M.M.: A characterization of the closed linear spans of two families of incomplete systems of analytic functions. Mat. Sb. (N.S.) **114(156)**(1), 3–84 (1981) (Russian). Math. USSR Sb. **42**(1), 1–70 (1982) (English translation)
4. Martirosyan, M.S.: On a representation in an incomplete system of rational functions. Izv. Nats. Akad. Nauk Arm. Mat. **32**(6), 30–38 (1997) (Russian). J. Contemp. Math. Anal. **32**(6), 26–34 (1997) (English translation)
5. Musoyan, V.Kh.: Summation of biorthogonal expansions in incomplete systems of exponentials and rational functions. Izv. Akad. Nauk Arm. SSR Ser. Mat. **21**(2), 163–186 (1986), also see p. 208 (Russian). Soviet, J. Contemp. Math. Anal. **21**(2), 59–83 (1986) (English translation)
6. Shen, X.C.: A remark on two theorems of Musoyan. Approx. Theory Appl. **3**(2–3), 84–90 (1987)

Blaschke Products as Solutions of a Functional Equation

Javad Mashreghi

Abstract In this paper, a family of Blaschke products, as non-trivial inner solutions of Schröder's equation, is introduced. This observation leads to the construction of a surjective composition operator on an infinite dimensional model subspace of H^2.

Keywords Schröder's equation · Blaschke products

Mathematics Subject Classification Primary 30D55 · Secondary 30D05 · 47B33

1 Introduction

The functional equation

$$\psi\big(\varphi(z)\big) = \lambda\,\psi(z), \tag{1}$$

where the function φ is given but ψ and λ are unknown, is known as *Schröder's equation* and has a very long and rich history [3, 4]. In classifying the composition operators with inner symbols on model subspaces of H^2, we faced with a generalized version of this equation in [6, 7]. More precisely, we needed to find inner functions φ, ϕ and ψ which satisfy the functional equation

$$\psi\big(\varphi(z)\big) \times \phi(z) = \psi(z) \quad (z \in \mathbb{D}). \tag{2}$$

We thoroughly discussed the case when ϕ is not a unimodular constant. However, it was shown that the general case (2), simplifies to (1) with $\lambda = 1$, i.e.

$$\psi\big(\varphi(z)\big) = \psi(z). \tag{3}$$

We did not provide an explicit non-trivial solution to (3). As a matter of fact, a complete characterization of the solutions of this case is still an open question. In this paper, as the first step in this direction, we introduce a family of Blaschke products which serve as non-trivial solutions of (3). Then, we will study the model subspaces created by these Blaschke products and study some composition operators on them.

J. Mashreghi (✉)
Département de mathématiques et de statistique, Université Laval, Québec, QC, Canada G1K 7P4
e-mail: javad.mashreghi@mat.ulaval.ca

J. Mashreghi, E. Fricain (eds.), *Blaschke Products and Their Applications*,
Fields Institute Communications 65, DOI 10.1007/978-1-4614-5341-3_5,

2 A Two-Sided Blaschke Sequence

Let t be a fixed parameter in the interval $(-1,1)$. Then the function

$$\varphi_t(z) = \frac{z+t}{1+tz}$$

is an automorphism of the open unit disc with a simple zero at $-t$ [5, p. 155]. One can easily check that the members of this family satisfy the composition rule

$$\varphi_s \circ \varphi_t = \varphi_{\frac{s+t}{1+st}}. \tag{4}$$

In particular,

$$\varphi_t \circ \varphi_{-t} = \varphi_0 = id. \tag{5}$$

The identity (4) is the main key to study the iterations $\varphi_\alpha^{[n]}$, where $n \in \mathbb{Z}$ and $0 < \alpha < 1$. As a matter of fact, if we write

$$\varphi_\alpha^{[n]} = \varphi_{\alpha_n} \quad (n \in \mathbb{Z}),$$

then, by (4) and (5), the two-sided sequence $(\alpha_n)_{n\in\mathbb{Z}}$ satisfies the recursive relation

$$\alpha_n = \frac{\alpha + \alpha_{n-1}}{1+\alpha\alpha_{n-1}} \quad (n \in \mathbb{Z}), \tag{6}$$

with $\alpha_0 = 0$ and $\alpha_{\pm 1} = \pm\alpha$. Note that $\alpha_n \in (0,1)$ if $n \geq 1$, and $\alpha_n \in (-1,0)$ if $n \leq -1$.

To proceed, write

$$1 - \alpha_n = \frac{(1-\alpha)(1-\alpha_{n-1})}{1+\alpha\alpha_{n-1}}.$$

Hence,

$$1 - \alpha_n \leq (1-\alpha)(1-\alpha_{n-1}),$$

which, by induction, gives

$$1 - \alpha_n \leq (1-\alpha)^n \quad (n \geq 1).$$

Similarly, we can show that

$$1 + \alpha_n \leq (1-\alpha)^{|n|} \quad (n \leq -1).$$

Therefore, $(\alpha_n)_{n\in\mathbb{Z}}$ is a two-sided Blaschke sequence. This fact allows us to define the two-sided Blaschke product

$$B_\alpha = \prod_{n\leq -1} (-1)\varphi_{\alpha_n} \times z \times \prod_{n\geq 1} \varphi_{\alpha_n}.$$

Note that to ensure the convergence of product, the factor -1 is needed when $n \leq -1$. The one-sided version of this Blaschke product was first introduced in [6].

Clearly, since $\varphi_\alpha^{[n]} = \varphi_{\alpha_n}$, $n \in \mathbb{Z}$, we can also write

$$B_\alpha = \prod_{n \leq -1} (-1)\varphi_\alpha^{[n]} \times \prod_{n \geq 0} \varphi_\alpha^{[n]}.$$

This format has an advantage. It reveals that B fulfils a functional equation. To detect this functional equation, write

$$\begin{aligned} B_\alpha \circ \varphi_\alpha &= \prod_{n \leq -1} (-1)\varphi_\alpha^{[n]} \circ \varphi_\alpha \times \prod_{n \geq 0} \varphi_\alpha^{[n]} \circ \varphi_\alpha \\ &= \prod_{n \leq -1} (-1)\varphi_\alpha^{[n+1]} \times \prod_{n \geq 0} \varphi_\alpha^{[n+1]} \\ &= \prod_{n \leq 0} (-1)\varphi_\alpha^{[n]} \times \prod_{n \geq 1} \varphi_\alpha^{[n]} \\ &= -z \times \prod_{n \leq -1} (-1)\varphi_{\alpha_n} \times \prod_{n \geq 1} \varphi_{\alpha_n}. \end{aligned}$$

Thus,

$$B_\alpha \circ \varphi_\alpha = -B_\alpha. \tag{7}$$

In other words, given $\varphi = \varphi_\alpha$, then B_α is a solution of the functional equation (1) with $\lambda = -1$. Moreover, if we apply (7) twice, we get

$$B_\alpha \circ \varphi_{\alpha_2} = B_\alpha. \tag{8}$$

This means that, with $\varphi = \varphi_{\alpha_2}$, the Blaschke product B_α is also a solution of the functional equation (3).

We end this section with an interesting open question. To explain the situation, note that the fixed points of φ_α are ± 1, and the point 1 is the Denjoy–Wolff point of φ_α. But

$$\varphi_\alpha'(1) = \frac{1-\alpha}{1+\alpha},$$

which is never equal to 1. Let us recall that the value of derivative at the Denjoy–Wolff point is always restricted to real numbers in the interval $(0, 1]$.

Open Question Can we construct an inner function φ whose Denjoy–Wolff point is 1 and, moreover, $\varphi'(1) = 1$, and for which the functional equation (3) has a non-trivial solution (preferably, a Blaschke product)?

3 A Surjective Composition Operator on K_B

A. Beurling [1] characterized the closed subspaces of H^2 which are invariant under the forward shift operator

$$\begin{aligned} S:\quad H^2 &\longrightarrow H^2 \\ f &\longmapsto zf. \end{aligned}$$

In this classical work, he show that they are precisely of the form ΘH^2 where Θ is an inner function. The corresponding *model subspace* K_Θ is, by definition, the orthogonal complement of ΘH^2, i.e.,

$$K_\Theta = H^2 \ominus \Theta H^2.$$

Therefore, model subspaces are in fact the closed subspaces of H^2 which are invariant under the backward shift operator S^*. For a detailed treatment of S^* and its invariant subspaces see [2] and [8]. There are several equivalent ways to describe model subspaces. In particular, if Θ is a Blaschke product, there is an easy way to characterize K_Θ, which suites this work. Write

$$\Theta(z) = \prod_n \frac{|z_n|}{z_n} \frac{z_n - z}{1 - \bar{z}_n z}.$$

Then K_Θ is precisely the closure of linear span of

$$f_n(z) = \frac{1}{1 - \bar{z}_n z}$$

in H^2.

We now study the model subspace created by the Blaschke product introduced in Sect. 2. A part of the following theorem follows from the results in [6]. However, for the sake of completeness, we provide a more direct and different proof.

Theorem 1 *Let* $0 < \alpha < 1$, *and let*

$$\mathbf{B}_\alpha = z^2 \times \prod_{n \le -1} (-1)\varphi_{\alpha_n} \times \prod_{n \ge 1} \varphi_{\alpha_n}.$$

Then the mapping

$$\begin{aligned} C_{\varphi_\alpha}:\quad K_{\mathbf{B}_\alpha} &\longrightarrow K_{\mathbf{B}_\alpha} \\ f &\longmapsto f \circ \varphi_\alpha \end{aligned}$$

is well-defined, bounded and surjective.

Proof Let

$$f_n(z) = \begin{cases} \frac{1}{1+\alpha_n z} & \text{if } n \neq 0, \\ z & \text{if } n = 0 \end{cases}$$

and $f_{00} = 1$. The reason for this indexing is clarified below. For simplicity, let $\mathcal{R}$ denote the range of C_{φ_α}.

We know that the span of the set $\{f_{00},\ f_n : n \in \mathbb{Z}\}$ is precisely $K_{\mathbf{B}_\alpha}$. Hence, to show that C_{φ_α} is well-defined and bounded, in the light of Littlewood's theorem and that model subspaces are closed in H^2, it is enough to verify that $C_{\varphi_\alpha} f_{00} \in K_{\mathbf{B}_\alpha}$ and that

$$C_{\varphi_\alpha} f_n \in K_{\mathbf{B}_\alpha} \quad (n \in \mathbb{Z}).$$

We start with the trivial, but important, identity $C_{\varphi_\alpha} f_{00} = f_{00}$ to deduce that $C_{\varphi_\alpha} f_{00} \in K_{\mathbf{B}_\alpha}$ and

$$f_{00} \in \mathcal{R}. \tag{9}$$

Since

$$C_{\varphi_\alpha} f_0 = C_{\varphi_\alpha} z = \varphi_\alpha$$

and we can write $\varphi_\alpha = f_{00}/\alpha - (1-\alpha^2)\, f_1/\alpha$, we have

$$C_{\varphi_\alpha} f_0 = \frac{1}{\alpha}\, f_{00} + \frac{\alpha^2 - 1}{\alpha}\, f_1.$$

Therefore, on one hand, we immediately see that

$$C_{\varphi_\alpha} f_0 \in K_{\mathbf{B}_\alpha},$$

and, on the other hand since $C_{\varphi_\alpha} f_{00} = f_{00}$, we rewrite the preceding identity as

$$C_{\varphi_\alpha}\left(f_0 - \frac{1}{\alpha}\, f_{00} \right) = \frac{\alpha^2 - 1}{\alpha}\, f_1$$

to deduce

$$f_1 \in \mathcal{R}. \tag{10}$$

For $n \neq 0$ and $n \neq -1$, by direct evaluation and using (6), we get

$$C_{\varphi_\alpha} f_n = \frac{\alpha}{\alpha_{n+1}}\, f_{00} + \left(1 - \frac{\alpha}{\alpha_{n+1}} \right) f_{n+1}.$$

But, for $n = -1$, we have

$$C_{\varphi_\alpha} f_{-1} = \frac{1}{1-\alpha^2}\, f_{00} + \frac{\alpha}{1-\alpha^2}\, f_0.$$

The last two relations imply

$$C_{\varphi_\alpha} f_n \in K_{\mathbf{B}_\alpha} \quad \bigl(n \in \mathbb{Z} \setminus \{0\}\bigr),$$

and if we properly rewrite those identities, we get

$$f_n \in \mathcal{R} \quad \bigl(n \in \mathbb{Z} \setminus \{1\}\bigr). \tag{11}$$

Finally, the relations (9), (10) and (11) ensure that C_{φ_α} is surjective. □

Acknowledgements This work was supported by NSERC (Canada) and FQRNT (Québec).

References

1. Beurling, A.: On two problems concerning linear transformations in Hilbert space. Acta Math. **81**, 239–255 (1949)
2. Cima, J., Ross, W.: The Backward Shift on the Hardy Space. Mathematical Surveys and Monographs, vol. 79. Am. Math. Soc., Providence (2000)
3. Cowen, C.: Iteration and the solution of functional equations for functions analytic in the unit disk. Trans. Am. Math. Soc. **265**, 69–95 (1981)
4. Kuczma, M.: Note on the Schröder's functional equation. J. Aust. Math. Soc. **4**, 149–151 (1963)
5. Mashreghi, J.: Representation Theorems in Hardy Spaces. London Mathematical Society Student Text Series, vol. 74. Cambridge University Press, Cambridge (2009)
6. Mashreghi, J., Shabankhah, M.: Composition of inner functions, submitted
7. Mashreghi, J., Shabankhah, M.: Composition operators on finite rank model subspaces. Glasgow Math. J., to appear
8. Ross, W.: The Backward Shift on Hp. Selected Topics in Complex Analysis. Oper. Theory Adv. Appl., vol. 158, pp. 191–211. Birkhäuser, Basel (2005)

Cauchy Transforms and Univalent Functions

Joseph A. Cima and John A. Pfaltzgraff

Abstract We use a formula of Pommerenke relating the primitives of functions which are the Cauchy transforms of measures on the unit circle to their behavior in the space of functions of bounded mean oscillation. This is a linear process and it has some smoothness. Further, there is a non-linear map from the Cauchy transforms into the normalized univalent functions. We show that for the subspace H^1 of Cauchy transforms the univalent functions so obtained have quasi-conformal extensions to all of the plane.

Keywords Cauchy transform · BMOA · Bloch

Mathematics Subject Classification Primary30H35 · Secondary47G10

1 Introduction

Classical Banach spaces have played a key role in the development of modern analysis. The principal thrust of many studies is to understand various properties of individual vectors in the space as well as to define and study bounded (usually linear) operators on the various spaces. In this paper we restrict ourselves to some classical spaces of analytic functions on the unit disc. The goal of our work is to consider the Banach space CT (Cauchy Transforms of measures on the unit circle) and relate it to the set $\mathbb{U}$ of univalent functions h on D normalized by $h(0) = 0$, $h'(0) = 1$. Some of these spaces that become involved along the way are the Bloch spaces, $\mathbb{B}$ and $\mathbb{B}_0$, the space of analytic functions of bounded mean oscillation, *BMOA*, and the classical Hardy spaces, H^p all on the unit disc. The term "Cauchy Transforms" appearing in the title of this print (and written for brevity as CT) appears in different settings. The one of primary interest in this work is the following.

J.A. Cima (✉) · J.A. Pfaltzgraff
Department of Mathematics, University of North Carolina, Chapel Hill, NC 27599-3250, USA
e-mail: cima@email.unc.edu

J.A. Pfaltzgraff
e-mail: jap@math.unc.edu

J. Mashreghi, E. Fricain (eds.), *Blaschke Products and Their Applications*, Fields Institute Communications 65, DOI 10.1007/978-1-4614-5341-3_6,

Definition 1 For μ a finite Borel measure on the unit circle $\mathbb{T}$ and z in the unit disc D, set

$$f(z) \equiv \int_{\mathbb{T}} \frac{d\mu(\zeta)}{(1-\bar{z}\zeta)} = C * d\mu(z).$$

The norm and further properties of CT are given in Sect. 1 below. The reader will find copious materials on these spaces in the books of Cima, Matheson and Ross [3], Garnett [5], Pommerenke [8], and Mashreghi [7] in the references.

One of the most important operators on such spaces is the operator of forming the primitive,

$$f^*(z) = P(f)(z) = \int_o^z f(w)dw, \quad |z| < 1.$$

For example it is known that if $f \in H^1$, then f^* is in the disc algebra A and has absolutely continuous boundary values, and if $f \in H^p$ with $0 < p < 1$, then $f^* \in H^q$ for $q = \frac{p}{1-p}$ (see Duren [4]). We limit our study of this primitive operator to the space CT. In this case it is easily checked that $f^* \in \mathbb{B}$. We show that P is compact from CT into CT. Further, there is an interesting non-linear mapping between CT and $\mathbb{U}$. This leads us to a class of such univalent h that have quasi-conformal extensions to all of $\mathbb{C} \equiv R^2$. Moreover, there is a natural measure (see definitions below) associated to the complex dilatation of this h which is shown to be a vanishing Carleson measure. We include the necessary definitions and a few important facts concerning earlier results in Sect. 1.

2 Definitions and Known Results

The unit disc in the complex plane is denoted by D and its boundary (the unit circle) is written as $\mathbb{T}$. The notations H^p, $\mathbb{B}$, $\mathbb{B}_o$, *BMOA* and CT (on the unit disc D) are all well known and can be found in the text material in the references. We mention a few necessary facts about the CT space. The space CT can be realized as the dual of the disc algebra and the duality yields the following

$$CT \cong (A)^* \cong \frac{L^1}{\overline{H}^1_o} + \mathbb{M}_s,$$

where the term $\mathbb{M}_s$ denotes the set of singular Borel measures on the circle.

This space is realized as a Banach space of analytic functions on the unit disc in the following way. For μ a finite Borel measure on $\mathbb{T}$ define

$$f(z) = C * d\mu(z) \equiv \int_{\mathbb{T}} \frac{d\mu(\zeta)}{(1-z\overline{\zeta})} \quad |z| < 1, \ \zeta = \exp(\imath\theta).$$

Note that there are many measures which produce the same analytic f. One uses the coset norm on CT.

$$\|f\|_{CT} = \inf\big(\|d\mu\| \mid f(z) = C * d\mu(z)\big),$$

where the inf is taken over all measures producing the given f and where $\|d\mu\|$ is the total variation norm for the measure.

Given $f \in CT$ there is a measure, say $d\mu_f$ with $f(z) = C * d\mu_f(z)$ and $\|d\mu_f\| = \|f\|_{CT}$ and it is useful to use this measure in our work. See [3] for more detail.

Definition 2 Assume $F(Z)$ is an orientation preserving homeomorphism of a planar domain E into $\mathbb{C} \simeq \mathbb{R}^2$. f is a quasiconformal mapping on E if f satisfies:

(i) f is ACL (absolutely continuous on line (segments) in E) and
(ii) there exists a constant k with $0 \le k < 1$ such that

$$|f_{\bar{z}}| \le k|f_z| \quad \text{a.e. in } E.$$

If h is a quasiconformal mapping of E into $\mathbb{C}$ and $\tilde{h}$ is a mapping from $\mathbb{C}$ into $\mathbb{C}$ which is quasiconformal and if $\tilde{h}|_W = h$, then h is a quasiconformal extension of h.

Results of Becker (Pommerenke [8]) show that if a univalent h on D satisfies certain conditions then h admits a quasiconformal extension to all of $\mathbb{C}$ given by

$$\tilde{h}(z) = h\left(\frac{1}{\bar{z}}\right) - \left(z - \frac{1}{\bar{z}}\right) h'\left(\frac{1}{\bar{z}}\right)$$

and with complex dilatation μ defined by

$$\mu(z) = -\frac{1}{(\bar{z})^2}\left(z - \frac{1}{\bar{z}}\right)\left(\frac{h''}{h'}\right)\left(\frac{1}{\bar{z}}\right).$$

In some sense the dilatation is a measure of how far the map h is from being conformal.

In the paper [1] they address the following two problems.

Problem 1 Suppose that f is a local homeomorphism defined on the domain $E \subset \mathbb{C}$. What additional conditions on f and E allow one to conclude that f is injective?

The second problem they study involves the Bers universal Teichmuller space T. With S_f the Schwarzian of f on E and

$$\mathcal{M} \equiv \left\{S_g : g \text{ is conformal on } D \text{ with } g(B) \subseteq \bar{\mathbb{R}}^2\right\}$$

define

$$T \equiv \left\{g \in \mathcal{M} \mid g \text{ has a quasiconformal extension to } \bar{\mathbb{R}}^2\right\}.$$

The set $\mathcal{M}$ can be given a norm (see [1]). T is known as the universal Teichmuller space.

The second problem is a follows.

Problem 2 Describe the closure of the Bers universal space in S.

In their paper [1], Astala and Gehring give a partial answer to Problem 1. They use their solution of Problem 1, in the locally quasiconformal case, to obtain new useful information concerning Problem 2.

Define a set S as follows.

Definition 3 S is the set defined by the functions $\log h'$, where h is conformal on D.

By the Koebe distortion theorem S is a norm bounded subset of $\mathbb{B}$. Let T denote the interior of S in $\mathbb{B}$. T is a model for the universal Teichmuller space. In this setting the following defined quantities are significant in certain characterizations.

Definition 4 A Jordan curve Γ is called a quasi-circle if there is a constant $C > 0$ such that for every z and ζ in Γ,

$$\text{diameter}(\gamma) \leq C|z - \zeta|$$

where γ is the shortest sub arc of Γ joining z to ζ.

Definition 5 A Jordan curve Γ is said to be a Lavrentiev curve if there is a constant $C > 0$ such that for every z and ζ in Γ,

$$l(\gamma) \leq C|z - \zeta|$$

where γ is the shortest sub arc of Γ joining z to ζ and $l(\gamma)$ is the arc length of γ.

We will use the following two important theorems relating T to quasiconformal extensions (see [2] Astala and Zinsmeister). They also give the flavor of results relating to our work.

Theorem 1 *The function* $\log h'$ *belongs to* T *if and only if* h *has a quasiconformal extension from* D *to all of* R^2.

Theorem 2 *The function* $\log h'$ *belongs to* T *if and only if* $\Gamma = \partial(h(\mathbb{T}))$ *is a quasi-circle (a Jordan curve for which there exists a constant* $C > 0$ *such that* $\forall z, \zeta \in \Gamma$, $\text{diam}(\gamma) \leq C|z - \zeta|$ *where* γ *is the smaller of the two subarcs of* Γ *joining* z *to* ζ).

Assume $\Omega \subseteq \mathbb{C}$ is a simply connected domain. We require the following definition (see Zinsmeister [9]).

Definition 6 A measure $d\nu$ on Ω is said to be a Carleson measure if and only if there is a constant $C > 0$ such that

$$|d\nu|\big(D(z,r)\big) \leq Cr$$

for every $z \in \partial(\Omega)$ and $r \leq$diameter $(\partial\Omega)$.

$D(z,r)$ is the disc with center z and radius r.

In the context of Definition 5 above, there are conditions on the complex dilatation of a quasiconformal mapping h on D, admitting a quasiconformal extension to $\mathbb{C}$, so that by using the Fefferman-Stein characterization of BMO one can prove that the measure

$$d\nu(z) = \mu_h(z)\big(|z|^2 - 1\big)^{-1} dxdy$$

is a Carleson measure on $(\mathbb{C}\backslash D)$. In the case we are considering for h univalent on D and $\log h'$ in the Bloch space, we will obtain a subset of CT for which Theorem 2 holds. In this case the form of the complex dilatation becomes simpler in that one can replace the term $\frac{h''}{h'}$ by a function in H'. One of the strengths of producing Carleson measures $d\nu$ is to show that the identity map $f \to f$ from H' into $L'(d\nu)$ is continuous. This has many applications. If the "big oh" estimate required on measure $d\nu$ is strengthened to a "little oh" condition, then frequently the operator is compact.

3 The Primitive of a CT Function

For $z, w \in \mathbb{D}$ and $\zeta = \exp(\imath\theta)$ let P denote the primitive map

$$f \to P(f)(z) = \int_o^z f(w)dw \equiv f^*(z)$$

from the space CT. For $f = C * \mu \in CT$ by a change of variables we may write

$$f^*(z) = \int_{\mathbb{T}} \log\left(\frac{1}{(1 - z\overline{\zeta})}\right)(-\overline{\zeta})d\mu(\zeta) \equiv \int_{\mathbb{T}} \log\left(\frac{1}{(1 - z\overline{\zeta})}\right)d\nu(\zeta). \qquad (1)$$

Proposition 1 *The map P from CT to the Bloch space $\mathbb{B}$ is linear, one to one and continuous.*

Proof Clearly P is linear and one to one into $\mathbb{B}$. We check the continuity. Given $f(z) = C * d\mu(z)$ where we may choose $\|f\|_{CT} = \|d\mu\|$ (the total variation norm) we have

$$\big(1 - |z|^2\big)\big|\big(f^*\big)'(z)\big| = \big(1 - |z|^2\big)\left|\int_{\mathbb{T}} \frac{d\mu(\zeta)}{(1 - z\overline{\zeta})}\right| \leq 2\|\mu\|. \qquad \square$$

This is sharp in the sense that $CT \subset \bigcap H^p$, $p < 1$ and $f(z) = \frac{(\log(1-z))}{(1-z)} \in H^p \setminus CT$, $p < 1$ but f^* is not in $\mathbb{B}$.

A bit more is true.

Theorem 3 *The map P maps $\frac{L^1}{H^1_o}$ into the small Bloch space $\mathbb{B}_o$.*

Proof Given $F \in L^1(\mathbb{T})$ it is known that the Cesaro sums $(\sigma_n(\zeta, F))$ tend to F in $L^1(\mathbb{T})$. As above there is a unique measure $d\mu$ (absolutely continuous with respect to the normalized Lebesgue measure $\frac{d\theta}{2\pi}$) for which

$$f(z) = \int_{\mathbb{T}} \frac{F(\zeta)dm(\zeta)}{(1 - z\overline{\zeta})} = \int_{\mathbb{T}} \frac{d\mu(\zeta)}{(1 - z\overline{\zeta})}$$

and with $\|f\|_{CT} = \|\mu\| \leq \|F\|_{L^1}$.

On the unit circle we write $\sigma_n(\theta) = A_n(\theta) + \overline{B}_n(\theta)$ where A_n and B_n are analytic polynomials in $\zeta = \exp(\imath\theta)$ and B_n has zero constant term.

Since $P(A_n)(z) = A_n^*(z)$ is in $\mathbb{B}_o$, and $\mathbb{B}_o$, is closed in $\mathbb{B}$, it suffices to show that A_n^* converges in $\mathbb{B}$, to f^*. To this end consider

$$\begin{aligned}(1 - |z|^2)\left|(A_n^*)'(z) - (f^*)'(z)\right| &= (1 - |z|^2)\left| \int_{\mathbb{T}} \frac{\sigma_n(\zeta)dm(\theta)}{(1 - z\overline{\zeta})} - \int_{\mathbb{T}} \frac{F(\zeta)dm(\theta)}{(1 - z\overline{\zeta})} \right| \\ &\leq C \int_{\mathbb{T}} \left|\sigma_n(\zeta) - F(\zeta)\right| dm(\theta) \to 0\end{aligned}$$

as $|z| \to 1$. □

4 The Connection to Univalent Functions

We begin this section by recalling a result of Pommerenke [8].

Theorem 4 *A function F analytic on the unit disc is in the Bloch space $\mathbb{B}$ if and only if there exist constants α, and β and a function h in $\mathbb{U}$ so that F satisfies the equality*

$$F(z) = \alpha \log h'(z) + \beta,$$

for $z \in D$.

We are considering $F(z) = P(f)(z) = f^*(z) \in \mathbb{B}$ for $f \in CT$. Note $F(0) = 0 = \alpha \log 1 + \beta$ so we have $\beta = 0$, and we may choose any number $\alpha \geq \|F\|_{\mathbb{B}}$. In studying the proof of this result one can choose the constant α larger than or equal to the Bloch norm of F but this changes the choice of the univalent mapping. Solving for h in this equation with $\gamma = \frac{1}{\alpha} > 0$ we find

$$h(z) = \int_o^z \exp\bigl(\gamma f^*(w)\bigr)dw = \int_o^z \exp\left(\gamma \int_o^w f(u)du\right)dw. \tag{2}$$

As an easy example assume we choose $f(z) = \frac{1}{(1-z)}$, $\alpha = 2$ and calculate to produce $h(z) = (-2)(\sqrt{(1-z)} - 1)$.

Proposition 2 *The nonlinear mapping given in Theorem* 4 *is continuous from the space CT with its topology into the space $\mathbb{U}$ in the compact open topology.*

The proof is straight forward using normal family arguments a few times.

The map above considered from the Banach CT into the Banach space *BMOA* is analytic and its Frechet Derivative can be computed by expanding the exponential in the integral. Since this has no applications to our work we omit the computation.

5 Connection to *BMOA* and Teichmuller Space

In [6] Hallenbeck and Samotij have proven the following result (which deserves to be more will known).

Theorem 5 *For $f \in CT$, $f^* \in BMOA$.*

We offer two proofs of this result. The first is our proof. In their paper [4], they state a key lemma and leave it to the reader to prove the lemma. We feel this result is so significant in its own right we offer a proof of this as well.

Proof 1 We use the duality statement

$$(H^1)^* \simeq BMOA.$$

To this end it suffices to show that for every analytic polynomial p, the inequality

$$|\langle p, f^* \rangle| \leq C \|p\|_{L^1(\mathbb{T})}$$

holds where C is independent of p. The integral above reads as

$$|\langle p, f^* \rangle| \leq \left| \int_{\mathbb{T}} p(t) \overline{\left(\int_{\mathbb{T}} \log(1 - t\overline{\zeta})\overline{\zeta} \, d\mu(\zeta) \right)} d|t| \right|.$$

Since $\log(1 - t\overline{\zeta})$ is in *BMOA* and its norm is independent of $|t| = 1$ we have

$$|\langle p, f^* \rangle| \leq \|d\mu\| \left(\|p\|_{L^1(\mathbb{T})} \left\| \log(1 - \bar{t}\zeta) \right\|_{BMOA} \right)$$

and this establishes the theorem and ends the first proof. □

Note that the example given earlier shows that one can not improve this inclusion (i.e. $f^*(z) = (\log(1-z))^2$ is not in BMOA).

For the second proof we establish their mean oscillation (MO) inequality on intervals in $\mathbb{T}$. For $g \in L'(\mathbb{T})$ and μ a finite Borel measure on $\mathbb{T}$ and I an interval in $\mathbb{T}$, set the following notation:

$$MO(g, I) = 1/|I| \int_I \left(\left| g(w) - 1/|I| \int_I g(\zeta) d\zeta \right| \right) |dw|,$$

$$g * dv(w) \equiv \int_T g(w\overline{x}) dv(x),$$

with $\mathbb{T}$ the unit circle, $w, x \in \mathbb{T}$.

Lemma 1 *Let $d\mu$ be a finite Borel measure on $\mathbb{T}$, and let $g \in L^1(\mathbb{T})$. With I an arc in $\mathbb{T}$ one has*

$$MO(g * \mu, I) \leq \|\mu\| \Big(\sup_{w \in \text{ supp } \mu} MO(g, \overline{w}I) \Big).$$

Proof We wish to prove

$$MO(g * \mu, I) \leq \|\mu\| \Big(\sup_{w \in \text{ supp } \mu} MO(g, \overline{w}I) \Big).$$

where $w \in$ supp $d\mu$.

$$MO(g * d\mu, I) = 1/|I| \int_I \left| \left(g * d\mu(w) - 1/|I| \int_I g * d\mu(u)|du| \right) \right| |dw|.$$

We compute with the integrand in the parenthesis.

$$\left| \int_T \left(g(w\overline{x}) d\mu(x) - \int_T 1/|I| \int_I g(u\overline{x})|du| \right) d\mu(x) \right|.$$

This is less than or equal to

$$\int_T \left| g(w\overline{x}) - 1/|I| \int_I g(u\overline{x}) du \right| \, |d\mu(x)|.$$

Now do the integration with respect to the remaining w variable.

$$\int_T \left(1/|I| \int_I \left| \left(g(w\overline{x}) - 1/|I| \int_I g(u\overline{x}) du \right) \right| \right) \, |d\mu(x)|.$$

Replace the term $(w\overline{x}) = m$ and note $|dw| = |dm|$ and the interval I is replaced by the interval $\overline{x}I$. Do the same computation in the u integration, with $u\overline{x} = n$. Again $|du| = |dn|$ and the interval I is replaced by the interval $\overline{x}I$.

Now rewrite the last integral with these changes to obtain

$$\int_T |d\mu(x)| MO(g, \overline{x}I).$$

Take the sup over the $x \in$ supp $d\nu$ and obtain

$$MO(g * d\nu, I) \leq \|\nu\| \Big(\sup_{x \in \text{ supp } \mu} MO(g, \overline{x} I) \Big),$$

where the x is in the support of the measure $d\nu$.

Using this key lemma it is straightforward to complete the proof of Theorem 5. □

At times it is easier to check sequential convergence of specific sequences in CT rather than in the image of CT in *BMOA* (e.g. our example with Cesaro means). So it would be useful to prove the linear map P is compact from CT with its topology into *BMOA* with its topology. Counter examples have not been forth coming and we have been unable to prove this. However, the following holds.

Theorem 6 *The linear map P is compact from CT into CT.*

Proof Recall that $BMOA \subset H^2 \subset CT$ so P maps CT into H^2. We also have the inequality $\|f^*\|_{H^2} \geq \|f^*\|_{CT}$. Thus it suffices to prove that P is compact into H^2. First, if $f(k)$ is a sequence in the unit ball of CT with $f(k)(z) = \sum_{n=o}^{\infty} f_n(k) z^n$ then

$$P\big(f(k)\big)(z) = \sum_n \frac{f_n(k)}{(n+1)} z^{n+1} \equiv F(k)(z).$$

The coefficients $|f_n(k)|$ are bounded by one (by assumption). The H^2 norm in this case is

$$\big\|F(k)\big\|_{L^2} = \sqrt{\sum_{n=o} \frac{|f_n(k)|^2}{(n+1)^2}} \leq \sqrt{\sum_{n=o} \frac{1}{(n+1)^2}} \leq C.$$

These bounds imply $\{F(k)\}$ is a normal family and hence there is a subsequence (which we again label as $F(k)$) and a holomorphic function $g(z)$ on D with

$$\lim_{k \to \infty} F(k)(z) = g(z)$$

uniformly on compacta of D.

As noted we may, with out loss of generality, choose measures μ_k with $\|\mu_k\| \leq 1$ and

$$f(k)(z) = C * d\mu_k(z).$$

By the Banach-Alaoglu theorem there is a subsequence of these measures (again written as μ_k) that converge weak* to a measure μ with $\|\mu\| \leq 1$. For each fixed $z \in D$ the kernel $\frac{1}{(1-\overline{z}\zeta)}$ is continuous on T and we have

$$f(k)(z) = C * d\mu_k(z) \to C * d\mu(z) \equiv h(z)$$

with $k \to \infty$. Hence, $h \in CT$, $P(h)(z) = g(z) \in H^2$ and

$$\lim_{k\to\infty} F(k)(z) = g(z)$$

uniformly on compacta in D.

It remains to prove the convergence in H^2,

$$\lim_{k\to\infty} \left\| F(k) - g \right\|_{H^2} = 0.$$

To simplify the notation we define

$$M_k(z) = F(k)(z) - g(z)$$

and show $\|M_k\|_{H^2} \to 0$ as $k \to \infty$.

Let $\epsilon > 0$ be given. Choose K so large that

$$\sum_{k=K+1} \frac{1}{(k+1)^2} \leq \epsilon.$$

Then writing $M_k(z) = \sum_{p=o}^{\infty} B_p(k) z^p$ we have

$$\begin{aligned} \|M_k\|_{H^2}^2 &= \sum_{p=o}^{K} \left|B_p(k)\right|^2 + \sum_{K+1}^{\infty} \left|B_p(k)\right|^2 \\ &\leq \sum_{p=o}^{K} \left|B_p(k)\right|^2 + 2\left(\sum_{p=K+1}^{\infty} \frac{|f_p(k)|^2}{(p+1)^2} + \sum_{p=K+1}^{\infty} \frac{|g_p|^2}{(p+1)^2} \right) \\ &\leq \sum_{p=0}^{K} \left|B_p(k)\right|^2 + 2\epsilon. \end{aligned} \tag{3}$$

Now M_k is a normal family so using the Cauchy integral formula all the derived families $(M_k)^j$, $j = 1, 2, \ldots, K$ are normal families. Some subsequence of each of these families tend to zero uniformly on compacta (and using the same notation as the original sequence) we may assume that there is one sequence (keeping the same sequence notation) (M_k), $k = 0, 1, 2, \ldots, K$ converging uniformly on compacta to zero. Hence, for each fixed integer $0 \leq p \leq K$,

$$\frac{M^p(k)(0)}{p!} = B_p(k) \to 0$$

as $k \to \infty$. Thus there is a positive integer $Q > K$ with

$$\left|B_p(k)\right| \leq \frac{\epsilon}{K}$$

for all $p = 0, 1, \ldots, K$ with $k > Q$. Thus from (3) it follows that

$$\lim_{k\to\infty} \|M_k\|_{H^2} = 0.$$

This shows that P is compact on CT into H^2 and consequently into CT. □

A natural question in the light of Theorem 1 and Proposition 1 is the following. "For which $f \in CT$ does the associated normalized conformal map h associated to f^* by Theorem 4 have a quasiconformal extension to R^2?"

We have the following result.

Theorem 7 *Let $f \in H^1(\subset CT)$ and let h be the associated normalized univalent map on the disc satisfying Theorem 4. Then h has a quasiconformal extension to R^2. Further, the complex dilatation $D(h)(z)$ of the extension is given by the formula*

$$D(h)(z) = \frac{-1}{\bar{z}^2}\left(z - \frac{1}{\bar{z}}\right)\left(\frac{1}{\alpha} f\left(\frac{1}{\bar{z}}\right)\right).$$

Proof Assume $f \in H^1$ and h is the univalent map associated to f in Theorem 4.

By Theorem 2 it is sufficient to prove that h satisfies the Ahlfors condition. The following stronger result is true. Namely, $\partial(h(D))$ satisfies the Lavrentiev condition.

Consider, as earlier, the integral

$$\int_T \bar{\zeta}\big(\log(1 - u\bar{\zeta})\big) f(\zeta) dm(\theta), \quad u = \exp(\imath\phi),\ \zeta = \exp(\imath\theta).$$

The function $\bar{\zeta}(\log(1-\overline{u}\zeta))$ is in BMOA and by the H^1-*BMOA* duality the above integral is bounded by the number $M = \|\log(1 - \bar{u}\zeta)\|_{BMOA}\|f\|_{H^1}$.

Hence, in the arclength integral for h (with $z' = h(\exp(\imath\phi_1)), \zeta' = h(\exp(\imath\phi_2))$ and γ the shorter arc of $\partial(h(D))$ (joining these two points) we have

$$\begin{aligned} l(\gamma) = \left|\int_{\phi_1}^{\phi_2} h'(u) d\phi\right| &\leq \int_{\phi_1}^{\phi_2} \exp\left(\gamma\Re\left(\int_T \bar{\zeta}\big(\log(1 - u\bar{\zeta})\big) f(\zeta) dm(\theta)\right)\right) d\phi \\ &\leq C\int_{\phi_1}^{\phi_2} d\phi = C|\phi_2 - \phi_1|. \end{aligned}$$

□

Note that in our earlier example with $f(z) = \frac{1}{(1-z)}$ and $h'(z) = \frac{1}{\sqrt{(1-z)}}$ the arc length estimate for the points $\exp(-\imath\delta)$ to $\exp(\imath\delta)$ is approximately $\log(\delta)$, whereas the integral is order δ so that the inequality

$$l(\gamma) \leq C(\delta)$$

fails.

The natural measure induced by the complex dilatation in the above theorem

$$\nu(z) = \frac{|D(z)|^2 dxdy}{(|z|^2 - 1)}$$

will satisfy a vanishing Carleson condition.

Theorem 8 *The measure ν indicated above is a vanishing Carleson measure.*

Proof Our domain $\Omega = \{|z| > 1\}$ and we consider $|s| = 1$. Without loss of generality (and for simplicity of notation), we may assume $s = 1$. Let $r > 0$ and $\epsilon > 0$ be given. The coefficient of our measure $d\nu$ (or $d|\nu|$) can be estimated by the quantity

$$\frac{|D(z)|^2}{(|z|^2 - 1)} \simeq \left| f\left(\frac{1}{\bar{z}}\right)\right| \left[\left| f\left(\frac{1}{\bar{z}}\right)\right| (|z|^2 - 1)\right].$$

For $|z| > 1$ there is $r_0 > 0$ so that if $0 < |z|^2 - 1 < r \le r_0$ then

$$\left| f\left(\frac{1}{\bar{z}}\right)\right| (|z|^2 - 1) < \epsilon.$$

Consider the domain

$$A \equiv D(1, r) \cap \Omega$$

and, with $z = 1 + w$ where $w = \rho e^{i\theta}$, $|\theta| \le \frac{\pi}{2}$, $\rho < r$, we have

$$|z|^2 - 1 \cong \rho^2 + 2\rho\cos\theta.$$

Evaluating the integral

$$\int_A d|\nu|(z) \le C \cdot \epsilon \int_A \left| f\left(\frac{1}{\bar{z}}\right)\right| dxdy,$$

we may rewrite the last integral

$$\int_0^r \int_{|\theta| \le \frac{\pi}{2}} \frac{\rho d\rho d\theta}{\rho^2 + 2\rho\cos\theta} = \int_0^r \left(\int_0^{\frac{\pi}{2}} \frac{d\theta}{((\sqrt{\rho})^2 + 2\cos\theta)} \right) d\rho.$$

Using tables we evaluate

$$\int_0^{\frac{\pi}{2}} \frac{d\theta}{((\sqrt{\rho})^2 + 2\cos\theta)} = \frac{-1}{\sqrt{4-\rho^2}} (\tanh)^{-1} \frac{(2-\rho)\tan(\frac{\theta}{2})}{\sqrt{4-\rho^2}} \Bigg|_0^{\frac{\pi}{2}}$$

$$= \frac{2(\tanh)^{-1}(\frac{2-\rho}{\sqrt{4-\rho^2}})}{\sqrt{4-\rho^2}} \le \frac{M}{\sqrt{4-\rho^2}}.$$

Hence, the integral

$$\int_A d|\nu|(z) \leq \epsilon \cdot M \to 0$$

as $r \to 0$.

Thus the measure

$$d|\nu|(z) = \frac{|D(z)|^2 dxdy}{(|z|^2 - 1)}$$

is a vanishing Carleson measure. □

6 Questions

In concluding we raise the following questions.

Q.1. If $f \in \frac{L^1}{H_o^1}$ does the associated h have a quasiconformal extension to $\mathbb{C}$?

Q.2. Is the span of the range of P on CT dense in BMOA?

References

1. Astala, K., Gehring, F.: Injectivity, the BMO norm and the universal Teichmuller space. J. Anal. Math. **46** (1986)
2. Astala, K., Zinsmeister, M.: Teichmuller spaces and BMOA. Math. Ann. **289**, 613–625 (1991)
3. Cima, J., Matheson, A., Ross, W.: The Cauchy Transform. Math. Surveys and Monographs, vol. 125. Am. Math. Soc. Providence (2006)
4. Duren, P.: Theory of H^p Spaces. Academic Press, New York (1970)
5. Garnett, J.: Bounded Analytic Functions. Academic Press, New York (1981)
6. Hallenbeck, D.J., Samotij, K.: On Cauchy integrals of logarithmic potentials and their multipliers. J. Math. Anal. Appl. **174**, 614–634 (1993)
7. Mashreghi, J.: Representation Theorems in Hardy Spaces. Cambridge University Press, Cambridge (2009)
8. Pommerenke, Ch.: Boundary Behavior of Conformal Maps. Springer, Berlin (1992)
9. Zinsmeister, M.L.: Domaines de Carleson. Michigan Math. J. **36**, 213–220 (1989)

Critical Points, the Gauss Curvature Equation and Blaschke Products

Daniela Kraus and Oliver Roth

Abstract In this survey paper we discuss the problem of characterizing the critical sets of bounded analytic functions in the unit disk of the complex plane. This problem is closely related to the Berger–Nirenberg problem in differential geometry as well as to the problem of describing the zero sets of functions in Bergman spaces. It turns out that for any non-constant bounded analytic function in the unit disk there is always a (essentially) unique "maximal" Blaschke product with the same critical points. These maximal Blaschke products have remarkable properties similar to those of Bergman space inner functions and they provide a natural generalization of the class of finite Blaschke products.

Keywords Blaschke products · Elliptic PDEs · Bergman spaces

Mathematics Subject Classification 30H05 · 30J10 · 35J60 · 30H20 · 30F45 · 53A30

1 Critical Points of Bounded Analytic Functions

A sequence of points (z_j) in a subdomain Ω of the complex plane $\mathbb{C}$ is called the zero set of an analytic function $f : \Omega \to \mathbb{C}$, if f vanishes precisely on this set. This means that if the point $\xi \in \Omega$ occurs m times in the sequence, then f has a zero at ξ of precise order m, and $f(z) \neq 0$ for $z \in \Omega \backslash (z_j)$. The following classical theorem due to Jensen [32], Blaschke [7] and F. and R. Nevanlinna [53] characterizes completely the zero sets of bounded analytic functions defined in the open unit disk $\mathbb{D} := \{z \in \mathbb{C} : |z| < 1\}$.

Theorem A *Let (z_j) be a sequence in $\mathbb{D}$. Then the following statements are equivalent.*

D. Kraus · O. Roth (✉)
Department of Mathematics, University of Würzburg, Emil–Fischer Straße 40, 97074 Würzburg, Germany
e-mail: roth@mathematik.uni-wuerzburg.de

D. Kraus
e-mail: dakraus@mathematik.uni-wuerzburg.de

J. Mashreghi, E. Fricain (eds.), *Blaschke Products and Their Applications*, Fields Institute Communications 65, DOI 10.1007/978-1-4614-5341-3_7,

(a) *There is an analytic self-map of* $\mathbb{D}$ *with zero set* (z_j).
(b) *There is a Blaschke product with zero set* (z_j).
(c) *The sequence* (z_j) *fulfills the Blaschke condition, i.e.* $\sum_{j=1}^{\infty}(1-|z_j|) < +\infty$.

We call a sequence of points (z_j) in a domain $\Omega \subseteq \mathbb{C}$ the critical set of an analytic function $f : \Omega \to \mathbb{C}$, if (z_j) is the zero set of the first derivative f' of the function f. There is an extensive literature on critical sets. In particular, there are many interesting results on the relation between the zeros and the critical points of analytic and harmonic functions. A classical reference for all this is the book of Walsh [63].

The first aim of this survey paper is to point out the following analogue of Theorem A for the critical sets of bounded analytic functions instead of their zeros sets.

Theorem 1 *Let* (z_j) *be a sequence in* $\mathbb{D}$. *Then the following statements are equivalent.*

(a) *There is an analytic self-map of* $\mathbb{D}$ *with critical set* (z_j).
(b) *There is a Blaschke product with critical set* (z_j).
(c) *There is a function in the weighted Bergman space*

$$\mathcal{A}_1^2 = \left\{ g : \mathbb{D} \to \mathbb{C} \text{ analytic} : \iint_{\mathbb{D}} \left(1-|z|^2\right)|g(z)|^2 \, d\sigma_z < +\infty \right\}$$

with zero set (z_j).

Here, and throughout, σ_z denotes two-dimensional Lebesgue measure with respect to z.

For the special case of *finite* sequences, a result related to Theorem 1 can be found in work of Heins [27, §29], Wang & Peng [64], Zakeri [68] and Stephenson [60]. They proved that for every finite sequence (z_j) in $\mathbb{D}$ there is always a *finite* Blaschke product whose critical set coincides with (z_j), see also Remark 2 below. A recent generalization of this result to *infinite* sequences is discussed in [41]. There it is shown that every Blaschke sequence (z_j) is the critical set of an infinite Blaschke product. However, the converse to this, known as the Bloch–Nevanlinna conjecture [14], is false. According to a result of Frostman, there do exist Blaschke products whose critical sets fail to satisfy the Blaschke condition, see [13, Theorem 3.6]. Thus the critical sets of bounded analytic functions are not just the Blaschke sequences and the situation for critical sets is much more complicated than for zero sets.

Theorem 1 identifies the critical sets of bounded analytic functions as the zeros sets of functions in the Bergman space $\mathcal{A}_1^2$. The simple geometric characterization of the zero sets of bounded analytic functions via the Blaschke condition (c) in Theorem A has not found an explicit counterpart for critical sets yet. However, condition (c) of Theorem 1 might be seen as an implicit substitute. The zero sets of (weighted) Bergman space functions have intensively been studied in the 1970's and 1990's by Horowitz [29, 30], Korenblum [37], Seip [56, 57], Luecking [46] and many others. As a result quite sharp necessary conditions and sufficient conditions for the zero sets of Bergman space functions are available. In view of Theorem 1 all results about zero sets of Bergman space functions carry now over to the critical sets of bounded analytic functions and vice versa. Unfortunately, a *geometric* characterization of the zero sets of (weighted) Bergman space functions is still unknown, "and it is well known that this problem is very difficult", cf. [25, p. 133].

The implications "(b) $\Longrightarrow$ (a)" and "(a) $\Longrightarrow$ (c)" of Theorem 1 are easy to prove. In fact, the statement "(a) $\Longrightarrow$ (c)" follows directly from the Littlewood–Paley identity (see [58, p. 178]), which says that for any holomorphic function $f : \mathbb{D} \to \mathbb{D}$ the derivative f' belongs to $\mathcal{A}_1^2$. Hence the critical set of any non-constant bounded analytic function is trivially the zero set of a function in $\mathcal{A}_1^2$. It is however not true that any $\mathcal{A}_1^2$ function is the derivative of a bounded analytic function, so "(c) $\Longrightarrow$ (a)" is more subtle.

In the following sections, we describe some of the ingredients of the proofs of the implications "(a) $\Longrightarrow$ (b)" and "(c) $\Longrightarrow$ (a)" in Theorem 1 as well as some further results connected to it. In particular, we explain the close relation of Theorem 1 to conformal differential geometry and the solvability of the Gauss curvature equation. This paper is expository, so there are essentially no proofs. For the proofs we refer to [39–41, 43]. Background material on Hardy spaces and Bergman spaces can be found e.g. in the excellent books [15, 19, 22, 25, 36, 47].

Finally, we draw attention to the recent paper [20] of P. Ebenfelt, D. Khavinson and H.S. Shapiro and the references therein, where the difficult problem of constructing a finite Blaschke product by its *critical values* is discussed.

2 Conformal Metrics and Maximal Blaschke Products

An essential characteristic of the proof of the implication "(a) $\Longrightarrow$ (b)" in Theorem 1 is the extensive use of negatively curved conformal pseudometrics. We give a short account of some of their properties and refer to [5, 27, 35, 38, 42, 59] for more details. In the following, G and D always denote domains in the complex plane $\mathbb{C}$.

2.1 Conformal Metrics and Developing Maps

We recall that a nonnegative upper semicontinuous function λ on G, $\lambda : G \to [0, +\infty)$, $\lambda \not\equiv 0$, is called conformal density on G. The corresponding quantity $\lambda(z)\,|dz|$ is called conformal pseudometric on G. If $\lambda(z) > 0$ for all $z \in G$, we say $\lambda(z)\,|dz|$ is a conformal metric on G. We call a conformal pseudometric $\lambda(z)\,|dz|$ regular on G, if λ is of class C^2 in $\{z \in G : \lambda(z) > 0\}$.

Example 1 (The Hyperbolic Metric) The ubiquitous example of a conformal metric is the Poincaré metric or hyperbolic metric

$$\lambda_{\mathbb{D}}(z)\,|dz| := \frac{|dz|}{1-|z|^2}$$

for the unit disk $\mathbb{D}$. Clearly, $\lambda_{\mathbb{D}}(z)\,|dz|$ is a regular conformal metric on $\mathbb{D}$.

The (Gauss) curvature κ_λ of a regular conformal pseudometric $\lambda(z)\,|dz|$ on G is defined by

$$\kappa_\lambda(z) := -\frac{\Delta(\log\lambda)(z)}{\lambda(z)^2}$$

for all points $z \in G$ where $\lambda(z) > 0$. Note that if $\lambda(z)\,|dz|$ is a regular conformal metric with curvature $\kappa_\lambda = \kappa$ on G, then the function $u := \log\lambda$ satisfies the partial differential equation (Gauss curvature equation)

$$\Delta u = -\kappa(z)\,e^{2u} \tag{1}$$

on G. If, conversely, a real-valued C^2-function u is a solution of the Gauss curvature equation (1) on G, then $\lambda(z) := e^{u(z)}$ induces a regular conformal metric $e^{u(z)}\,|dz|$ on G with curvature κ.

Example 2 The hyperbolic metric $\lambda_{\mathbb{D}}(z)\,|dz|$ on $\mathbb{D}$ has constant negative curvature[1] -4.

Using analytic maps, conformal metrics can be transferred from one domain to another as follows. Let $\lambda(w)\,|dw|$ be a conformal pseudometric on a domain D and let $w = f(z)$ be a non-constant analytic map from another domain G to D. Then the conformal pseudometric

$$\big(f^*\lambda\big)(z)\,|dz| := \lambda\big(f(z)\big)\,\big|f'(z)\big|\,|dz|,$$

defined on G, is called the pullback of $\lambda(w)\,|dw|$ under the map f. Now, Gauss curvature is important since it is a conformal invariant in the following sense.

Theorem B (Theorema Egregium) *For every analytic map $w = f(z)$ and every regular conformal pseudometric $\lambda(w)\,|dw|$ the relation*

$$\kappa_{f^*\lambda}(z) = \kappa_\lambda\big(f(z)\big)$$

is satisfied provided $\lambda(f(z))\,|f'(z)| > 0$.

---

[1] *Warning:* A rival school of thought calls $\frac{2\,|dz|}{1-|z|^2}$ the hyperbolic metric of $\mathbb{D}$. This metric has constant curvature -1.

Hence pullbacks of conformal pseudometrics can be used for constructing conformal pseudometrics with various prescribed properties while controlling their curvature. For instance, if $\lambda(w)\,|dw|$ is a conformal metric on D (without zeros) and $w = f(z)$ is a non-constant analytic map from G to D, then $(f^*\lambda)(z)\,|dz|$ is a pseudometric on G with zeros exactly at the critical points of f. In order to take the multiplicity of the zeros into account, it is convenient to introduce the following formal definition.

Definition 1 (Zero Set of a Pseudometric) Let $\lambda(z)\,|dz|$ be a conformal pseudometric on G. We say $\lambda(z)\,|dz|$ has a zero of order $m_0 > 0$ at $z_0 \in G$ if

$$\lim_{z \to z_0} \frac{\lambda(z)}{|z - z_0|^{m_0}} \quad \text{exists and } \neq 0.$$

We will call a sequence $\mathcal{C} = (\xi_j) \subset G$

$$(\xi_j) := (\underbrace{z_1, \ldots, z_1}_{m_1\text{-times}}, \underbrace{z_2, \ldots, z_2}_{m_2\text{-times}}, \ldots), \quad z_k \neq z_n \text{ if } k \neq n,$$

the zero set of a conformal pseudometric $\lambda(z)\,|dz|$, if $\lambda(z) > 0$ for $z \in G \backslash \mathcal{C}$ and if $\lambda(z)\,|dz|$ has a zero of order $m_k \in \mathbb{N}$ at z_k for all k.

Now, every conformal pseudometric of constant curvature -4 can locally be represented by a holomorphic function. This is the content of the following result.

Theorem C (Liouville's Theorem) *Let $\mathcal{C}$ be a sequence of points in a simply connected domain G and let $\lambda(z)\,|dz|$ be a regular conformal pseudometric on G with constant curvature -4 on G and zero set $\mathcal{C}$. Then $\lambda(z)\,|dz|$ is the pullback of the hyperbolic metric $\lambda_{\mathbb{D}}(z)\,|dz|$ under some analytic map $f : G \to \mathbb{D}$, i.e.*

$$\lambda(z) = \frac{|f'(z)|}{1 - |f(z)|^2}, \quad z \in G. \tag{2}$$

If $g : G \to \mathbb{D}$ is another analytic function, then $\lambda(z) = (g^*\lambda_{\mathbb{D}})(z)$ for $z \in G$ if and only if $g = T \circ f$ for some conformal automorphism T of $\mathbb{D}$.

A holomorphic function f with property (2) will be called a *developing map* for $\lambda(z)\,|dz|$. Note that the critical set of each developing map coincides with the zero set of the corresponding conformal pseudometric.

Example 3 The developing maps of the hyperbolic metric $\lambda_{\mathbb{D}}(z)\,|dz|$ are precisely the conformal automorphisms of $\mathbb{D}$, i.e., the finite Blaschke products of degree 1.

For later applications we wish to mention the following variant of Liouville's Theorem.

Theorem 2 *Let G be a simply connected domain and let $h : G \to \mathbb{C}$ be an analytic map. If $\lambda(z)\,|dz|$ is a regular conformal metric with curvature $-4\,|h(z)|^2$, then there exists a holomorphic function $f : G \to \mathbb{D}$ such that*

$$\lambda(z) = \frac{1}{|h(z)|}\frac{|f'(z)|}{1-|f(z)|^2}, \qquad z \in G.$$

Moreover, f is uniquely determined up to postcomposition with a unit disk automorphism.

Liouville [45] stated Theorem C for the special case that $\lambda(z)\,|dz|$ is a regular conformal metric. We therefore refer to Theorem C as well as to Theorem 2 as Liouville's theorem. Theorem C and in particular the special case that $\lambda(z)\,|dz|$ is a conformal metric has a number of different proofs, see for instance [8, 11, 12, 49, 55, 65]. Theorem 2 is discussed in [41].

Liouville's theorem plays an important rôle in this paper. It provides a bridge between the world of bounded analytic functions and the world of conformal pseudometrics with constant negative curvature. The critical points on the one side correspond to the zeros on the other side. Unfortunately, this bridge only works for simply connected domains, see Remark 7.

2.2 Maximal Conformal Pseudometrics and Maximal Blaschke Products

Apart from having constant negative curvature the hyperbolic metric on $\mathbb{D}$ has another important property: it is *maximal* among all regular conformal pseudometrics on $\mathbb{D}$ with curvature bounded above by -4. This is the content of the following result.

Theorem D (Fundamental Theorem) *Let $\lambda(z)\,|dz|$ be a regular conformal pseudometric on $\mathbb{D}$ with curvature bounded above by -4. Then $\lambda(z) \le \lambda_{\mathbb{D}}(z)$ for every $z \in \mathbb{D}$.*

Theorem D is due to Ahlfors [1] and it is usually called Ahlfors' lemma. However, in view of its relevance Beardon and Minda proposed to call Ahlfors' lemma the *fundamental theorem.* We will follow their suggestion in this paper. As a result of the fundamental theorem we have

$$\lambda_{\mathbb{D}}(z) = \max\big\{\lambda(z) \,:\, \lambda(z)\,|dz| \text{ is a regular conformal pseudometric on } \mathbb{D} \text{ with curvature } \le -4\big\}$$

for any $z \in \mathbb{D}$.

Remark 1 (Developing Maps and Universal Coverings) Let G be a hyperbolic subdomain of the complex plane $\mathbb{C}$, i.e., the complement $\mathbb{C}\backslash G$ consists of more than one point. In analogy with the Poincaré metric for the unit disk, a regular conformal metric $\lambda_G(z)\,|dz|$ of constant curvature -4 on G is said to be the hyperbolic metric for G, if it is the maximal regular conformal pseudometric with curvature ≤ -4 on G, i.e. $\lambda(z) \leq \lambda_G(z)$, $z \in G$, for all regular conformal pseudometrics $\lambda(z)\,|dz|$ on G with curvature bounded above by -4. Then $\lambda_G(z)\,|dz|$ and $\lambda_{\mathbb{D}}(z)\,|dz|$ are connected via the universal coverings $\pi : \mathbb{D} \to G$ of G by the formula $(\pi^*\lambda_G)(z)\,|dz| = \lambda_{\mathbb{D}}(z)\,|dz|$, see for example [27, §9] and [42, 48]. Hence, every branch of the inverse of a universal covering map $\pi : \mathbb{D} \to G$ is *locally* the developing map of the hyperbolic metric $\lambda_G(z)\,|dz|$ of G. In particular, if G is a hyperbolic simply connected domain, then the developing maps for $\lambda_G(z)\,|dz|$ are precisely the conformal mappings from G onto $\mathbb{D}$.

We now consider prescribed zeros.

Theorem E *Let $\mathcal{C} = (\xi_j)$ be a sequence of points in G and*

$$\Phi_{\mathcal{C}} := \big\{\lambda : \lambda(z)\,|dz| \text{ is a regular conformal pseudometric in } G \\ \text{with curvature } \leq -4 \text{ and zero set } \mathcal{C}^* \supseteq \mathcal{C}\big\}.$$

If $\Phi_{\mathcal{C}} \neq \emptyset$, then

$$\lambda_{max}(z) := \sup\big\{\lambda(z) : \lambda \in \Phi_{\mathcal{C}}\big\}, \quad z \in G,$$

induces the unique maximal regular conformal pseudometric $\lambda_{max}(z)\,|dz|$ on G with constant curvature -4 and zero set $\mathcal{C}$.

Thus, if $\Phi_{\mathcal{C}} \neq \emptyset$, i.e., if there exists at least one regular conformal pseudometric $\lambda(z)\,|dz|$ on G with curvature ≤ -4 whose zero set contains the given sequence $\mathcal{C}$, then there exists a (maximal) regular conformal pseudometric $\lambda(z)\,|dz|$ on G with *constant* curvature -4 whose zero set is *exactly* the sequence $\mathcal{C}$. In particular, Theorem E can be applied, if there exists a non-constant holomorphic function $f : G \to \mathbb{D}$ with critical set $\mathcal{C}^* \supseteq \mathcal{C}$ since then the pseudometric $(f^*\lambda_{\mathbb{D}})(z)\,|dz|$ belongs to $\Phi_{\mathcal{C}}$. The proof of Theorem E relies on a modification of Perron's method and can be found in [27, §12 & §13] and [39].

Example 4 If G is a hyperbolic domain and $\mathcal{C} = \emptyset$, then the maximal regular conformal pseudometric $\lambda_{max}(z)\,|dz|$ on G with constant curvature -4 and zero set $\mathcal{C}$ is exactly the hyperbolic metric $\lambda_G(z)\,|dz|$ for G.

Thus maximal pseudometrics are generalizations of the hyperbolic metric and their developing maps are therefore of special interest.

Definition 2 (Maximal Functions) Let $\mathcal{C}$ be a sequence of points in $\mathbb{D}$ such that there exists a maximal regular conformal pseudometric $\lambda_{max}(z)\,|dz|$ on $\mathbb{D}$ with constant curvature -4 and zero set $\mathcal{C}$. Then every developing map for $\lambda_{max}(z)\,|dz|$ is called maximal function for $\mathcal{C}$.

Some remarks are in order. First, in view of Theorem C every maximal function is uniquely determined by its critical set $\mathcal{C}$ up to postcomposition with a unit disk automorphism and, conversely, the postcomposition of any maximal function with a unit disk automorphism is again a maximal function. Second, if $\mathcal{C} = \emptyset$, then the maximal functions for $\mathcal{C}$ are precisely the unit disk automorphisms, i.e., the finite Blaschke products of degree 1.

Now, we have the following result, see [39].

Theorem 3 *Every maximal function is a Blaschke product.*

It is to emphasize that since the postcomposition of any maximal function with a unit disk automorphism is again a maximal function, every maximal function is an *indestructible* Blaschke product.

We note that Theorem E combined with Theorem 3 immediately gives the implication "(a) $\Longrightarrow$ (b)" in Theorem 1. In fact, if f is a non-constant analytic self-map of $\mathbb{D}$ with critical set $\mathcal{C}$, then $\lambda(z)\,|dz| := (f^*\lambda_{\mathbb{D}})(z)\,|dz|$ is a regular conformal pseudometric on $\mathbb{D}$ with zero set $\mathcal{C}$. Thus Theorem E guarantees the existence of a maximal conformal pseudometric on $\mathbb{D}$ with zero set $\mathcal{C}$. Now, Theorem 3 says that the corresponding maximal function for $\mathcal{C}$ is a Blaschke product.

In the special case that the maximal function has finitely many critical points, the statement of Theorem 3 follows from the next result which is due to Heins [27, §29].

Theorem F *Let $\mathcal{C} = (z_1, \ldots, z_n)$ be a finite sequence in $\mathbb{D}$ and $f : \mathbb{D} \to \mathbb{D}$ analytic. Then the following statements are equivalent.*

(a) *f is a maximal function for $\mathcal{C}$.*
(b) *f is Blaschke product of degree $n+1$ with critical set $\mathcal{C}$.*

We shall give a quick proof of Theorem F in Remark 5 below.

Remark 2 (Constructing Finite Blaschke Products with Prescribed Critical Points) In his proof of Theorem F, Heins showed that for any finite sequence $\mathcal{C} = (z_1, \ldots, z_n)$ in $\mathbb{D}$ there is always a finite Blaschke product B of degree $n+1$ with critical set $\mathcal{C}$. The essential step is nonconstructive and consists in showing that the set of critical points of all finite Blaschke products of degree $n+1$, which is clearly closed, is also open in the poly disk $\mathbb{D}^n$ by applying Brouwer's fixed point theorem. The same result was later obtained by Wang & Peng [64] and Zakeri [68] by using similar arguments. A completely different approach via Circle Packing is due to Stephenson, see Lemma 13.7 and Theorem 21.1 in [60]. Stephenson builds discrete finite Blaschke products with prescribed branch set and shows that under refinement these discrete Blaschke products converge locally uniformly in $\mathbb{D}$ to the desired classical Blaschke product.

We are not aware of any *efficient* constructive method for computing a (nondiscrete) finite Blaschke product from its critical points.

Maximal functions form a particular class of Blaschke products. It is therefore convenient to make the following definition.

Definition 3 (Maximal Blaschke Products) A non-constant Blaschke product is called a maximal Blaschke product, if it is a maximal function for its critical set.

As was mentioned earlier, every maximal Blaschke product is indestructible and every finite Blaschke product is a maximal Blaschke product. Moreover, if $\mathcal{C}$ is the critical set of any non-constant analytic function $f : \mathbb{D} \to \mathbb{D}$, then there is maximal Blaschke product with critical set $\mathcal{C}$. A maximal Blaschke product is uniquely determined by its critical set $\mathcal{C}$ up to postcomposition with a unit disk automorphism.

Geometrically, the finite maximal Blaschke products are just the finite branched coverings of $\mathbb{D}$. One is therefore inclined to consider maximal Blaschke products for infinite branch sets as "infinite branched coverings":

Critical set	Maximal Blaschke product	Mapping properties
$\mathcal{C} = \emptyset$	Automorphism of $\mathbb{D}$	Unbranched covering of $\mathbb{D}$; Conformal self-map of $\mathbb{D}$
$\mathcal{C}$ finite	Finite Blaschke product	Finite branched covering of $\mathbb{D}$
$\mathcal{C}$ infinite	Indestructible infinite maximal Blaschke product	"Infinite branched covering of $\mathbb{D}$"

The class of maximal conformal pseudometrics and their corresponding maximal functions have already been studied by Heins in [27, §25 & §26]. Heins obtained some necessary conditions as well as sufficient conditions for maximal functions (see Theorem G and Theorem H below), but he did not prove that maximal functions are always Blaschke products. He also posed the problem of characterizing maximal functions, cf. [27, §26 & §29].

3 Some Properties of Maximal Blaschke Products

It turns out that maximal Blaschke products do have remarkable properties and provide in some sense a fairly natural generalization of the class of finite Blaschke products. In this section we take a closer look at some of the properties of maximal Blaschke products. In the following, we denote by H^∞ the set of bounded analytic functions $f : \mathbb{D} \to \mathbb{C}$ and set $\|f\|_\infty = \sup_{z \in \mathbb{D}} |f(z)|$. This makes $(H^\infty, \|\cdot\|_\infty)$ a Banach space.

3.1 Schwarz' Lemmas

Theorem 4 (Maximal Blaschke Products as Extremal Functions) *Suppose $\mathcal{C}$ is a sequence of points in $\mathbb{D}$ such that*

$$\mathcal{F}_{\mathcal{C}} := \left\{ f \in H^\infty \, : \, f'(z) = 0 \, for \; z \in \mathcal{C} \right\}$$

contains at least some non-constant function and let

$$m := \min\left\{ n \in \mathbb{N} \, : \, f^{(n)}(0) \neq 0 \, for \; some \; f \in \mathcal{F}_{\mathcal{C}} \right\} \geq 1.$$

Then the unique extremal function to the extremal problem

$$\max\left\{ \operatorname{Re} f^{(m)}(0) \, : \, f \in \mathcal{F}_{\mathcal{C}}, \; \|f\|_\infty \leq 1 \right\} \tag{3}$$

is the maximal Blaschke product F for $\mathcal{C}$ normalized by $F(0) = 0$ and $F^{(m)}(0) > 0$.

We refer to [43] for the proof of Theorem 4. To put Theorem 4 in perspective, note that if $m = 1$, then the extremal problem (3) is exactly the problem of maximizing the derivative at a point, i.e., exactly the character of Schwarz' lemma.

Remark 3 (The Nehari–Schwarz Lemma) If $\mathcal{C} = \emptyset$, then $\mathcal{F}_{\mathcal{C}} = H^\infty$, i.e., the set of all bounded analytic functions in $\mathbb{D}$. In this case, Theorem 4 is of course just the statement of Schwarz' lemma. If $\mathcal{C}$ is a finite sequence, then Theorem 4 is exactly Nehari's 1947 generalization of Schwarz' lemma (Nehari [52], in particular the Corollary to Theorem 1). Hence Theorem 4 can be considered as an extension of the Nehari–Schwarz Lemma.

Remark 4 (The Riemann Mapping Theorem and the Ahlfors Map) We consider a domain $\Omega \subseteq \mathbb{C}$ containing 0. Let $\mathcal{C} = (z_j)$ be the critical set of a non-constant function f in $H^\infty(\Omega)$, where $H^\infty(\Omega)$ denotes the set of all functions analytic and bounded in Ω. We let N denote the number of times that 0 appears in the sequence $\mathcal{C}$ and set

$$\mathcal{F}_{\mathcal{C}}(\Omega) := \left\{ f \in H^\infty(\Omega) \, : \, f'(z) = 0 \text{ for any } z \in \mathcal{C} \right\}.$$

Then, by a normal family argument, there is always at least one extremal function for the extremal problem

$$\max\left\{ \operatorname{Re} f^{(N+1)}(0) \, : \, f \in \mathcal{F}_{\mathcal{C}}(\Omega), \; \|f\|_\infty \leq 1 \right\}. \tag{$*$}$$

In the following three cases there is a unique extremal function to the extremal problem $(*)$.

(i) *$\Omega \neq \mathbb{C}$ is simply connected and $\mathcal{C} = \emptyset$ (conformal maps)*:
In this case, $N = 0$ and the extremal problem $(*)$ has exactly one extremal function, the normalized Riemann map Ψ for Ω, that is, the unique conformal map Ψ from Ω onto $\mathbb{D}$ normalized such that $\Psi(0) = 0$ and $\Psi'(0) > 0$.

(ii) *$\Omega \neq \mathbb{C}$ is simply connected and $\mathcal{C} \neq \emptyset$ (prescribed critical points)*:
Let Ψ be the normalized Riemann map for Ω. Then $\Psi(\mathcal{C})$ is the critical set of a non-constant function in $H^\infty = H^\infty(\mathbb{D})$. If $B_{\Psi(\mathcal{C})}$ is the extremal function in $\mathcal{F}_{\Psi(\mathcal{C})}$ according to Theorem 4, then $B_{\Psi(\mathcal{C})} \circ \Psi$ is the unique extremal function for $(*)$.

(iii) *$\Omega \neq \mathbb{C}$ is not simply connected and $\mathcal{C} = \emptyset$ (Ahlfors' maps)*:
If Ω has connectivity $n \geq 2$, none of whose boundary components reduces to a point, then the extremal problem $(*)$ has exactly one solution, namely the Ahlfors map $\Psi : \Omega \to \mathbb{D}$. It is a $n:1$ map from Ω onto $\mathbb{D}$ such that $\Psi(0) = 0$ and $\Psi'(0) > 0$, see Ahlfors [2] and Grunsky [23].

The Schwarz lemma (i.e., the case $\mathcal{C} = \emptyset$ of Theorem 4) can be stated in an invariant form, the Schwarz–Pick lemma which says that

$$\frac{|f'(z)|}{1-|f(z)|^2} \leq \frac{1}{1-|z|^2}, \qquad z \in \mathbb{D}, \tag{4}$$

for any analytic map $f : \mathbb{D} \to \mathbb{D}$, with equality for some point $z \in \mathbb{D}$ if and only if f is a conformal disk automorphism. Hence maximal Blaschke products without critical points serve as extremal functions. In a similar way, the more general statement of Theorem 4 admits an invariant formulation as follows (see [39]).

Theorem 5 (Sharpened Schwarz–Pick Inequality) *Let $f : \mathbb{D} \to \mathbb{D}$ be a non-constant analytic function with critical set $\mathcal{C}$ and let $\mathcal{C}^*$ be a subsequence of $\mathcal{C}$. Then there exists a maximal Blaschke product F with critical set $\mathcal{C}^*$ such that*

$$\frac{|f'(z)|}{1-|f(z)|^2} \leq \frac{|F'(z)|}{1-|F(z)|^2}, \qquad z \in \mathbb{D}.$$

If $\mathcal{C}^*$ is finite, then F is a finite Blaschke product.

Furthermore, $f = T \circ F$ for some automorphism T of $\mathbb{D}$ if and only if

$$\lim_{z \to w} \frac{|f'(z)|}{1-|f(z)|^2} \, \frac{1-|F(z)|^2}{|F'(z)|} = 1$$

for some $w \in \mathbb{D}$.

If $\mathcal{C}^* = \mathcal{C} \neq \emptyset$, this gives the sharpening

$$\frac{|f'(z)|}{1-|f(z)|^2} \leq \frac{|F'(z)|}{1-|F(z)|^2} < \frac{1}{1-|z|^2}$$

for all $f \in \mathcal{F}_{\mathcal{C}}$ of the Schwarz–Pick inequality (4), which is best possible in some sense.

3.2 Related Extremal Problems in Hardy and Bergman Spaces

Let $\mathcal{C}$ be a sequence in $\mathbb{D}$, assume that $\mathcal{C}$ is the critical set of a bounded analytic function $f:\mathbb{D}\to\mathbb{D}$ and let N denote the multiplicity of the point 0 in $\mathcal{C}$. Then according to Theorem 4 the maximal Blaschke product F for $\mathcal{C}$ normalized by $F(0)=0$ and $F^{(N+1)}(0)>0$ is the unique solution to the extremal problem

$$\max\left\{\operatorname{Re} f^{(N+1)}(0) : f\in H^\infty,\ \|f\|_\infty\le 1 \text{ and } f'(z)=0 \text{ for } z\in\mathcal{C}\right\}.$$

This extremal property of a maximal Blaschke product is reminiscent of the well-known extremal property of

(i) Blaschke products in the Hardy spaces H^∞ and

$$H^p := \left\{f:\mathbb{D}\to\mathbb{C} \text{ analytic} : \|f\|_p<+\infty\right\},$$

where $1\le p<+\infty$ and

$$\|f\|_p := \left(\lim_{r\to 1}\frac{1}{2\pi}\int_0^{2\pi}\left|f\left(re^{it}\right)\right|^p dt\right)^{1/p};$$

and

(ii) canonical divisors in the (weighted) Bergman spaces

$$\mathcal{A}^p_\alpha = \left\{f:\mathbb{D}\to\mathbb{C} \text{ analytic} : \|f\|_{p,\alpha}<+\infty\right\},$$

where $-1<\alpha<+\infty$ and $1\le p<+\infty$ and

$$\|f\|_{p,\alpha} := \left(\frac{1}{\pi}\iint_{\mathbb{D}}\left(1-|z|^2\right)^\alpha \left|f(z)\right|^p d\sigma_z\right)^{1/p}.$$

Note that in (i) and (ii) the prescribed data are not the critical points, but the zeros.

More precisely, let the sequence $\mathcal{C}=(z_j)$ in $\mathbb{D}$ be the zero set of an H^p function and let N be the multiplicity of the point 0 in $\mathcal{C}$. Then the (unique) solution to the extremal problem

$$\max\left\{\operatorname{Re} f^{(N)}(0) : f\in H^p,\ \|f\|_p\le 1 \text{ and } f(z)=0 \text{ for } z\in\mathcal{C}\right\}$$

is a Blaschke product B with zero set $\mathcal{C}$ which is normalized by $B^{(N)}(0)>0$, see [19, §5.1].

In searching for an analogue of Blaschke products for Bergman spaces, Hedenmalm [24] (see also [17, 18]) had the idea of posing an appropriate counterpart of the latter extremal problem for Bergman spaces. As before, let $\mathcal{C} = (z_j)$ be a sequence in $\mathbb{D}$ where the point 0 occurs N times and assume that $\mathcal{C}$ is the zero set of a function in $\mathcal{A}^p_\alpha$. Then the extremal problem

$$\max\left\{\operatorname{Re} f^{(N)}(0) : f \in \mathcal{A}^p_\alpha,\ \|f\|_{p,\alpha} \le 1 \text{ and } f(z) = 0 \text{ for } z \in \mathcal{C}\right\}$$

has a unique extremal function $\mathcal{G} \in \mathcal{A}^p_\alpha$, which vanishes precisely on $\mathcal{C}$ and is normalized by $\mathcal{G}^{(N)}(0) > 0$. The function $\mathcal{G}$ is called the canonical divisor for $\mathcal{C}$. It plays a prominent rôle in the modern theory of Bergman spaces.

In summary, we have the following situation:

Prescribed data	Function space	Extremal function
Critical set $\mathcal{C}$	H^∞	Maximal Blaschke product for $\mathcal{C}$
Zero set $\mathcal{C}$	H^p	Blaschke product
Zero set $\mathcal{C}$	$\mathcal{A}^p_\alpha$	Canonical divisor for $\mathcal{C}$

In light of this strong analogy, one expects that maximal Blaschke products enjoy similar properties as finite Blaschke products and canonical divisors. An example is their analytic continuability. It is a familiar result that a Blaschke product has a holomorphic extension across every open arc of $\partial\mathbb{D}$ that does not contain any limit point of its zero set, see [22, Chap. II, Theorem 6.1]. The same is true for a canonical divisor in the Bergman spaces $\mathcal{A}^p_0$. This was proved by Sundberg [62] in 1997, who improved earlier work of Duren, Khavinson, Shapiro and Sundberg [17, 18] and Duren, Khavinson and Shapiro [16]. Now following the model that critical points of maximal functions correspond to the zeros of Blaschke products and canonical divisors respectively, one hopes that a maximal Blaschke product has an analytic continuation across every open arc of $\partial\mathbb{D}$ which does not meet any limit point of its critical set. This in fact turns out to be true:

Theorem 6 (Analytic Continuability, [43]) *Let $F : \mathbb{D} \to \mathbb{D}$ be a maximal Blaschke product with critical set $\mathcal{C}$. Then F has an analytic continuation across each arc of $\partial\mathbb{D}$ which is free of limit points of $\mathcal{C}$. In particular, the limit points of the critical set of F coincide with the limit points of the zero set of F.*

Another rather strong property of finite Blaschke products is their semigroup property with respect to composition. In contrast, the composition of two infinite Blaschke products does not need to be a Blaschke product (just consider destructible Blaschke products). However, in the case of maximal Blaschke products the following result holds.

Theorem 7 (Semigroup Property, [43]) *The set of maximal Blaschke products is closed under composition.*

It would be interesting to get some information about the critical *values* of maximal Blaschke products and to explore the possibility of factorizing maximal Blaschke products in a way similar to the recent extension of Ritt's theorem for finite Blaschke products due to Ng and Wang (see [51]).

3.3 Boundary Behaviour of Maximal Blaschke Products

We now shift attention to the boundary behaviour of maximal Blaschke products. Ideally, one should be able to determine whether a bounded analytic function $F : \mathbb{D} \to \mathbb{D}$ is a maximal Blaschke product either from the behaviour of

$$\frac{|F'(z)|}{1-|F(z)|^2} \quad \text{as } |z| \to 1$$

or from the behaviour of

$$\int_0^{2\pi} \log \frac{|F'(re^{it})|}{1-|F(re^{it})|^2}\, dt \quad \text{as } r \to 1.$$

We only have some partial results in this connection and we begin our account with the case of finite Blaschke products.

Theorem 8 (Boundary Behaviour of Finite Blaschke Products) *Let $I \subset \partial\mathbb{D}$ be some open arc and let $f : \mathbb{D} \to \mathbb{D}$ be an analytic function. Then the following statements are equivalent.*

(a) $\lim_{z\to\zeta}(1-|z|^2)\frac{|f'(z)|}{1-|f(z)|^2} = 1$ *for every* $\zeta \in I$,
(b) $\lim_{z\to\zeta} \frac{|f'(z)|}{1-|f(z)|^2} = +\infty$ *for every* $\zeta \in I$,
(c) *f has a holomorphic extension across the arc I with $f(I) \subset \partial\mathbb{D}$.*

In particular, if $I = \partial\mathbb{D}$, then f is in either case a finite Blaschke product.

The equivalence of conditions (a) and (b) in Theorem 8 for the special case $I = \partial\mathbb{D}$ is due to Heins [28]; the general case is proved in [40]. We now extend Theorem 8 beyond the class of finite Blaschke products and start with the following auxiliary result.

Proposition 1 (See [39]) *Let $f : \mathbb{D} \to \mathbb{D}$ be an analytic function and I some subset of $\partial\mathbb{D}$.*

(1) *If*

$$\angle \lim_{z\to\zeta}\left(1-|z|^2\right)\frac{|f'(z)|}{1-|f(z)|^2} = 1 \quad \textit{for every } \zeta \in I,$$

then f has a finite angular derivative[2] *at a.e. $\zeta \in I$. In particular,*

$$\angle \lim_{z\to\zeta} \left|f(z)\right| = 1 \quad \textit{for a.e. } \zeta \in I.$$

(2) *If f has a finite angular derivative (and $\angle \lim_{z\to\zeta} |f(z)| = 1$) at some $\zeta \in I$, then*

$$\angle \lim_{z\to\zeta} \left(1-|z|^2\right)\frac{|f'(z)|}{1-|f(z)|^2} = 1.$$

In particular, when $I = \partial\mathbb{D}$, we obtain the following corollary.

Corollary 1 (See [39]) *Let $f : \mathbb{D} \to \mathbb{D}$ be an analytic function. Then the following statements are equivalent.*

(a) $\angle \lim_{z\to\zeta}(1-|z|^2)\frac{|f'(z)|}{1-|f(z)|^2} = 1$ *for a.e. $\zeta \in \partial\mathbb{D}$.*
(b) *f is an inner function with finite angular derivative at almost every point of $\partial\mathbb{D}$.*

We further note that conditions (1) and (2) in Proposition 1 do not complement each other. Therefore we may ask if an analytic self-map f of $\mathbb{D}$ which satisfies

$$\angle \lim_{z\to 1} \frac{|f'(z)|}{1-|f(z)|^2}\left(1-|z|^2\right) = 1$$

does have an angular limit at $z = 1$; this might then be viewed as a converse of the Julia–Wolff–Carathéodory theorem, see [58, p. 57].

For maximal Blaschke products whose critical sets satisfy the Blaschke condition one can show that condition (a) in Corollary 1 holds:

Theorem 9 (See [39]) *Let $C = (z_j)$ be a Blaschke sequence in $\mathbb{D}$.*

(a) *The maximal conformal pseudometric $\lambda_{max}(z)\,|dz|$ on $\mathbb{D}$ with constant curvature -4 and zero set C satisfies*

$$\angle \lim_{z\to\zeta} \frac{\lambda_{max}(z)}{\lambda_{\mathbb{D}}(z)} = 1 \quad \textit{for a.e. } \zeta \in \partial\mathbb{D}.$$

(b) *Every maximal function for C has a finite angular derivative at almost every point of $\partial\mathbb{D}$.*

Remark 5 The boundary behaviour of a maximal Blaschke product contains useful information. For instance, it leads to a quick proof of Theorem F. To see this, let $F : \mathbb{D} \to \mathbb{D}$ be a maximal function for a finite sequence C and $\lambda_{max}(z)\,|dz| = (F^*\lambda_{\mathbb{D}})(z)\,|dz|$ be the maximal conformal metric with constant curvature -4 and

[2] See [58, p. 57].

zero set $\mathcal{C}$. By Theorem 3, the maximal function F is an indestructible Blaschke product. Now, let B be a finite Blaschke product with *zero set* $\mathcal{C}$. Then

$$\lambda(z)\,|dz| := \left|B(z)\right| \lambda_{\mathbb{D}}(z)\,|dz|$$

is a regular conformal pseudometric on $\mathbb{D}$ with curvature $-4/|B(z)|^2 \leq -4$ and zero set $\mathcal{C}$. Thus, by the maximality of $\lambda_{max}(z)\,|dz|$,

$$\lambda(z) \leq \lambda_{max}(z) \quad \text{for } z \in \mathbb{D}$$

and consequently

$$\left|B(z)\right| \leq \frac{\lambda_{max}(z)}{\lambda_{\mathbb{D}}(z)} \quad \text{for all } z \in \mathbb{D}. \tag{5}$$

Since B is a finite Blaschke product, we deduce from (5) and the Schwarz–Pick lemma (4) that

$$\lim_{z\to\zeta} \frac{\lambda_{max}(z)}{\lambda_{\mathbb{D}}(z)} = \lim_{z\to\zeta} \frac{|F'(z)|}{1-|F(z)|^2}\left(1-|z|^2\right) = 1 \quad \text{for all } \zeta \in \partial\mathbb{D}.$$

Applying Theorem 8 shows that F is a finite Blaschke product. The branching order of F is clearly $2n$. Thus, according to the Riemann–Hurwitz formula, see [21, p. 140], the Blaschke product F has degree $m = n+1$.

On the other hand, assume that F is a finite Blaschke product of degree $n+1$. Then, by Theorem 8,

$$\lim_{z\to\zeta} \frac{|F'(z)|}{1-|F(z)|^2}\left(1-|z|^2\right) = 1 \quad \text{for all } \zeta \in \partial\mathbb{D}.$$

Theorem 10 below shows that F is a maximal Blaschke product.

The next result gives a sufficient condition for maximality of a Blaschke product F in terms of the boundary behaviour of the *integral means* of the quantity

$$\left(1-|z|^2\right)\frac{|F'(z)|}{1-|F(z)|^2} = \frac{\lambda(z)}{\lambda_{\mathbb{D}}(z)}.$$

Theorem 10 (See [39]) *Let $\lambda(z)\,|dz|$ be a conformal pseudometric on $\mathbb{D}$ with constant curvature -4 and zero set $\mathcal{C}$ such that*

$$\lim_{r\to 1} \int_0^{2\pi} \log \frac{\lambda(re^{it})}{\lambda_{\mathbb{D}}(re^{it})}\, dt = 0. \tag{6}$$

Then $\lambda(z)\,|dz|$ is the maximal conformal pseudometric $\lambda_{max}(z)\,|dz|$ on $\mathbb{D}$ with constant curvature -4 and zero set $\mathcal{C}$.

If $\mathcal{C}$ is a Blaschke sequence, then the corresponding maximal conformal pseudometric $\lambda_{max}(z)\,|dz|$ on $\mathbb{D}$ with constant curvature -4 and zero set $\mathcal{C}$ satisfies condition (6):

Theorem 11 (See [39]) *Let $\mathcal{C}$ be a Blaschke sequence in $\mathbb{D}$. A conformal pseudometric $\lambda(z)\,|dz|$ on $\mathbb{D}$ with constant curvature -4 and zero set $\mathcal{C}$ is the maximal conformal pseudometric $\lambda_{max}(z)\,|dz|$ on $\mathbb{D}$ with constant curvature -4 and zero set $\mathcal{C}$ if and only if* (6) *holds.*

We don't know whether this result is true for any sequence $\mathcal{C}$ for which there is a non-constant bounded analytic function with critical set $\mathcal{C}$.

3.4 *Heins' Results on Maximal Functions*

For completeness, we close this section with a discussion of Heins' results on maximal functions, cf. [27, §25 & §26]. A first observation is that every maximal function is surjective. In fact more is true; a maximal function is "locally" surjective. Here is the precise definition.

Definition 4 Let $f : \mathbb{D} \to \mathbb{D}$ be an analytic function. A point $q \in \mathbb{D}$ is called locally omitted by f provided that either $q \in \mathbb{D} \backslash f(\mathbb{D})$ or else $q \in f(\mathbb{D})$ and there exists a domain Ω, $q \in \Omega$, such that for some component U of $f^{-1}(\Omega)$ the restriction of f to U omits q, i.e. $q \notin f(U)$.

Theorem G (Heins [27]) *A maximal function has no locally omitted point.*

So far, the results about maximal functions leave an important question unanswered, namely, how to tell whether a given function in H^∞ is a maximal function? Heins' second result gives a topological sufficient criterion and provides therefore a source of "examples" of maximal functions. It is based on the following concept.

Definition 5 A holomorphic function $f : \mathbb{D} \to \mathbb{D}$ is called locally of island type if f is onto and if for each $w \in \mathbb{D}$ there is an open disk $K(w)$ about w such that each component of $f^{-1}(K(w))$ is compactly contained in $\mathbb{D}$.

Obviously, every surjective analytic self-map of $\mathbb{D}$ with constant finite valence, that is, every finite Blaschke product is locally of island type.

Theorem H (Heins [27]) *Every function locally of island type is a maximal function.*

4 The Gauss Curvature PDE and the Berger–Nirenberg Problem

We return to a discussion of Theorem 1. The results of Sect. 2 (Theorem E and Theorem 3) provide a proof of implication "(a) $\Longrightarrow$ (b)" in Theorem 1. In this section, we discuss implication "(c) $\Longrightarrow$ (a)". The key idea is the following.

Theorem 12 *Let $h : \mathbb{D} \to \mathbb{C}$ be a non-constant holomorphic function with zero set $\mathcal{C}$. Then the following statements are equivalent.*

(a) *There exists a holomorphic function $f : \mathbb{D} \to \mathbb{D}$ with critical set $\mathcal{C}$.*
(b) *There exists a C^2-solution $u : \mathbb{D} \to \mathbb{R}$ to the Gauss curvature equation*

$$\Delta u = 4\left|h(z)\right|^2 e^{2u}. \tag{7}$$

Let us sketch a proof here. If $f : \mathbb{D} \to \mathbb{D}$ is a holomorphic function with critical set $\mathcal{C}$, then a quick computation shows that

$$u(z) := \log\left(\frac{1}{|h(z)|} \frac{|f'(z)|}{1-|f(z)|^2}\right)$$

is a C^2-solution to (7). This proves "(a) $\Longrightarrow$ (b)". Conversely, if there is a C^2-solution $u : \mathbb{D} \to \mathbb{R}$ to the curvature equation (7), then

$$\lambda(z)\,|dz| := e^{u(z)}\,|dz|$$

is a regular conformal metric with curvature $-4|h(z)|^2$ on $\mathbb{D}$. Hence, by Theorem 2,

$$u(z) = \log\left(\frac{1}{|h(z)|} \frac{|f'(z)|}{1-|f(z)|^2}\right)$$

with some analytic self-map f of $\mathbb{D}$. Thus the zero set of h agrees with the critical set of f.

In view of Theorem 12, the task is now to characterize those holomorphic functions $h : \mathbb{D} \to \mathbb{C}$ for which the PDE (7) has a solution. In fact this problem is a special case of the Berger–Nirenberg problem from differential geometry:

Berger–Nirenberg Problem *Given a function $\kappa : R \to \mathbb{R}$ on a Riemann surface R. Is there a conformal metric on R with Gauss curvature κ?*

The Berger–Nirenberg problem is well-understood for the projective plane, see [50] and has been extensively studied for compact Riemannian surfaces, see [3, 9, 33, 61] as well as for the complex plane [4, 10, 54].[3] However much less is known for proper domains D of the complex plane, see [6, 31, 34]. In this situation the Berger–Nirenberg problem reduces to the question if for a given function $k : D \to \mathbb{R}$ the Gauss curvature equation

$$\Delta u = k(z)\,e^{2u} \tag{8}$$

has a solution on D. We just note that k is the negative of the curvature κ of the conformal metric $e^{u(z)}\,|dz|$.

In the next theorem we give some necessary conditions as well as sufficient conditions for the solvability of the Gauss curvature equation (8) only in terms of the curvature function k and the domain D.

[3]These are just some of the many references.

Theorem 13 (See [39]) *Let D be a bounded and regular domain*[4] *and let k be a nonnegative locally Hölder continuous function on D.*

(1) *If for some (and therefore for every) $z_0 \in D$*

$$\iint_D g_D(z_0, \xi)\, k(\xi)\, d\sigma_\xi < +\infty,$$

then (8) *has a C^2-solution $u : D \to \mathbb{R}$, which is bounded from above.*

(2) *If* (8) *has a C^2-solution $u : D \to \mathbb{R}$ which is bounded from below and if this solution has a harmonic majorant on D, then*

$$\iint_D g_D(z, \xi)\, k(\xi)\, d\sigma_\xi < +\infty$$

for all $z \in D$.

(3) *There exists a bounded C^2-solution $u : D \to \mathbb{R}$ to* (8) *if and only if*

$$\sup_{z \in D} \iint_D g_D(z, \xi)\, k(\xi)\, d\sigma_\xi < +\infty.$$

If we choose $D = \mathbb{D}$ and $z_0 = 0$, then $g_{\mathbb{D}}(0, \xi) = -\log|\xi|$. Hence as a consequence of the inequality

$$\frac{1 - |\xi|^2}{2} \le \log\frac{1}{|\xi|} \le \frac{1 - |\xi|^2}{|\xi|}, \qquad 0 < |\xi| < 1,$$

we obtain the following equivalent formulation of Theorem 13.

Corollary 2 (See [39]) *Let k be a nonnegative locally Hölder continuous function on $\mathbb{D}$.*

(1) *If*

$$\iint_{\mathbb{D}} (1 - |\xi|^2)\, k(\xi)\, d\sigma_\xi < +\infty, \tag{9}$$

then (8) *has a C^2-solution $u : \mathbb{D} \to \mathbb{R}$, which is bounded from above.*

(2) *If* (8) *has a C^2-solution $u : \mathbb{D} \to \mathbb{R}$ which is bounded from below and if this solution has a harmonic majorant on $\mathbb{D}$, then*

$$\iint_{\mathbb{D}} (1 - |\xi|^2)\, k(\xi)\, d\sigma_\xi < +\infty.$$

(3) *There exists a bounded C^2-solution $u : \mathbb{D} \to \mathbb{R}$ to* (8) *if and only if*

$$\sup_{z \in \mathbb{D}} \iint_{\mathbb{D}} \log\left|\frac{1 - \overline{\xi} z}{z - \xi}\right| k(\xi)\, d\sigma_\xi < +\infty.$$

[4] I.e. there exists Green's function g_D for D which vanishes continuously on ∂D.

Both, Theorem 13 and Corollary 2, are not best possible, because (8) may indeed have solutions, even if

$$\iint_D g_D(z,\xi)\, k(\xi)\, d\sigma_\xi = +\infty$$

for some (and therefore for all) $z \in D$. Here is an explicit example.

Example 5 For $\alpha \geq 3/2$ define

$$h(z) = \frac{1}{(z-1)^\alpha}$$

for $z \in \mathbb{D}$ and set $k(z) = 4\,|h(z)|^2$ for $z \in \mathbb{D}$. Then an easy computation yields

$$\iint_{\mathbb{D}} \left(1 - |z|^2\right) k(z)\, d\sigma_z = +\infty$$

and a straightforward check shows that the function

$$u_f(z) := \log\left(\frac{1}{|h(z)|}\, \frac{|f'(z)|}{1 - |f(z)|^2}\right)$$

is a solution to (8) on $\mathbb{D}$ for every locally univalent analytic function $f : \mathbb{D} \to \mathbb{D}$.

Remark 6 The sufficient condition (9) improves earlier results of Kalka & Yang in [34]. In fact, Kalka & Yang give explicit examples for the function k which tend to $+\infty$ at the boundary of $\mathbb{D}$ such that the existence of a solution to (8) can be guaranteed. All these examples are radially symmetric and satisfy (9). Kalka & Yang also supplement their existence results by nonexistence results. They find explicit lower bounds for the function k in terms of radially symmetric functions which grow to $+\infty$ at the boundary of $\mathbb{D}$, such that (8) has no solution.

We wish to emphasize that the necessary conditions and the sufficient conditions for the solvability of the curvature equation (8) of Kalka & Yang do not complement each other. In particular, the case when the function k oscillates is not covered. For the proof of their nonexistence results Kalka & Yang needed to use Yau's celebrated Maximum Principle [66, 67], which is an extremely powerful tool. In [39], an almost elementary proof of these nonexistence results is given, which has the additional advantage that Ahlfors' type lemmas for conformal metrics with variable curvature and explicit formulas for the corresponding maximal conformal metrics are obtained. In [39] the nonexistence theorems of Kalka & Yang are further extended by allowing the function k to oscillate.

It turns out that although condition (9) is not necessary for the existence of a solution to (8) it is strong enough to deduce a necessary and sufficient condition for the solvability of the Gauss curvature equation of the particular form (7):

Theorem 14 (See [39]) *Let $h : \mathbb{D} \to \mathbb{C}$ be a holomorphic function. Then the Gauss curvature equation* (7) *has a solution if and only if h has a representation as a product of an $\mathcal{A}_1^2$ function and a nonvanishing analytic function.*

Note that Theorem 12 combined with Theorem 14 shows that the class of all holomorphic functions $h : \mathbb{D} \to \mathbb{C}$ whose zero sets coincide with the critical sets of the class of bounded analytic function is exactly the Bergman space $\mathcal{A}_1^2$. This proves implication "(c) $\Longrightarrow$ (a)" in Theorem 1.

A further remark is that Theorem 13 (c) characterizes those curvature functions k for which (8) has at least one bounded solution. For the case of the unit disk $\mathbb{D}$, this result can be stated as follows.

Theorem 15 *Let $\varphi : \mathbb{D} \to \mathbb{C}$ be analytic and $k(z) = 4\,|\varphi'(z)|^2$. Then there exists a bounded solution to the Gauss curvature equation* (8) *if and only if $\varphi \in BMOA$, where*

$$BMOA = \left\{ \varphi : \mathbb{D} \to \mathbb{C} \textit{ analytic} : \sup_{z \in \mathbb{D}} \iint_{\mathbb{D}} g_{\mathbb{D}}(z,\xi)\, \left|\varphi'(\xi)\right|^2 d\sigma_\xi < +\infty \right\}$$

is the space of analytic functions of bounded mean oscillation on $\mathbb{D}$, see [44, *p.* 314].

Finally, we note that in Theorem 13 and Corollary 2, condition (1) does not imply condition (3). The Gauss curvature equation (8) may indeed have solutions none of which is bounded. For example, choose $\varphi \in H^2 \backslash BMOA$ and set $k(z) = 4\,|\varphi'(z)|^2$. Then, according to Theorem 15, every solution to (8) must be unbounded.

The following result of Heins adds another item to the list of equivalent statements in Theorem 1.

Theorem I (Heins [27]) *Let $\mathcal{C} = (z_j)$ be a sequence in $\mathbb{D}$. Then the following conditions are equivalent.*

(a) *There is an analytic function $f : \mathbb{D} \to \mathbb{D}$ with critical set (z_j).*
(b) *There is a function in the Nevanlinna class $\mathcal{N}$ with critical set (z_j).*

Here, a function f analytic in $\mathbb{D}$ is said to belong to the Nevanlinna class $\mathcal{N}$ if the integrals

$$\int_0^{2\pi} \log^+ \left| f\left(re^{it}\right) \right| dt$$

remain bounded as $r \to 1$.

Let us indicate how the results of the present survey allow a quick proof of Theorem I.

Proof (b) $\Rightarrow$ (a): Let $\varphi \in \mathcal{N}$. Then $\varphi = \varphi_1/\varphi_2$ is the quotient of two analytic self-maps of $\mathbb{D}$, see for instance [15, Theorem 2.1]. W.l.o.g. we may assume φ_2 is zerofree. Differentiation of φ yields

$$\varphi'(z) = \frac{1}{\varphi_2(z)^2}\left(\varphi_1'(z)\,\varphi_2(z) - \varphi_1(z)\,\varphi_2'(z)\right).$$

Since $\varphi_1', \varphi_2' \in \mathcal{A}_1^2$ and $\mathcal{A}_1^2$ is a vector space, it follows that the function $\varphi_1'\,\varphi_2 - \varphi_1\,\varphi_2'$ belongs to $\mathcal{A}_1^2$. Thus Theorem 14 ensures the existence of a solution $u : \mathbb{D} \to \mathbb{R}$ to

$$\Delta u = \left|\varphi'(z)\right|^2 e^{2u}.$$

Hence Theorem 12 gives the desired result. □

The results of this section about the solvability of the Gauss curvature equation (8) do have further consequences for the critical sets of bounded analytic functions. For instance, one can give an answer to a question of Heins [27, §31]. In order to state Heins' question properly, we recall the well-known Jensen formula which connects the rate of growth of an analytic function with the density of its zeros. Thus the restriction on the growth of the derivative of an analytic self-map of $\mathbb{D}$ imposed by the Schwarz–Pick lemma (4) forces an upper bound for the number of critical points of non-constant analytic self-maps of $\mathbb{D}$. More precisely,

$$\sup_{\substack{f \in H^\infty \\ \|f\|_\infty \le 1,\, f \not\equiv \text{const.}}} \left(\limsup_{r \to 1} \frac{N(r; f')}{\log \frac{1}{1-r^2}} \right) \le 1, \tag{10}$$

where

$$N\left(r; f'\right) := \int_0^1 \frac{n(t; f')}{t}\, dt$$

and $n(r; f')$ denotes the number of zeros of f' counted with multiplicity in the disk $\mathbb{D}_r := \{z \in \mathbb{C} : |z| < r\}$, $0 < r < 1$.

Heins showed that equality holds in (10), cf. [26]. More precisely, using his solution of the Schwarz–Picard problem, he proved that there exists for every $p = 2, 3, \ldots$ a non-constant analytic function $f_p : \mathbb{D} \to \mathbb{D}$ such that

$$\limsup_{r \to 1} \frac{N(r; f_p')}{\log(\frac{1}{1-r^2})} = \frac{2p-3}{2p-2}.$$

Letting $p \to +\infty$ shows that (10) is best possible. In this way, Heins [27, §31] was led to ask whether there is a bounded analytic function which realizes the supremum in (10). This question is answered in our next theorem—even with some additional information.

Theorem 16 *Let $\beta \in [0,1]$. Then there exists a non-constant analytic function (and even a maximal Blaschke product) $f : \mathbb{D} \to \mathbb{D}$ such that*

$$\limsup_{r\to 1} \frac{N(r; f')}{\log(\frac{1}{1-r^2})} = \beta.$$

For the proof of Theorem 16 we refer the reader to [39].
We close with the following remark.

Remark 7 With the help of the Riemann mapping theorem, the results of this paper about critical sets of bounded analytic functions $f : \mathbb{D} \to \mathbb{D}$ can easily be transferred to the class $H^\infty(\Omega)$ of bounded analytic functions $f : \Omega \to \mathbb{D}$, when $\Omega \neq \mathbb{C}$ is a simply connected domain. The critical sets of bounded analytic functions on multiply connected domains are much more difficult to fathom.

Acknowledgements The authors received support from the Deutsche Forschungsgemeinschaft (Grants: Ro 3462/3–1 and Ro 3462/3–2).

This paper is based on lectures given at the workshop on *Blaschke products and their Applications* (Fields Institute, Toronto, July 25–29, 2011). The authors would like to thank the organizers of this workshop, Javad Mashreghi and Emmanuel Fricain, as well as the Fields Institute and its staff, for their generous support and hospitality.

References

1. Ahlfors, L.: An extension of Schwarz's lemma. Trans. Am. Math. Soc. **43**, 359–364 (1938)
2. Ahlfors, L.: Bounded analytic functions. Duke Math. J. **14**, 1–11 (1947)
3. Aubin, T.: Some Nonlinear Problems in Riemannian Geometry. Springer, Berlin (1998)
4. Aviles, P.: Conformal complete metrics with prescribed non-negative Gaussian curvature in $\mathbb{R}^2$. Invent. Math. **83**, 519–544 (1986)
5. Beardon, A., Minda, D.: The hyperbolic metric and geometric function theory. In: Ponnusamy, S., Sugawa, T., Vuorinen, M. (eds.) Quasiconformal Mappings and Their Applications. Narosa, New Delhi (2007)
6. Bland, J., Kalka, M.: Complete metrics conformal to the hyperbolic disc. Proc. Am. Math. Soc. **97**(1), 128–132 (1986)
7. Blaschke, W.: Eine Erweiterung des Satzes von Vitali über Folgen analytischer Funktionen. S.-B. Sächs. Akad. Wiss. Leipz. Math.-Natur. Kl. **67**, 194–200 (1915)
8. Bieberbach, L.: $\Delta u = e^u$ und die automorphen Funktionen. Math. Ann. **77**, 173–212 (1916)
9. Chang, S.Y.A.: Non-linear Elliptic Equations in Conformal Geometry. Eur. Math. Soc., Zurich (2004)
10. Cheng, K.S., Ni, W.M.: On the structure of the conformal Gaussian curvature equation on $\mathbb{R}^2$. Duke Math. J. **62**(3), 721–737 (1991)
11. Chou, K.S., Wan, T.: Asymptotic radial symmetry for solutions of $\Delta u + e^u = 0$ in a punctured disc. Pac. J. Math. **163**(2), 269–276 (1994)
12. Chou, K.S., Wan, T.: Correction to "Asymptotic radial symmetry for solutions of $\Delta u + e^u = 0$ in a punctured disc". Pac. J. Math. **171**(2), 589–590 (1995)
13. Colwell, P.: Blaschke Products. University of Michigan Press, Ann Arbor (1985)
14. Duren, P.: On the Bloch–Nevanlinna conjecture. Colloq. Math. **20**, 295–297 (1969)
15. Duren, P.: Theory of H^p Spaces. Dover, New York (2000)

16. Duren, P., Khavinson, D., Shapiro, H.S.: Extremal functions in invariant subspaces of Bergman spaces. Ill. J. Math. **40**, 202–210 (1996)
17. Duren, P., Khavinson, D., Shapiro, H.S., Sundberg, C.: Contractive zero-divisors in Bergman spaces. Pac. J. Math. **157**(1), 37–56 (1993)
18. Duren, P., Khavinson, D., Shapiro, H.S., Sundberg, C.: Invariant subspaces in Bergman spaces and the biharmonic equation. Mich. Math. J. **41**(2), 247–259 (1994)
19. Duren, P., Schuster, A.: Bergman Spaces. Am. Math. Soc., Providence (2004)
20. Ebenfelt, P., Khavinson, D., Shapiro, H.S.: Two–dimensional shapes and lemniscates. Complex Anal. Dyn. Syst. IV, Contemp. Math., **553**, 45–59 (2011)
21. Forster, O.: Lectures on Riemann Surfaces. Springer, Berlin (1999)
22. Garnett, J.B.: Bounded Analytic Functions, revised 1st edn. Springer, Berlin (2007)
23. Grunsky, H.: Lectures on Theory of Functions in Multiply Connected Domains. Vandenhoeck & Rupprecht, Gortingen (1978)
24. Hedenmalm, H.: A factorization theorem for square area-integrable analytic functions. J. Reine Angew. Math. **442**, 45–68 (1991)
25. Hedenmalm, H., Korenblum, B., Zhu, K.: Theory of Bergman Spaces. Springer, Berlin (2000)
26. Heins, M.: A class of conformal metrics. Bull. Am. Math. Soc. **67**, 475–478 (1961)
27. Heins, M.: On a class of conformal metrics. Nagoya Math. J. **21**, 1–60 (1962)
28. Heins, M.: Some characterizations of finite Blaschke products of positive degree. J. Anal. Math. **46**, 162–166 (1986)
29. Horowitz, C.: Zeros of functions in the Bergman spaces. Duke Math. J. **41**, 693–710 (1974)
30. Horowitz, C.: Factorization theorems for functions in the Bergman spaces. Duke Math. J. **44**, 201–213 (1977)
31. Hulin, D., Troyanov, M.: Prescribing curvature on open surfaces. Math. Ann. **293**(2), 277–315 (1992)
32. Jensen, J.: Sur un nouvel et important théorème de la théorie des fonctions. Acta Math. **22**, 359–364 (1899)
33. Kazdan, J.: Prescribing the Curvature of a Riemannian Manifold. CMBS Regional Conf. Ser. in Math., vol. 57 (1985)
34. Kalka, M., Yang, D.: On conformal deformation of nonpositive curvature on noncompact surfaces. Duke Math. J. **72**(2), 405–430 (1993)
35. Keen, L., Lakic, N.: Hyperbolic Geometry from a Local Viewpoint. Cambridge University Press, Cambridge (2007)
36. Koosis, P.: Introduction to H^p Spaces. Cambridge Tracts, 2nd edn. (1998)
37. Korenblum, B.: An extension of the Nevanlinna theory. Acta Math. **135**, 187–219 (1975)
38. Krantz, St.: Complex Analysis—The Geometric Viewpoint, 2nd edn. Math. Assoc. of America, Washington (2004)
39. Kraus, D.: Critical sets of bounded analytic functions, zero sets of Bergman spaces and nonpositive curvature. Proc. Lond. Math. Soc., to appear
40. Kraus, D., Roth, O., Ruscheweyh, St.: A boundary version of Ahlfors' lemma, locally complete conformal metrics and conformally invariant reflection principles for analytic maps. J. Anal. Math. **101**, 219–256 (2007)
41. Kraus, D., Roth, O.: Critical points of inner functions, nonlinear partial differential equations, and an extension of Liouville's theorem. J. Lond. Math. Soc. **77**(1), 183–202 (2008)
42. Kraus, D., Roth, O.: Conformal metrics. In: Topics in Modern Function Theory, Ramanujan Math. Soc., 41 pp., to appear
43. Kraus, D., Roth, O.: Maximal Blaschke products, submitted
44. Laine, I.: Complex differential equations. In: Battelli, F., Fečkan, M. (eds.) Handbook of Differential Equations: Ordinary Differential Equations, vol. IV. Elsevier, Amsterdam (2008)
45. Liouville, J.: Sur l'équation aux différences partielles $\frac{d^2 \log \lambda}{du dv} \pm \frac{\lambda}{2a^2} = 0$. J. Math. **16**, 71–72 (1853)
46. Luecking, D.H.: Zero sequences for Bergman spaces. Complex Var. Theory Appl. **30**(4), 345–362 (1996)
47. Mashreghi, J.: Representation Theorems in Hardy Spaces. LMS Student Texts, vol. 74 (2009)

48. Minda, C.D.: The hyperbolic metric and coverings of Riemann surfaces. Pac. J. Math. **84**(1), 171–182 (1979)
49. Minda, D.: Conformal metrics. Unpublished notes
50. Moser, J.: On a nonlinear problem in differential geometry. In: Dynamical Syst., Proc. Sympos., Univ. Bahia, Salvador, vol. 1971, pp. 273–280 (1973)
51. Ng, T.W., Wang, M.-X.: Ritt's theory on the unit disk. Preprint (2011)
52. Nehari, Z.: A generalization of Schwarz' lemma. Duke Math. J. **14**, 1035–1049 (1947)
53. Nevanlinna, R.: Über die Eigenschaften analytischer Funktionen in der Umgebung einer singulären Stelle oder Linie. Acta Soc. Sci. Fenn. **50**(5) (1922), 46 pp.
54. Ni, W.-M.: Recent progress on the elliptic equation $\Delta u + K\, e^{2u} = 0$ on $\mathbb{R}^2$. Rend. Semin. Mat., Torino Fasc. Spec., 1–10 (1989)
55. Nitsche, J.: Über die isolierten Singularitäten der Lösungen von $\Delta u = e^u$. Math. Z. **68**, 316–324 (1957)
56. Seip, K.: On a theorem of Korenblum. Ark. Mat. **32**, 237–243 (1994)
57. Seip, K.: On Korenblum's density condition for zero sequences of $A^{-\alpha}$. J. Anal. Math. **67**, 307–322 (1995)
58. Shapiro, J.H.: Composition Operators and Classical Function Theory. Springer, Berlin (1993)
59. Smith, S.J.: On the uniformization of the n-punctured disc. Ph.D. Thesis, University of New England (1986)
60. Stephenson, K.: Introduction to Circle Packing: The Theory of Discrete Analytic Functions. Cambridge University Press, Cambridge (2005)
61. Struwe, M.: A flow approach to Nirenberg's problem. Duke Math. J. **128**(1), 19–64 (2005)
62. Sundberg, C.: Analytic continuability of Bergman inner functions. Mich. Math. J. **44**(2), 399–407 (1997)
63. Walsh, J.: The Location of Critical Points of Analytic and Harmonic Functions. Am. Math. Soc., Providence (1950)
64. Wang, Q., Peng, J.: On critical points of finite Blaschke products and the equation $\Delta u = e^{2u}$. Kexue Tongbao **24**, 583–586 (1979) (Chinese)
65. Yamada, A.: Bounded analytic functions and metrics of constant curvature on Riemann surfaces. Kodai Math. J. **11**(3), 317–324 (1988)
66. Yau, S.T.: Harmonic functions on complete Riemannian manifolds. Commun. Pure Appl. Math. **28**, 201–228 (1975)
67. Yau, S.T.: A general Schwarz lemma for Kähler manifolds. Am. J. Math. **100**, 197–203 (1978)
68. Zakeri, S.: On critical points of proper holomorphic maps on the unit disk. Bull. Lond. Math. Soc. **30**(1), 62–66 (1996)

Growth, Zero Distribution and Factorization of Analytic Functions of Moderate Growth in the Unit Disc

Igor Chyzhykov and Severyn Skaskiv

Abstract We give a survey of results on zero distribution and factorization of analytic functions in the unit disc in classes defined by the growth of $\log|f(re^{i\theta})|$ in the uniform and integral metrics. We restrict ourself to the case of finite order of growth. For a Blaschke product B we obtain a necessary and sufficient condition for the uniform boundedness of all p-means of $\log|B(re^{i\theta})|$, where $p > 1$.

Keywords Analytic function · Factorization · Zero distribution · Canonical product · Order of growth

Mathematics Subject Classification Primary 30J99 · Secondary 30D35 · 30H15 · 37A45

1 Introduction

Let $D(z,t) = \{\zeta \in \mathbb{C} : |\zeta - z| < t\}$, $z \in \mathbb{C}$, $t > 0$, and $\mathbb{D} = D(0,1)$. Denote by $H(\mathbb{D})$ the class of analytic functions in $\mathbb{D}$. For $f \in H(\mathbb{D})$ we define the *maximum modulus* $M(r,f) = \max\{|f(z)| : |z| = r\}$, $0 \le r < 1$. The zero sequence of a function $f \in H(\mathbb{D})$ will be denoted by Z_f. In the sequel, the symbol C with indices stands for positive constants which depend on parameters indicated. We write $a(r) \sim b(r)$ if $\lim_{r\uparrow 1} a(r)/b(r) = 1$, $x^+ = \max\{x, 0\}$. Throughout this paper, by $(1-w)^\alpha$, $w \in \mathbb{D}$, $\alpha \in \mathbb{R}$, we mean the branch of the power function such that $(1-w)^\alpha|_{w=0} = 1$. $BV[a,b]$ stands for the class of functions of bounded variation on $[a,b]$.

We are primarily interested in zero distribution of analytic functions from classes defined by growth conditions in the unit disc. The topic is closely related to the problem of factorization of such classes.

I. Chyzhykov (✉) · S. Skaskiv
Faculty of Mechanics and Mathematics, Ivan Franko National University of Lviv, Universytets'ka 1, 79000, Lviv, Ukraine
e-mail: chyzhykov@yahoo.com

J. Mashreghi, E. Fricain (eds.), *Blaschke Products and Their Applications*, Fields Institute Communications 65, DOI 10.1007/978-1-4614-5341-3_8,

Usually, the orders of growth of an analytic function f in $\mathbb{D}$ are defined as

$$\rho_M[f] = \limsup_{r\uparrow 1} \frac{\log^+\log^+ M(r,f)}{-\log(1-r)}, \qquad \rho_T[f] = \limsup_{r\uparrow 1} \frac{\log^+ T(r,f)}{-\log(1-r)},$$

where $T(r,f) = \frac{1}{2\pi}\int_0^{2\pi} \log^+|f(re^{i\theta})|\,d\theta$. It is well known that

$$\rho_T[f] \le \rho_M[f] \le \rho_T[f] + 1, \tag{1}$$

and all admissible values of the orders are possible ([1, 2, 15]).

The paper is organized in the following way. In Sects. 1 and 2 we give a survey of results on zero distribution and factorization in subclasses of $H(\mathbb{D})$ defined by the growth conditions on $T(r,f)$ and $\log M(r,f)$, respectively. In Sect. 3 we consider the concept of ρ_∞-order, which goes back to work of C.N. Linden [16, 17]. This notion allows us to prove several new results for functions f with $\rho_M[f] < 1$. Finally, in Sect. 4 we prove a criterion of uniform boundedness of the integral means of $\log|B(re^{i\theta})|$, where B is a Blaschke product.

We do not consider zero distribution and factorization either of functions of infinite order or meromorphic functions. We refer the reader who is interested in factorization of meromorphic functions to [13].

2 Classes Defined by the Growth of $T(r,f)$

2.1 Growth of Nevanlinna Characteristic and Zero Distribution

To be more precise we start with canonical products. Let $Z = (z_n)$ be a sequence of complex numbers in $\mathbb{D}$ without accumulation points in $\mathbb{D}$. We define the *exponent of convergence* of Z by

$$\mu[Z] = \inf\left\{\mu \ge 0 : \sum_{z_n\in Z} \left(1 - |a_n|\right)^{\mu+1} < \infty\right\},$$

with the convention that $\inf\emptyset = +\infty$. It is well known [6, 7, 19, 21] that the Djrbashian–Naftalevich–Tsuji *canonical product*,

$$P(z,Z,q) = \prod_{n=1}^{\infty} E\left(\frac{1-|z_n|^2}{1-\bar{z}_n z}, q\right), \tag{2}$$

where $E(w,0) = 1 - w$,

$$E(w,q) = (1-w)\exp\left\{w + w^2/2 + \cdots + w^q/q\right\}, \quad q \in \mathbb{N},$$

is an analytic function with the zero sequence Z provided that $\sum_{z_n \in Z}(1 - |a_n|)^{q+1} < \infty$. We note that if $q = 0$ then $P(z, Z, 0) = CB(z, Z)$, where $C = \prod_{z_n \in Z} |z_n|$,

$$B(z, Z) = \prod_{z_n \in Z} \frac{\bar{z}_n(z_n - z)}{|z_n|(1 - \bar{z}_n z)}$$

is the Blaschke product constructed by the sequence Z.

Let $n(r, Z_P)$ be the number of zeros in $\overline{D}(0, r)$,

$$\rho_n[P] = \limsup_{r \uparrow 1} \frac{\log^+ n(r, P)}{-\log(1 - r)}, \tag{3}$$

be the order of the counting function of Z_P. Under the technical assumption that $0 \notin Z_f$ we also consider the *Nevanlinna counting function* $N(r, Z_f) = \int_0^r \frac{n(t, Z_f)}{t} dt$. Note that $N(r, Z_f) \le T(r, f) + O(1)$ due to the first fundamental theorem of R. Nevanlinna [12].

In 1953 Naftalevich [19], and in 1956 Tsuji [21] proved that

$$\rho_T[P] = \big(\rho_n[P] - 1\big)^+. \tag{4}$$

Moreover, $(\rho_n[P] - 1)^+$ is equal to the convergence exponent $\mu(Z_P)$, and the order of $N(r, Z_P)$.

This result was improved by F. Shamoyan in [22, 23].

Let $\omega \in C^1[0, 1)$ be positive, monotone and such that

$$\int_0^1 \omega(t)\, dt < +\infty, \qquad \sup_{r \in [r_0, 1)} \left| \frac{(1 - r)\omega'(r)}{\omega(r)} \right| < q_\omega < +\infty, \tag{5}$$

where $r_0 \in (0, 1)$. If ω is an increasing function we assume in addition that $0 < q_\omega < 1$. The class A^*_ω consists of analytic functions f in the unit disc satisfying

$$\int_0^1 \omega(r) T(r, f)\, dr < +\infty. \tag{6}$$

If $\omega(r) = (1 - r)^{\alpha - 1}$, $\alpha > 0$, the A^*_ω coincides with Djrbashian's class A^*_α which consists of analytic functions f in the unit disc satisfying

$$\int_0^1 (1 - r)^{\alpha - 1} T(r, f)\, dr < +\infty. \tag{7}$$

Remark that (7) implies $\rho_T[f] \le \alpha$. On the other hand $f \in A^*_\alpha$ provided that $\alpha > \rho_T[f]$.

Theorem 1 (F.A. Shamoyan, [23, Theorem 1]) *Let ω be a monotone positive function satisfying* (5), $Z = (z_k) \subset \mathbb{D}$. *In order that Z be a sequence of zeros of a function*

$f \in A^*_\omega$, $f \not\equiv 0$ it is necessary and sufficient that

$$\sum_{z_k \in Z} \left(1 - |z_k|\right)^2 \omega\left(|z_k|\right) < +\infty. \tag{8}$$

Moreover, under condition (8) *Djrbashian's canonical product $P(z, Z, \alpha)$ (see* (10) *below) is convergent in $\mathbb{D}$ and belongs to A^*_ω for $\alpha > q_\omega$.*

2.2 Factorization of Classes Defined by the Growth of $T(r, f)$

Canonical and parametric representations of functions analytic in $\mathbb{D}$ and of finite order of the growth were obtained [6–8] in the 1940s by M.M. Djrbashian using the Riemann–Liouville fractional integral.

Theorem 2 (M.M. Djrbashian) *If $f \in A^*_\alpha$, $\alpha > 0$ then f admits a representation*

$$f(z) = C_\lambda z^\lambda P(z, Z_f, \alpha) \exp\left\{ \frac{\alpha}{\pi} \int_{\mathbb{D}} \frac{\log|f(\zeta)|(1 - |\zeta|^2)\, dm_2(\zeta)}{(1 - \bar{\zeta} z)^{\alpha+2}} \right\}, \tag{9}$$

where C_λ is a complex constant, $\lambda \in \mathbb{Z}_+$, m_2 is the planar Lebesgue measure, $P(z, Z_f, \alpha)$ is a canonical product with the zeros Z_f, and of the form

$$P(z, Z_f, \alpha) = \prod_k \left(1 - \frac{z}{z_k}\right) \exp\left\{-U_\alpha(z, z_k)\right\}, \tag{10}$$

where

$$U_\alpha(z, z_k) = \frac{2\alpha}{\pi} \int_{\mathbb{D}} \frac{\log|1 - \frac{w}{\zeta}|(1 - |w|^2)\, dm_2(w)}{(1 - \bar{w} z)^{\alpha+2}}, \qquad z \in \mathbb{D}.$$

Moreover, $P(z, Z_f, \alpha)$ converges in $\mathbb{D}$ if and only if

$$\sum_{z_k \in Z_f} \left(1 - |z_k|\right)^{\alpha+1} < +\infty.$$

M.M. Djrbashian [7] noted that $P(z, Z_f, \alpha)$ has the form (2) if $\alpha \in \mathbb{N}$.
Besides the class A^*_α, which can be defined by the condition

$$\sup_{0<r<1} \int_0^{2\pi} \left(\int_0^r (r - t)^{\alpha-1} \log^+ \left| f\left(te^{i\varphi}\right) \right| dt \right) d\varphi < +\infty,$$

M.M. Djrbashian considered the class A_α defined by

$$\sup_{0<r<1} \int_0^{2\pi} \left(\int_0^r (r - t)^{\alpha-1} \log \left| f\left(te^{i\varphi}\right) \right| dt \right)^+ d\varphi < +\infty.$$

Obviously, $A^*_\gamma \subset A^*_\alpha \subset A_\alpha \subset A_\beta$, $\gamma < \alpha < \beta$. Moreover, the function $g_\alpha(z) = \exp\{\frac{1}{(1-z)^{\alpha+1}}\}$ belongs to $A_\alpha \setminus A^*_\alpha$.

Theorem 3 (M.M. Djrbashian) *The class A_α, $\alpha > -1$, coincides with the class of functions represented in the form*

$$f(z) = C_\lambda z^\lambda B_\alpha(z) \exp\left\{\int_0^{2\pi} \frac{d\psi(\theta)}{(1-e^{-i\theta}z)^{\alpha+1}}\right\} \equiv C_\lambda z^\lambda B_\alpha(z) \exp\{g_\alpha(z)\}, \quad (11)$$

where $\psi \in BV[0,2\pi]$, $\sum_{z_k \in Z_f} (1-|z_k|)^{\alpha+1} < +\infty$; $B_\alpha(z) = \prod_k (1-\frac{z}{z_k}) \times \exp\{-W_\alpha(z,z_k)\}$ is Djrbashian's product

$$W_\alpha(z,\zeta) = \int_{|\zeta|}^1 \frac{(1-x)^\alpha}{x} dx + \sum_k \frac{\Gamma(\alpha+k+1)}{\Gamma(\alpha+1)\Gamma(1+k)}$$
$$\times \left((\bar\zeta z)^k \int_{|\zeta|}^1 \frac{(1-x)^\alpha}{x^{k+1}} dx - \left(\frac{z}{\zeta}\right)^k \int_0^{|\zeta|} (1-x)^\alpha x^{k-1}\, dx \right).$$

More general results for arbitrary growth are obtained in [9].

3 Classes Defined by the Growth $\log M(r,f)$

3.1 Growth of the Maximum Modulus and Zero Distribution

B. Khabibullin [13] considered the following problem.

Problem 1 *Given a sequence Z in $\mathbb{D}$ without accumulation points in $\mathbb{D}$, find the lowest possible growth of* $\log M(r,f)$ *in the class of analytic functions $f \not\equiv 0$ vanishing on Z.*

An increasing continuous function $d\colon [a,1) \to [0,1)$, where $a \in [0,1)$ is called [13] a *shift function* if $t < d(t) < 1$ for $t \in [a,1)$.

Theorem 4 (B.N. Khabibullin [13, Theorem 1]) *Let Z be a sequence in $\mathbb{D}$, d be convex or concave shift function. Then there exists a function $f \in H(\mathbb{D})$, $f \not\equiv 0$ such that $Z_f \supset Z$ and* $\log M(r,f) \le \frac{C}{d(r)-r} N(d(r),Z)$ *for some positive constant C.*

Another approach was used by C.N. Linden. In 1964 [14] he established a connection between $\rho_M[P]$ and the zero distribution of P, where P is of the form (2). To clarify this connection we need some definitions.

Let

$$\square(re^{i\varphi}) = \left\{\rho e^{i\theta} : r \le \rho \le \frac{1+r}{2}, |\theta-\varphi| \le \frac{\pi(1-r)}{2}\right\},$$

and $\nu(re^{i\varphi})$ be the number of zeros of P in $\square(re^{i\varphi})$. We define

$$\nu_1(r,P)=\max_{\varphi}\nu\big(re^{i\varphi}\big),\qquad \nu[P]=\limsup_{r\uparrow 1}\frac{\log^+\nu_1(r,P)}{-\log(1-r)}. \tag{12}$$

Theorem 5 (C.N. Linden, [14, Theorem V]) *With the notation above we have*

$$\rho_M[P]\begin{cases}=\nu[P], & \rho_M[P]\ge 1,\\ \le \nu[P]\le 1, & \rho_M[P]<1.\end{cases} \tag{13}$$

This result was improved and generalized by F. Shamoyan in [22, 23]. We follow the notation of [22]. Let φ be nonnegative increasing function on $(0,+\infty)$. Set

$$X_{\varphi}^{\infty}=\left\{f\in H(\mathbb{D}):\log\big|f(z)\big|\le C(f)\varphi\left(\frac{1}{1-|z|}\right)\right\}. \tag{14}$$

Assume that for

$$\beta_{\varphi}=\liminf_{x\to+\infty}\frac{x\varphi'(x)}{\varphi(x)},\qquad \alpha_{\varphi}=\limsup_{x\to+\infty}\frac{x\varphi'(x)}{\varphi(x)}$$

we have $\beta_{\varphi}\le\alpha_{\varphi}<+\infty$.

Theorem 6 (F.A. Shamoyan, [22, Theorem 1]) *Suppose that φ satisfies the above conditions.*

(i) *Let $\beta_{\varphi}>1$. If $f\in X_{\varphi}^{\infty}$, $f(0)=1$, then $\nu_1(r,Z_f)\le C\varphi(\frac{1}{1-r})$ for some positive constant C;*
(ii) *let $\beta_{\varphi}>0$. If Z be an arbitrary sequence in $\mathbb{D}$ such that $\nu_1(r,Z)\le C\varphi(\frac{1}{1-r})$ for some positive constant C, then $P(z,Z,\alpha)\in X_{\varphi}^{\infty}$ for every $\alpha>\alpha_{\varphi}+1$.*

As we see, this theorem gives a description of zeros for functions $f\in H(\mathbb{D})$ of finite order $\rho_M[f]>1$. A counterpart of this result for functions of infinite order is obtained in [23, Theorem 2].

3.2 *Factorization of Classes Defined by the Growth of* $\log M(r,f)$

In [14] Linden proved the following result.

Theorem 7 (Linden, [14, Theorem I]) *Let f be analytic in $\mathbb{D}$ and of order $\rho_M[f]\ge 1$. Then*

$$f(z)=z^pP(z)g(z),$$

where P is a canonical product displaying the zeros of f, p is nonnegative integer, g is non-zero and both P and g are analytic and of ρ_M-order at most $\rho_M[f]$.

Further, in Theorem IV [14], Linden showed that if $\rho_M[f] < 1$ one has

$$\max\{\rho_M[P], \rho_M[g]\} \le \max\{\rho_M[f], \nu[f]\}.$$

For $\varphi(x) = x^\rho$, $\rho > 0$ we denote $X_\rho = X_\varphi^\infty$.

V.I. Matsaev and Ye.Z. Mogulski [18] established that if we take $P(z) = P(z, Z_f, s)$, $s \ge [\rho] + 1$, $s \in \mathbb{N}$, in the representation of Theorem 7, then the function g has the form

$$g(z) = \exp \int_0^{2\pi} S_q\big(ze^{-i\theta}\big)\gamma(\theta)\,d\theta, \quad z \in \mathbb{D}, \tag{15}$$

where $q = [\rho] + 1$, $S_q(z) = \Gamma(q+1)(\frac{2}{(1-z)^q} - 1)$ is the generalized Schwarz kernel, γ is a real valued function such that $\gamma \in \mathrm{Lip}(q - \rho)$ for noninteger ρ, and γ satisfies Zygmund's condition $|\gamma(\theta + h) - 2\gamma(\theta) + \gamma(\theta - h)| \le Ch$ for integer ρ.

In [24] F. Shamoyan showed that non-zero factor $U_\alpha(z)$ in Djrbashian' representation (9) can be written in the form (15) with q not necessarily integer such that $q > \alpha$, and $(k = [q - \alpha])$

$$\int_0^{2\pi}\int_0^{2\pi} \frac{\gamma^{(k)}(t+\theta) - 2\gamma^{(k)}(\theta) + \gamma^{(k)}(\theta - t)}{|t|^{1+q-\alpha}}\,dt\,d\theta < +\infty.$$

In view of relation (1) the following problem arises naturally.

Problem 2 *Given* $0 \le \sigma \le \rho \le \sigma + 1$, *describe the class* $\mathcal{H}_{\sigma,\rho}$ *of analytic functions in* $\mathbb{D}$ *such that* $\rho_T[f] = \sigma$, $\rho_M[f] = \rho$.

In [15] Linden constructed canonical products form $\mathcal{H}_{\sigma,\rho}$ when $\rho \ge 1$, and $\rho - 1 \le \sigma \le \rho$. In [2] Problem 2 was solved by the first author under the restriction that $\rho \ge 1$. A solution is given in terms of so called *complete measure* of an analytic function in the sense of Grishin (see [10, 11]).

Let $f \in H(\mathbb{D})$ be of the form

$$f(z) = C_q z^\lambda P(z, Z_f, q)\exp\left\{\int_0^{2\pi} S_q\big(ze^{-i\theta}\big)d\psi^*(\theta)\right\}, \tag{16}$$

where $\psi^* \in BV[0, 2\pi]$, $\sum_{z_k \in Z_f}(1 - |a_k|)^{q+1} < +\infty$, $\lambda \in \mathbb{Z}_+$, $C_q \in \mathbb{C}$.

Let M be Borel's subset of $\overline{\mathbb{D}}$. A *complete measure* λ_f *of genus* q *in the sense of Grishin* is defined as

$$\lambda_f(M) = \sum_{Z_f \cap M} \big(1 - |z_k|\big)^{q+1} + \psi(M \cap \partial\mathbb{D}), \tag{17}$$

where ψ is the Stieltjes measure associated with ψ^*.

A characterization of λ_f for $f \in \mathcal{H}_{\sigma,\rho}$ is given in [2, Theorem 4]. Another application of λ_f can be found in [3].

4 A Concept of ρ_∞-Order

Many theorems valid on analytic functions of finite order in $\mathbb{D}$ fail to hold when ρ_M-order is smaller than 1 (see e.g. [2, 14, 16]).

In particular, for a Blaschke product B we always have $0 \le \nu[B] \le 1$, so Theorems 5 and 6 give no new information on zero distribution of B.

The question arises:

Question 1 What kind of growth characteristic can describe zero distribution in the case when $\rho_M[f] \le 1$?

For a meromorphic function $f(z)$, $z \in \mathbb{D}$, and $p \ge 1$ we define

$$m_p(r, f) = \left(\frac{1}{2\pi}\int_0^{2\pi} \left|\log\left|f\left(re^{i\theta}\right)\right|\right|^p d\theta\right)^{\frac{1}{p}}, \quad 0 < r < 1.$$

We write

$$\rho_p[f] = \limsup_{r\uparrow 1} \frac{\log m_p(r, f)}{-\log(1-r)}.$$

A characterization of ρ_p-orders can be found in [17].

We define ρ_∞-*order* of f as

$$\rho_\infty[f] = \lim_{p\to\infty} \rho_p[f],$$

(existence of the limit follows from the fact that L_p-norms are monotone in p). It follows from the first fundamental theorem of Nevanlinna that $\rho_1[f] = \rho_T[f]$. Besides, it is known (e.g. [16]) that $\rho_M[f] \le \rho_p[f] + \frac{1}{p}$ $(p > 0)$, which generalizes (1). Consequently, $\rho_M[f] \le \rho_\infty[f]$. Moreover, Linden [16] proved that $\rho_\infty[f] = \rho_M[f]$ provided that $\rho_M[f] \ge 1$. Thus, the values $\rho_\infty[f]$ and $\nu[f]$ have similar behavior with respect to the maximum modulus order, when f is a canonical product.

Remark 1 To avoid confusion, we have to note that Linden used the notation $\lambda_\infty(f)$ for $\rho_M[f]$. But he did not consider the limit $\lim_{p\to\infty} \rho_p[f]$ when $\rho_M[f] < 1$.

For a sequence Z in $\mathbb{D}$ with finite convergence exponent we define $\nu[Z] = \nu[P(z, Z, q)]$ for an appropriate choice of q. It is clear that the definition does not depend on q.

The following theorem answers the question posed above.

Theorem 8 (I. Chyzhykov, [5, Theorem 1.1]) *Given a sequence Z in $\mathbb{D}$ such that $\nu = \nu[Z] < \infty$ and an integer s such that $s \ge [\nu] + 1$, we define the canonical product $P_s(z) = P(z, Z, s)$. Then $\rho_\infty[P_s] = \nu$.*

Corollary 1 ([5, Theorem 1.2]) *Let $f \in H(\mathbb{D})$. Then $\nu[f] \le \rho_\infty[f]$.*

Example 1 Let $z_k = 1 - 1/(k \log^2 k)$, $k \in \{3, \dots\}$. We consider the Blaschke product $B(z, Z)$. Since $|B|$ is bounded in $\mathbb{D}$, we have $\rho_M[B] = \rho_T[B] = 0$, and consequently $\rho_\infty[B] \le 1$.

On the other hand, it is easy to check that

$$n(r, B) \sim \frac{1}{(1-r)\log^2(1-r)}, \quad r \uparrow 1,$$

and

$$\frac{d_1}{(1-r)\log^2(1-r)} \le \nu(r) \le \frac{d_2}{(1-r)\log^2(1-r)}, \quad r \uparrow 1,$$

for some positive constants d_1, d_2. Hence, $\nu[B] = 1$, and by Theorem 1 $\rho_\infty[B] = 1$.

Taking into account Corollary 1 we deduce that $\max\{\rho_M[P], \rho_M[g]\} \le \rho_\infty[f]$ in Theorem 7. A counterpart of Theorem 7 is valid without restrictions on the value of order.

Theorem 9 (I. Chyzhykov, [5, Theorem 2.1]) *Let f be analytic in $\mathbb{D}$, and of finite order $\rho_\infty[f]$. Then*

$$f(z) = z^p P(z) g(z),$$

where P is a canonical product displaying the zeros of f, p is nonnegative integer, g is non-zero and both P and g are analytic and of ρ_∞-order at most $\rho_\infty[f]$.

Some other applications of the concept of ρ_∞-order such as logarithmic derivative estimates can be found in [4].

The proof of Theorem 8 relies on the inequality $s \ge [\nu] + 1$. Since the theorem is not applicable for Blaschke products one may ask what are relations between zero distribution of a Blaschke product and its ρ_∞-order.

Here we prove the following Carleson-type result. Let

$$S(\varphi, \delta) = \left\{\rho e^{i\theta} \in \overline{\mathbb{D}} : \rho \ge 1 - \delta, -\pi\delta < \theta - \varphi \le \pi\delta\right\}$$

be the Carleson square based on the arc $[e^{i(\varphi - \pi\delta)}, e^{i(\varphi + \pi\delta)}]$.

Theorem 10 *Let Z be a sequence in $\mathbb{D}$ such that $\sum_{z_k \in Z} (1 - |z_k|)^{s+1} < +\infty$ for some nonnegative integer s, $P_s(z) = P(z, Z, s)$.*

(i) *Let $\gamma \in (0, s+1]$. If*

$$\sum_{z_n \in S(\varphi,\delta)} \left(1 - |z_n|\right)^{s+1} \le C_1 \delta^\gamma, \quad \delta \in (0,1), \tag{18}$$

for some constant C_1 independent of φ and δ, then for all $p \ge 1$

$$m_p\big(r, \log|P_s|\big) \le \begin{cases} C_2 (1-r)^{\gamma - s - 1} (\log \frac{1}{1-r} + 1), & \gamma \in (0, s+1); \\ C_2 (\log^2(1-r) + 1), & \gamma = s+1. \end{cases} \tag{19}$$

(ii) *If* $s = 0$, *and for all* $p \geq 1$ *we have* $m_p(r, \log|B|) \leq K(1-r)^{1-\gamma}$ *for some constant* K *independent of* p *and* r *and* $\gamma \in (0, 1]$, *then* (18) *holds.*

For a Blaschke product we define $\lambda(\varphi, r) = \sum_{z_k \in Z_B \cap S(\varphi, \frac{1-r}{2})} (1 - |z_k|)$.

Corollary 2 *Let* B *be a Blaschke product. Set*

$$t[B] = \sup\Big\{\gamma \geq 0 : \max_{\varphi} \lambda(\varphi, r) = O\big((1-r)^{\gamma}\big)\Big\}.$$

Then $\rho_\infty[B] = 1 - t[B]$.

Corollary 3 *If* B *is an interpolating Blaschke product, then* $m_p(r, \log|B|) \leq C(\log^2 \frac{1}{1-r} + 1)$ *for all* $p \geq 1$.

5 Proof of Theorem 10

We start with proving (i). We write $E_m(re^{i\varphi}) = S(\varphi, (1-r)2^{m-1})$, $m \in \mathbb{N}$, $E_0(z) = \emptyset$. So $E_1(re^{i\varphi}) = S(\varphi, 1-r)$, and $E_m(re^{i\varphi}) = \overline{\mathbb{D}}$ for $m \geq m(r) = [\log_2 \frac{1}{1-r}]$.

Lemma 1 *Let* Z *be a sequence in* $\mathbb{D}$ *such that* $\sum_{z_k \in Z} (1 - |z_k|)^{s+1} < \infty$. *Suppose that for some* K *and* $\gamma \in (0, s+1]$ *condition* (18) *holds. Then*

$$\sum_{k=1}^{\infty} \left| \frac{1 - |z_k|^2}{1 - z\bar{z}_k} \right|^{s+1} \leq \begin{cases} \frac{C_3}{(1-|z|)^{s+1-\gamma}}, & \gamma \in (0, s+1), \\ C_3 \log \frac{1}{1-|z|}, & \gamma = s+1, \end{cases} \qquad z \in \mathbb{D}$$

for some constant $C_3 = C_3(s, \gamma) > 0$.

Proof of the lemma It is easy to see that $|1 - r\rho_k e^{i(\varphi - \theta_k)}| \geq C_4(1-r)2^m$ for $z_k = \rho_k e^{i\theta_k} \in \mathbb{D} \setminus E_m$ with some absolute constant C_4. Then

$$\begin{aligned} \sum_k \frac{(1 - |z_k|^2)^{s+1}}{|1 - re^{i\varphi}\bar{z}_k|^{s+1}} &= \sum_{m=1}^{m(r)} \sum_{z_k \in E_m \setminus E_{m-1}} \frac{(1 - \rho_k^2)^{s+1}}{|1 - r\rho e^{i(\varphi - \theta_k)}|} \\ &\leq \frac{2^{s+1}}{(C_4(1-r)2^{m-1})^{s+1}} \sum_{z \in E_m} (1 - \rho_k)^{s+1} \\ &\leq \frac{4^{s+1}}{(C_4(1-r))^{s+1}} \sum_{m=1}^{m(r)} \frac{C_1((1-r)2^m)^{\gamma}}{2^{m(s+1)}} \\ &\leq \frac{C_5(s)}{(1-r)^{s+1-\gamma}} \sum_{m=1}^{m(r)} 2^{m(\gamma - s - 1)}. \end{aligned}$$

The last sum is bounded by a constant depending on γ and s for $\gamma \in (0, s+1)$, and equals $m(r)$ in the case $\gamma = s+1$. This implies the assertion of the lemma. □

We shall need some known results.

Theorem 11 (See [20, Theorem V.24, p. 222; Theorem V.25, p. 224]) *For the canonical product* $P_s(z)$

$$\log^+ \big|P_s(z)\big| \le C_6(s) \sum_m \left| \frac{1-|z_m|^2}{1-z\bar{z}_m} \right|^{s+1}, \quad z \in \mathbb{D}, \sum_m \big(1-|z_m|\big) = +\infty; \tag{20}$$

if D_m *denotes the disc* $D(z_m, (1-|z_m|^2)^{s+4})$ *then*

$$\log^+ \frac{1}{|P_s(z)|} \le K \log \frac{1}{1-|z|} \sum_m \left| \frac{1-|z_m|^2}{1-z\bar{z}_m} \right|^{s+1}, \quad \frac{1}{2} \le |z| < 1,\ z \notin \bigcup_m D_m. \tag{21}$$

We first suppose that $\gamma < s+1$. Then, let $s \in \mathbb{N}$. We have to prove that

$$\int_0^{2\pi} \big|\log\big|P_s\big(re^{i\theta}\big)\big|\big|^p \, d\theta \le C_7^p \frac{\log^p \frac{1}{1-r}}{(1-r)^{p(s+1-\gamma)}}. \tag{22}$$

We deal with the integral in (22) by covering the range of integration by $[\pi/(1-r)]+1$ intervals of the form $[\tau + r - 1, \tau + 1 - r]$ for $\tau = 2k(1-r)$ and $k \in \{0, \ldots, [\pi/(1-r)]\}$, showing that

$$\int_{\tau+r-1}^{\tau+1-r} \big|\log\big|P_s\big(re^{i\theta}\big)\big|\big|^p \, d\theta \le C_8^p (1-r)^{-p(s+1-\gamma)+1} \log^p \frac{1}{1-r} \tag{23}$$

for each τ, where the constant C_8 is independent of τ. For convenience and without loss of generality, we may suppose that $\tau = 0$ and $\frac{3}{4} \le |z_m| < 1$. For given r, let $\gamma_r = \{z = re^{i\theta} : r-1 \le \theta \le 1-r\}$, and $F(r) = \{m : D_m \cap \gamma_r \ne \emptyset\}$, where D_m are the exceptional discs of Theorem 11. From the definition of the discs D_m and assumptions on (z_m) it follows that $1 - 4^{-3} \le \frac{1-r}{1-|z_m|} \le 1 + 4^{-3}$. Hence $\sum_{z_m \in F(r)} (1-|z_m|)^{s+1} \ge \frac{(1-r)^{s+1}}{2^{s+1}} |F(r)|$, where $|F(r)|$ denotes the number of elements in the set $F(r)$. Thus, by (18), we have

$$\big|F(r)\big| \le C_9 (1-r)^{\gamma-1-s}. \tag{24}$$

We consider the factorization $P_s = B_1 B_2 B_3$, where

$$B_1(z) = \prod_{m \notin F(r)} E\left(\frac{1-|z_m|^2}{1-\bar{z}_m z}, s\right),$$

$$B_2(z) = \prod_{m \in F(r)} \exp \sum_{j=1}^{s} \frac{1}{j} \left(\frac{1-|z_m|^2}{1-z\bar{z}_m}\right)^j,$$

$$B_3(z) = \prod_{m \in F(r)} \left(1 - \frac{1-|z_m|^2}{1-z\bar{z}_m}\right) = \prod_{m \in F(r)} \left(\frac{\bar{z}_m(z_m - z)}{1-z\bar{z}_m}\right).$$

First we note that Theorem 11 and Lemma 1 give

$$\begin{aligned}\int_{r-1}^{1-r} \bigl|\log\bigl|B_1\bigl(re^{i\theta}\bigr)\bigr|\bigr|^p d\theta &\le \int_{r-1}^{1-r} C_{10}^p \log^p \frac{1}{1-r} \left(\sum_m \left|\frac{1-|z_m|^2}{1-re^{i\theta}\bar{z}_m}\right|^{s+1}\right)^p d\theta \\ &\le C_{10}^p \log^p \frac{1}{1-r} \frac{1}{(1-r)^{p(s+1-\gamma)}} 2(1-r) \\ &= \frac{C_{11}^p(s,\gamma) \log^p \frac{1}{1-r}}{(1-r)^{p(s+1-\gamma)-1}}. \end{aligned} \tag{25}$$

Next, the inequality $|1 - z\bar{z}_m| > \frac{1}{2}(1-|z_m|^2)$ yields

$$\bigl|\log\bigl|B_2(z)\bigr|\bigr| < \sum_{m \in F(r)} \sum_{j=1}^{s} \frac{1}{j} \left|\frac{1-|z_m|^2}{1-z\bar{z}_m}\right|^j \le C_{12}\bigl|F(r)\bigr|.$$

Hence (24) implies

$$\int_{r-1}^{1-r} \bigl|\log\bigl|B_2\bigl(re^{i\theta}\bigr)\bigr|\bigr|^p d\theta \le C_{13}^p (1-r)^{1-p(s+1-\gamma)}. \tag{26}$$

Finally, in [16, p. 124] it is proved that

$$\int_{r-1}^{1-r} \bigl|\log\bigl|B_3\bigl(re^{i\theta}\bigr)\bigr|\bigr|^p d\theta \le C_{14}\bigl|F(r)\bigr|^p (1-r). \tag{27}$$

Inequality (23) now follows from (25)–(27).

In the case $s = 0$ the only difference in the proof is that there is no product B_2, and $|B_1(z)| \le (\prod_m |z_m|)^{-1}$.

We now suppose that $\gamma = s+1$. In this case $|F(r)|$ is bounded uniformly in r. Instead of (25), using Lemma 1, we obtain

$$\int_{r-1}^{1-r} \left|\log\left|B_1\left(re^{i\theta}\right)\right|\right|^p d\theta \le \int_{r-1}^{1-r} C_{10}^p \log^p \frac{1}{1-r} \left(\sum_m \left|\frac{1-|z_m|^2}{1-re^{i\theta}\bar{z}_m}\right|^{s+1}\right)^p d\theta$$

$$\le 2C_{10}^p \log^{2p} \frac{1}{1-r}(1-r). \tag{28}$$

Hence, $m_p(r,\ \log|P_s|) = O(\log^2(1-r))$ as $r \uparrow 1$.

We now prove (ii). Consider the function

$$K(z,\zeta) = \frac{1}{1-|\zeta|}\log\left|\frac{1-z\bar{\zeta}}{z-\zeta}\right|, \quad z \in \mathbb{D}, \zeta \in \overline{\mathbb{D}}.$$

This function has many nice properties. It is nonnegative. Moreover ($\zeta = \rho e^{i\theta}$, $z = re^{i\varphi}$),

$$K(z,\zeta) = \frac{1}{2(1-\rho)}\log\left(1 + \frac{(1-r^2)(1-\rho^2)}{r^2 - 2r\rho\cos(\varphi-\theta)+\rho^2}\right), \tag{29}$$

and therefore

$$\lim_{\rho\uparrow 1} K\left(z, \rho e^{i\theta}\right) = \frac{1-|z|^2}{|e^{i\theta}-z|^2}.$$

We need the following property:

$$\left|K(z,\zeta)\right| \ge \frac{1}{12}\frac{1-|z|^2}{|z-\zeta|^2}, \qquad 1-|\zeta| \le \frac{1}{2}\left(1-|z|\right). \tag{30}$$

Indeed, since $\log(1+x) \ge \frac{x}{1+x}$, $x \in (0,1)$, using (29), we deduce that

$$\left|K(z,\zeta)\right| \ge \frac{1+\rho}{2}\frac{1-r^2}{|z-\zeta|^2}\frac{1}{1+\frac{(1-r^2)(1-\rho^2)}{|z-\zeta|^2}}. \tag{31}$$

The condition $1-|\zeta| \le \frac{1}{2}(1-|z|)$ yields $|\zeta| \ge \frac{1}{2}$, and

$$|z-\zeta| \ge 1-|z| - \left(1-|\zeta|\right) \ge \frac{1-|z|}{2}.$$

Therefore

$$\frac{1+\rho}{2}\frac{1-r^2}{|z-\zeta|^2}\frac{1}{1+\frac{(1-r^2)(1-\rho^2)}{|z-\zeta|^2}}$$

$$\ge \frac{3}{2}\frac{1}{1+\frac{1-\rho^2}{1-r}4(1+r)} \ge \frac{3}{4}\frac{1}{1+2(1+\rho)(1+r)} \ge \frac{1}{12}.$$

Inequality (30) is proved.

Then we can write $\log|B(z)| = -\sum_{z_k\in Z} K(z,z_k)(1-|z_k|) + \sum_k \log|z_k|$. Using (30), we obtain

$$\begin{aligned}&\left|\log\left|B\left(re^{i\theta}\right)\right|\right| \\ &\quad \geq \sum_{z_k\in S(\varphi,\frac{1-r}{2})} K\left(re^{i\varphi},\zeta\right)\left(1-|z_k|\right) \geq \frac{1}{12}\sum_{z_k\in S(\varphi,\frac{1-r}{2})} \frac{(1-r^2)(1-|z_k|)}{|re^{i\varphi}-z_k|^2}.\end{aligned}$$

Elementary geometric arguments show that $|re^{i\varphi}-\rho e^{i\theta}| \leq |re^{i\varphi}-e^{i\theta}|$ for $1>\rho\geq r\geq 0$. It then follows that

$$\begin{aligned}\left|\log\left|B\left(re^{i\theta}\right)\right|\right| &\geq \frac{1}{12}\sum_{z_k\in S(\varphi,\frac{1-r}{2})} \frac{(1-r^2)(1-|z_k|)}{|re^{i\varphi}-e^{i\theta}|^2} \\ &\geq \frac{1}{3(\frac{\pi^2}{4}+1)}\frac{1-r^2}{(1-r)^2}\sum_{z_k\in S(\varphi,\frac{1-r}{2})}\left(1-|z_k|\right) \\ &\geq \frac{\sum_{z_k\in S(\varphi,\frac{1-r}{2})}(1-|z_k|)}{3(\frac{\pi^2}{4}+1)(1-r)}.\end{aligned}$$

Recall that $\lambda(\varphi,r) = \sum_{z_k\in S(\varphi,\frac{1-r}{2})}(1-|z_k|)$. From the definition of $S(\varphi,\delta)$ it follows that for fixed r the function $\lambda(\varphi,r)$ is piecewise constant and continuous from the right. Therefore it attains its maximum on some interval $[\varphi_1(r),\varphi_2(r))$, $\varphi_2(r)>\varphi_1(r)$. By the assumption of the theorem we deduce that

$$\frac{C_1}{(1-r)^{1-\gamma}} \geq \left(\int_0^{2\pi}\left|\log\left|B\left(re^{i\varphi}\right)\right|\right|^p d\varphi\right)^{\frac{1}{p}} \geq C_{15}\frac{(\int_0^{2\pi}(\lambda(\varphi,r))^p\,d\varphi)^{\frac{1}{p}}}{1-r}.$$

Hence

$$\max_{\varphi}\lambda(\varphi,r)\left(\varphi_2(r)-\varphi_1(r)\right)^{\frac{1}{p}} \leq \left(\int_0^{2\pi}\lambda^p(\varphi,r)\,d\varphi\right)^{\frac{1}{p}} \leq \frac{C_1}{C_{15}}(1-r)^{\gamma}.$$

Letting $p\to\infty$ we obtain the assertion of (ii).

Acknowledgements This paper was inspired by the "Conference on Blaschke Products and their Applications" (Fields Institute, Toronto, July 25–29, 2011) organized by Javad Mashreghi and Emmanuel Fricain. I would like to thank the organizers and the staff of the Fields Institute for hospitality and financial support.

References

1. Chyzhykov, I.E.: On a complete description of the class of functions without zeros analytic in a disk and having given orders. Ukr. Math. J. **59**(7), 1088–1109 (2007)

2. Chyzhykov, I.E.: Growth of analytic functions in the unit disc and complete measure in the sense of Grishin. Mat. Stud. **29**(1), 35–44 (2008)
3. Chyzhykov, I.: Argument of bounded analytic functions and Frostman's type conditions. Ill. J. Math. **53**(2), 515–531 (2009)
4. Chyzhykov, I., Heittokangas, J., Rättyä, J.: Sharp logarithmic derivative estimates with applications to differential equations in the unit disc. J. Aust. Math. Soc. **88**(2), 145–167 (2010)
5. Chyzhykov, I.: Zero distribution and factorization of analytic functions of slow growth in the unit disc. Proc. Am. Math. Soc. (accepted)
6. Djrbashian, M.M.: On canonical representation of functions meromorphic in the unit disk. Dokl. Akad. Nauk Arm. SSR **3**(1), 3–9 (1945)
7. Djrbashian, M.M.: On the representation problem of analytic functions. Soobshch. Inst. Math. Mekh. Akad. Nauk Arm. SSR **2**, 3–50 (1948)
8. Djrbashian, M.M.: Integral Transforms and Representations of Functions in the Complex Domain. Nauka, Moscow (1966) (in Russian)
9. Djrbashian, M.M.: Theory of factorization and boundary properties of functions meromorphic in the disc. In: Proceedings of the ICM, Vancouver, BC (1974)
10. Fedorov, M.A., Grishin, A.F.: Some questions of the Nevanlinna theory for the complex half-plane. Math. Phys. Anal. Geom. **1**(3), 223–271 (1998) (Kluwer Academic)
11. Grishin, A.: Continuity and asymptotic continuity of subharmonic functions. Math. Phys. Anal. Geom., ILPTE **1**(2), 193–215 (1994) (in Russian)
12. Hayman, W.K.: Meromorphic Functions. Clarendon Press, Oxford (1964)
13. Khabibullin, B.: Zero subsets, representation of meromorphic functions and Nevanlinna characteristics in a disc. Mat. Sb. **197**(2), 117–130 (2006)
14. Linden, C.N.: The representation of regular functions. J. Lond. Math. Soc. **39**, 19–30 (1964)
15. Linden, C.N.: On a conjecture of Valiron concerning sets of indirect Borel point. J. Lond. Math. Soc. **41**, 304–312 (1966)
16. Linden, C.N.: Integral logarithmic means for regular functions. Pac. J. Math. **138**(1), 119–127 (1989)
17. Linden, C.N.: The characterization of orders for regular functions. Math. Proc. Cambr. Philos. Soc. **111**(2), 299–307 (1992)
18. Matsaev, V.I., Mogulski, E.Z.: A division theorem for analytic functions with the given majorant and some of its applications. Zap. Nauč. Semin. POMI **56**, 73–89 (1976) (in Russian)
19. Naftalevich, A.G.: On interpolating of functions meromorphic in the unit disc. Dokl. Akad. Nauk SSSR **88**(2), 205–208 (1953) (in Russian)
20. Tsuji, M.: Potential Theory in Modern Function Theory. Chelsea, New York (1975). Reprinting of the 1959 edn.
21. Tsuji, M.: Canonical product for a meromorphic function in a unit circle. J. Math. Soc. Jpn. **8**(1), 7–21 (1956)
22. Shamoyan, F.A.: A factorization theorem of M.M. Dzhrbashian's and characteristic of the zeros of analytic functions in the disk with a majorant of finite growth. Izv. Akad. Nauk. Arm. SSR Mat. **XIII**(5–6), 405–422 (1978) (in Russian)
23. Shamoyan, F.A.: Zeros of functions analytic in a disk, that increase near the boundary. Izv. Akad. Nauk Arm. SSR Ser. Mat. **XVIII**(1), 15–27 (1983) (in Russian)
24. Shamoyan, F.A.: Several remarks on parametric representation of Nevanlinna-Dzhrbashian's classes. Math. Notes **52**(1), 7227–7237 (1992)

Hardy Means of a Finite Blaschke Product and Its Derivative

Alan Gluchoff and Frederick Hartmann

Abstract In this chapter we consider several topics related to finite Blaschke products $B_n(z) = \prod_{k=1}^{n} \frac{z_k - z}{1-\bar{z}_k z}$ in the unit disc of the complex plane and their Hardy means $M_p^p(r, B) = \frac{1}{2\pi}\int_0^{2\pi} |B(re^{i\theta})|^p d\theta$. We discuss two explicit formulae for $1 - M_2^2(r, B)$: when B has distinct zeroes or a single zero repeated n times. We relate the growth of the means $M_2^2(r, B)$ and $M_2^2(r, B')$ to "sampling means" $\sum_{k=1}^{n} |B(rz_k)|(1 - |z_k|^2)$ and $\sum_{k=1}^{n} |B'(rz_k)|(1 - |z_k|^2)$. It is shown, for products of degree two and three, that if the zeroes lie on the circle of radius $|z| = \rho < 1$ with constant angle ϕ between successive zeroes, then $1 - M_2^2(r, B)$ is an increasing function of ϕ. We conjecture that this holds true for products of arbitrary finite degree.

Keywords Mean modulus · Blaschke products

Mathematics Subject Classification Primary 30C45 · Secondary 30C10 · 30D50

1 Introduction

Let $\mathbb{D} = \{z : |z| < 1\}$ be the unit disc of the complex plane $\mathbb{C}$, and let $z_k \in \mathbb{D}$, $k = 1, \ldots, n$, $n \geq 1$. Then

$$B_n(z) = \prod_{k=1}^{n} \frac{z_k - z}{1 - \bar{z}_k z}$$

A. Gluchoff (✉) · F. Hartmann
Department of Mathematics and Statistics, Villanova University, Villanova, PA, 19085 USA
e-mail: alan.gluchoff@villanova.edu

F. Hartmann
e-mail: frederick.hartmann@villanova.edu

J. Mashreghi, E. Fricain (eds.), *Blaschke Products and Their Applications*,
Fields Institute Communications 65, DOI 10.1007/978-1-4614-5341-3_9,

is the finite Blaschke product with zero set $\{z_k\}_{k=1}^n$. For $0 < p < \infty$ and $0 < r < 1$ the Hardy p-mean is defined for f analytic in $\mathbb{D}$ by

$$M_p(r, f) = \left(\frac{1}{2\pi}\int_0^{2\pi} \left|f\left(re^{i\theta}\right)\right|^p d\theta\right)^{1/p},$$

and H^p is the space of all f analytic in $\mathbb{D}$ for which $\sup_{0<r<1} M_p(r, f) < \infty$. We recall that H^∞ is the space of all bounded analytic functions in $\mathbb{D}$ with $\|f\|_\infty = \sup_{z\in\mathbb{D}} |f(z)|$. We further define for any $B_n(z)$ the quantities $\Delta_1(r, B_n) = 1 - M_1(r, B_n)$ and $\Delta_2(r, B_n) = 1 - M_2^2(r, B_n)$. By the properties of the Hardy means and Blaschke products it follows that $\Delta_1(r, B_n)$ and $\Delta_2(r, B_n)$ are decreasing functions of r and $\lim_{r\to 1} \Delta_1(r, B_n) = \lim_{r\to 1} \Delta_2(r, B_n) = 0$; also we have $\Delta_1(r, B_n) \le \Delta_2(r, B_n) \le 2\Delta_1(r, B_n)$ for all r, $0 < r < 1$.

In this paper we investigate various problems related to $\Delta_k(r, B_n), k = 1, 2$ and $M_1(r, B_n')$. In [5] an explicit formula for $\Delta_2(r, B_n)$ in terms of r and $\{z_k\}_{k=1}^n$, was given in the case of distinct $\{z_k\}$, $z_k \neq 0$ for all k. In Sect. 2 of this paper some further comments on this expression are given and it is related to an identity of Sylvester. A new formula for $\Delta_2(r, B_n)$ in the case of repeated zeroes ($z_k = \rho e^{i\phi}$, for all k, for fixed $0 < \rho < 1$ and $0 \le \phi \le 2\pi$) is derived. In the third section we relate the quantities $\Delta_1(r, B_n)$ and $M_1(r, B_n')$ to the corresponding means $\sum_{k=1}^n |B_n(rz_k)|(1 - |z_k|)$ and $\sum_{k=1}^n |B_n'(rz_k)|(1 - |z_k|)$ respectively, and show how these corresponding quantities compare in growth for $0 < r < 1$. Thus the rate of growth of $|B_n|$ and $|B_n'|$ on circles $\{z : |z| = r\}$ is related to their growth on $\{rz_k\}_{k=1}^n$, $0 < r < 1$; the means $\Delta_1(r, B_n)$ and $M_1(r, B_n')$ are thus compared with what might be called "sampling means" on $\{rz_k\}_{k=1}^n$. The proofs of these comparisons use the formulae of Sect. 2.

In the fourth section we compare, for fixed r, $\Delta_2(r, B_n)$ for B_n with zeroes on $\{z : |z| = \rho\}$ both with repeated zeroes, $z_k = \rho e^{i\phi}$, $k = 1, \ldots, n$ and for equally spaced zeroes, $z_k = \rho e^{i[\phi + 2\pi k/n]}$, $k = 1, \ldots, n$. This comparison also uses the formulae developed in Sect. 2. In spite of the availability of expressions for these quantities, comparison of the means proves quite difficult, and results are proved for products of low degree which suggest a conjecture for higher degree products.

We conclude this section by noting that in the 1980's there was interest in $\Delta_2(r, B)$ and $M_1(r, B')$ for an infinite Blaschke product B; (see [3, 4]) in these papers the rate at which $\Delta_2(r, B)$ approaches zero was related to the distributional properties of the zeroes and their rate of approach to the boundary of $\mathbb{D}$. Results were obtained for products with zeroes on a radial axis and also for interpolating Blaschke products. Finite products present challenges of a different nature.

2 Derivations of Formulae for $\Delta_2(r, B_n)$

In this section we record a previously proved expression for $\Delta_2(r, B_n)$ where B_n has distinct zeroes and derive a new expression for $\Delta_2(r, B_n)$ where B_n has repeated zeroes.

Theorem 1 *Let*

$$B_n(z) = \prod_{k=1}^{n} \frac{z_k - z}{1 - \overline{z}_k z},$$

where $0 < |z_k| < 1$ *for* $k = 1, 2, \ldots, n$, $z_k \neq z_j$ *if* $k \neq j$, $z = re^{i\theta}$, $0 < r < 1$. *Then*

$$\Delta_2(r, B_n) = \sum_{k=1}^{n} \frac{(1-r^2)(1-|z_k|^2)}{1-r^2|z_k|^2} \prod_{\substack{j\neq k\\ j=1}}^{n} \frac{z_j - r^2 z_k}{1-\overline{z}_j r^2 z_k} \Big/ \prod_{\substack{j\neq k\\ j=1}}^{n} \frac{z_j - z_k}{1-\overline{z}_j z_k}.$$

Proof This result was proved by the first author in [5]. □

Remark 1 An elementary but lengthy proof of this theorem follows from the identity

$$\left|\frac{z_k - z}{1-\overline{z}_k z}\right|^2 = \frac{r_k^2 - 2rr_k\cos(\theta-\theta_k) + r^2}{1 - 2rr_k\cos(\theta-\theta_k) + r^2 {r_k}^2}$$

(where $z_k = r_k e^{i\theta_k}$) and the standard residue theorem substitutions $\cos(\theta) = \frac{1}{2}(z + z^{-1})$, $\sin(\theta) = \frac{1}{2i}(z - z^{-1})$. We omit the details.

Remark 2 Letting $r \to 0$ in Theorem 1 gives

$$1 - \prod_{k=1}^{n} |z_k|^2 = \sum_{k=1}^{n} (1-|z_k|^2) \prod_{\substack{j\neq k\\ j=1}}^{n} \left(\frac{1-\overline{z}_j z_k}{z_j - z_k}\right) z_j.$$

By choosing $z_k = y_k^{1/2} e^{i\theta_k}$ for arbitrary $0 < y_k < 1$, $\theta_k \in \mathbb{R}$ we obtain

$$1 - \prod_{k=1}^{n} y_k = \sum_{k=1}^{n} (1-y_k) \prod_{\substack{j\neq k\\ j=1}}^{n} \frac{y_j^{1/2} e^{i\theta_j} - y_j y_k^{1/2} e^{i\theta_k}}{y_j^{1/2} e^{i\theta_j} - y_k^{1/2} e^{i\theta_k}}.$$

This is a special case of Sylvester's identity [1, p. 132, (2.1)]

$$1 - \prod_{k=1}^{n} y_k = \sum_{k=1}^{n} (1-y_k) \prod_{\substack{j\neq k\\ j=1}}^{n} \frac{1 - x_j y_j / x_k}{1 - x_j / x_k}$$

with $x_k = [y_k^{1/2} e^{i\theta_k}]^{-1}$.

Theorem 2 *If* $z_k = \rho e^{i(2\pi k/n)}$, $n \geq 2$, $k = 1, 2, \ldots, n$, $0 < \rho < 1$, $0 < r < 1$, *then*

$$\Delta_2(r, B_n) = \frac{(1-r^{2n})(1-\rho^{2n})}{1 - r^{2n}\rho^{2n}}.$$

Proof This follows by substitution in Theorem 1 and some elementary manipulations. Also since in this case $B_n(z) = (z^n - \rho^n)/(1 - \rho^n z^n)$, $[M_2(r, B_n)]^2$ can be evaluated directly. □

Theorem 3 *Let*

$$B_n(z) = \left[\frac{z - \rho e^{i\phi}}{1 - \rho e^{-i\phi} z}\right]^n, \quad 0 < \rho < 1,\ n \geq 2.$$

Then $\Delta_2(r, B_n) = (1 - r^2)(1 - \rho^2)/(1 - r^2\rho^2)^{2n-1} S$, *where*

$$S = \sum_{\substack{k_1+k_2+k_3=0 \\ k_i \geq 0}}^{n-1} (-1)^{k_3} \left[\binom{n}{k_1}\left(1 - r^2\right)^{n-k_1-1}\left(r^2\right)^{k_2}\right]$$
$$\cdot \left[\binom{n}{k_2}\left(1 - \rho^2\right)^{n-k_2-1}\left(\rho^2\right)^{k_2}\right]$$
$$\cdot \left[\binom{n + k_3 - 1}{k_3}\left(1 - r^2\rho^2\right)^{n-k_3-1}\left(r^2\rho^2\right)^{k_3}\right].$$

Proof Without loss of generality we may take $\phi = 0$. Then by [5] we have

$$\begin{aligned}
&\int_0^{2\pi} \left|B_n\left(re^{i\theta}\right)\right|^2 \frac{d\theta}{2\pi} \\
&= \frac{1}{2\pi i} \int_{|z|=1} \frac{B_n(r^2 z)}{z B_n(z)} dz \\
&= \mathrm{Res}_{z=0} \frac{B_n(r^2 z)}{z B_n(z)} + \mathrm{Res}_{z=\rho} \frac{B_n(r^2 z)}{z B_n(z)} \\
&= 1 + \frac{1}{(n-1)!} \frac{d^{(n-1)}}{dz^{(n-1)}} \left[\left(r^2 z - \rho\right)^n (1 - \rho z)^n \left(1 - \rho r^2 z\right)^{-n} z^{-1}\right]\Big|_{z=\rho}.
\end{aligned}$$

Thus,

$$\Delta_2(r, B_n) = \frac{-1}{(n-1)!}\left[\sum_{k_1+k_2+k_3+k_4=n-1} (n-1)! \frac{f_1^{(k_1)}}{k_1!} \frac{f_2^{(k_2)}}{k_2!} \frac{f_3^{(k_3)}}{k_3!} \frac{f_4^{(k_4)}}{k_4!}\right],$$

where $f_1(z) = (r^2 z - \rho)^n$, $f_2(z) = (1 - \rho z)^n$, $f_3(z) = (1 - \rho r^2 z)^{-n}$, $f_4(z) = z^{-1}$. Calculation of the derivatives of the f_k's show that

$$\begin{aligned}
&\Delta_2(r, B_n) \\
&= - \sum_{k_1+k_2+k_3+k_4=n-1} \left[\frac{n!(r^2\rho - \rho)^{n-k_1} r^{2k_1}}{(n-k_1)! k_1!} \cdot \frac{n!(1-\rho^2)^{n-k_2}(-1)^{k_2}\rho^{k_2}}{(n-k_2)! k_2!}\right.
\end{aligned}$$

$$\cdot \frac{(n+k_3-1)!(1-\rho^2 r^2)^{-n-k_3}\rho^{k_3} r^{2k_3}}{(n-1)!k_3!} \cdot \frac{(-1)^{k_4} k_4! \rho^{-k_4-1}}{k_4!}\Bigg]$$

$$= - \sum_{k_1+k_2+k_3+k_4=n-1} (-1)^{k_3+1}\left[\binom{n}{k_1}\binom{n}{k_2}\binom{n+k_3-1}{k_3}\right.$$

$$\left.\cdot \frac{(1-r^2)^{n-k_1}(1-\rho^2)^{n-k_2}}{(1-r^2\rho^2)^{n+k_3}}\rho^{2(k_2+k_3)} r^{2(k_1+k_3)}\right]$$

$$= \sum_{k_1+k_2+k_3=0}^{n-1} (-1)^{k_3}\left[\binom{n}{k_1}\binom{n}{k_2}\binom{n+k_3-1}{k_3}\right.$$

$$\left.\cdot \frac{(1-r^2)^{n-k_1}(1-\rho^2)^{n-k_2}}{(1-r^2\rho^2)^{n+k_3}}\rho^{2(k_2+k_3)} r^{2(k_1+k_3)}\right]$$

$$= \frac{(1-r^2)(1-\rho^2)}{(1-r^2\rho^2)^{2n-1}} \cdot S,$$

where

$$S = \sum_{\substack{k_1+k_2+k_3=0 \\ k_i \geq 0}}^{n-1} (-1)^{k_3}\left[\binom{n}{k_1}\left(1-r^2\right)^{n-k_1-1}\left(r^2\right)^{k_2}\right]$$

$$\cdot\left[\binom{n}{k_2}\left(1-\rho^2\right)^{n-k_2-1}\left(\rho^2\right)^{k_2}\right]$$

$$\cdot\left[\binom{n+k_3-1}{k_3}\left(1-r^2\rho^2\right)^{n-k_3-1}\left(r^2\rho^2\right)^{k_3}\right].$$

□

Remark 3 It follows from this identity that $\Delta_2(r, B_n)$ is symmetric in r and ρ.

3 Comparison of Hardy and Sampling Means

Carleson's interpolation Theorem [2, p. 149] has focused attention on the comparison of the growth of f in H^p with f in $A^p(\mu) = \{f$ analytic in $\mathbb{D}$: $(\int_{|z|<1} |f(z)|^p d\mu(z))^{1/p}\} = \|f\|_{p(\mu)} < \infty$, where μ is a finite measure on $\mathbb{D}$. In particular, necessary and sufficient conditions guaranteeing a continuous inclusion of either $H^p \subset A^p(d\mu)$ or $A^p(d\mu) \subset H^p$ have been investigated, see [6] for a recent survey of results. Particular attention was paid to the measures of the form $\mu(z_k) = 1 - |z_k|$ where $\{z_k\}$ is a uniformly separated sequence, i.e., there is a constant $\delta > 0$ for which $\inf_k \prod_{j=1, j\neq k}^{\infty} |\frac{z_k - z_j}{1-\bar{z}_j z_k}| \geq \delta$.

In this section we show how as $r \to 1$ the rate of decay of $\Delta_1(r, B_n)$ compares with $\sum_{k=1}^{n} |B_n(rz_k)|(1 - |z_k|)$ and how, similarly, the rate of growth of $\int_0^{2\pi} |B_n'(re^{i\theta})|\frac{d\theta}{2\pi}$ compares with $\sum_{k=1}^{n} |B_n'(rz_k)|(1-|z_k|)$. Thus for these Blaschke products we see how the Hardy means compare with the means sampled along the set $\{rz_k, 0 < r < 1\}$, where $\{z_k\}_{k=1}^{\infty}$ is the zero set for the product.

We need the following lemma.

Lemma 1 *Let $f \in H^p$, $0 < p < \infty$, and $\|f\|_p = 1$. Then for any $\alpha > 0$ there are constants $c_0(\alpha)$, $c_1(\alpha) > 0$ such that, for all r, $0 < r < 1$ we have $c_0(\alpha)(1 - M_p(r^\alpha, f)) \le 1 - M_p(r, f) \le c_1(\alpha)(1 - M_p(r^\alpha, f))$.*

Proof If $\alpha = 1$ then there is nothing to prove, so assume $0 < \alpha < 1$. By Hardy's convexity theorem [2, p. 9] $\log M_p(r, f)$ is a convex function of $\log r$; it follows that $\log M_p(e^t, f)$ is also convex on $-\infty < t < 0$, hence $M_p(e^t, f)$ is convex on $-\infty < t < 0$. Thus if $0 < r_1 < r_2 < 1$ we have $M_p(r_1^\alpha r_2^{1-\alpha}, f) \le \alpha M_p(r_1, f) + (1-\alpha)M_p(r_2, f)$, thus $1 - M_p(r_1^\alpha r_2^{1-\alpha}) \ge \alpha + (1-\alpha) - \alpha M_p(r_1, f) - (1-\alpha)M_p(r_2, f) = \alpha(1 - M_p(r_1, f)) + (1-\alpha)(1 - M_p(r_2, f))$. Letting $r_2 \to 1$ gives the inequality in one direction; the other direction follows with $c_0 = 1$ since $M_p(r, f)$ is non-decreasing. The case $\alpha > 1$ follows from the case $\alpha < 1$ in the obvious way. □

Theorem 4 *Let $B_n(z) = \prod_{k=1}^{n} \frac{z_k - z}{1 - \bar{z}_k z}$ ($z_k \neq 0$ for all k, z_k distinct). Then there are constants c_0, $c_1 > 0$ depending only on $\{z_k\}_{k=1}^{n}$ for which*

$$c_0 \sum_{k=1}^{n} \left|B_n'(rz_k)\right|\left(1 - |z_k|\right) \le \int_0^{2\pi} \left|B_n'\left(re^{i\theta}\right)\right| \frac{d\theta}{2\pi}$$

$$\le c_1 \frac{1}{1-r} \int_r^1 \sum_{k=1}^{n} \left|B'(tz_k)\right|\left(1 - |z_k|\right) dt \qquad (1)$$

for all r, $0 < r < 1$.

Proof Since for a fixed r, $0 < r < 1$, we have $B_n'(rz) \in H^1$, then by Carleson's Theorem there is a constant $c_0 > 0$ for which $c_0 \sum_{k=1}^{n} |B_n'(rz_k)|(1 - |z_k|) \le \int_0^{2\pi} |B_n'(re^{i\theta})|\frac{d\theta}{2\pi}$ for all r, $0 < r < 1$. For the other direction, recall that for $f \in H^\infty$, $\|f\|_\infty = 1$ we have $|f'(z)| \le \frac{1-|f(z)|^2}{1-|z|^2}$, thus, for a fixed r, $0 < r < 1$, we have $\int_0^{2\pi} |B_n'(re^{i\theta})|\frac{d\theta}{2\pi} \le \frac{\Delta_2(r, B_n)}{1-r^2} \le c\frac{\Delta_2(r^{1/2}, B_n)}{1-r^2}$ by Lemma 1. But by the residue calculation in [5] we have $\frac{\Delta_2(r^{1/2}, B_n)}{1-r^2} = \frac{1}{1-r^2} \sum_{k=1}^{n} \frac{B_n(rz_k)(1-|z_k|^2)}{z_k \prod_{j=1, j\neq k}^{n} \frac{z_j - z_k}{1-\bar{z}_j z_k}}$. By the Fundamental Theorem of Calculus $B_n(z_k) - B_n(rz_k) = \int_{rz_k}^{z_k} B_n'(z)dz$, hence $-B_n(rz_k) =$

$\int_r^1 B_n'(tz_k)z_k dt$, so $|B_n(rz_k)| \leq \int_r^1 |B_n'(tz_k)||z_k|dt$. Putting this together we have

$$\int_0^{2\pi} \left|B_n'\left(re^{i\theta}\right)\right|\frac{d\theta}{2\pi} \leq \frac{c\Delta_2(r^{1/2}, B_n)}{1-r^2} \leq \frac{c}{1-r^2}\sum_{k=1}^n \frac{|B_n(rz_k)|(1-|z_k|^2)}{|z_k|\prod_{j=1, j\neq k}^n |\frac{z_j - z_k}{1-\bar{z}_j z_k}|}$$

$$\leq \frac{c}{1-r^2}\sum_{k=1}^n \frac{(\int_r^1 |B_n'(tz_k)|dt)(1-|z_k|^2)}{|z_k|\prod_{j=1, j\neq k}^n |\frac{z_j - z_k}{1-\bar{z}_j z_k}|}$$

$$\leq \frac{2c}{\delta(1-r)}\int_r^1 \sum_{k=1}^n \left|B_n'(tz_k)\right|\left(1-|z_k|\right)dt,$$

where $\delta = \min_k[|z_k|\prod_{j=1, j\neq k}^n |\frac{z_j - z_k}{1-\bar{z}_j z_k}|]$. This is the required result with $c_1 = \frac{2c}{\delta}$. □

For B_n itself we have the following:

Theorem 5 *Let $B_n(z) = \prod_{k=1}^n \frac{z_k - z}{1-\bar{z}_j z_k}$, $z_k \neq 0$, for all k, z_k distinct, then there are constants $c_0, c_1 > 0$ depending only on $\{z_k\}_{k=1}^n$ such that $c_0\ \Delta_1(r, B_n) \leq \sum_{k=1}^n |B_n(rz_k)|(1-|z_k|) \leq c_1 \int_r^1 \frac{\Delta_1(t, B_n)dt}{1-t}$ for all r, $0 < r < 1$.*

Proof As in the previous theorem, $\Delta_1(r, B_n) < \Delta_2(r, B_n) \leq c\ \Delta_2(r^{1/2}, B_n) = c|\sum_{k=1}^n \frac{B_n(rz_k)(1-|z_k|^2)}{z_k\prod_{j=1, j\neq k}^n \frac{z_k - z_j}{1-\bar{z}_j z_k}}| \leq \frac{2c}{\delta}\sum_{k=1}^n |B_n(rz_k)|(1-|z_k|)$, giving the first inequality with $c_0 = \delta/2c$.

For the other direction $B_n(z_k) - B_n(rz_k) = \int_r^1 B_n'(tz_k)z_k\ dt$, thus $|B_n(rz_k)| \leq \int_r^1 |B_n'(tz_k)|\ dt$, hence the Carleson theorem again gives $c_0^* \sum_{k=1}^n |B_n'(rz_k)|(1 - |z_k|) \leq \int_0^{2\pi} |B_n'(re^{i\theta})\frac{d\theta}{2\pi} \leq \frac{\Delta_2(r, B_n)}{1-r^2}$, thus

$$\sum_{k=1}^n \left|B_n(rz_k)\right|\left(1-|z_k|\right) \leq \sum_{k=1}^n \left(\int_r^1 \left|B_n'(tz_k)\right|\,dt\right)\left(1-|z_k|\right)$$

$$= \int_r^1 \sum_{k=1}^n \left|B_n'(tz_k)\right|\left(1-|z_k|\right)dt$$

$$\leq \frac{1}{c_0^*}\int_r^1 \frac{\Delta_2(t, B_n)}{1-t^2}\,dt \leq \frac{2}{c_0^*}\int_r^1 \frac{\Delta_1(t, B_n)}{1-t}\,dt.$$

This is the second inequality with $c_1 = 2/c_0^*$. □

Theorem 5 can be generalized as follows:

Theorem 6 *Let $B_n(z)$ be a Blaschke product, $\alpha \in \mathbb{D}$ be such that $S_\alpha = \{w_k : w_k \in B_n^{-1}(\alpha)\}$ has distinct values. Then there are constants c_0, $c_1 > 0$ depending only on*

S_α and α for which

$$c_0\Delta_1(r, B_n - \alpha) \le \sum_{k=1}^{n} \big|B_n(rw_k) - \alpha\big|\big(1 - |w_k|\big)$$

$$\le c_1 \int_r^1 \frac{\Delta_1(t, B_n - \alpha)}{1-t}\, dt, \quad \textit{for all } r,\ 0 < r < 1.$$

Proof Define $B_\alpha(z) = \frac{\alpha - B_n(z)}{1-\overline{\alpha}B_n(z)}$, then B_α is a finite Blaschke product with distinct zeroes at $\{w_k\}_{k=1}^n$. By a well-known identity [2, p. 150] we have $1 - |B_\alpha(z)|^2 = [\frac{1-|\alpha|^2}{1-\overline{\alpha}B_n(z)^2}](1 - |B_n(z)|^2)$. By applying Theorem 5 to $B_\alpha(z)$ and using the identity together with standard bounds the result follows. □

Remark 4 In Theorem 4 the term on the extreme right is an average of $\sum_{k=1}^n |B_n'(tz_k)|(1-|z_k|)$ over the interval $[r-1, 1]$, hence there is a "rough proportionality" of the quantity $M_1(r, B_n')$ and $\sum_{k=1}^n |B_n'(tz_k)|(1-|z_k|)$. The quantity on the extreme right in the statement of Theorem 5 exceeds the average of $\Delta_1(t, B_n)$ over the same interval, so the bound is in a sense weaker than that of Theorem 4.

4 Location of Zeroes

In this section we study the effect of the location of the zeroes, $z_k = \rho e^{i\phi_k}$, z_k in the unit disk, on the mean modulus of the Blaschke product. We are able to obtain results only in the cases of two or three zeroes, and even these cases involve considerable complexities. Let for z, z_k in the unit disk,

$$B_n\big(z, \{z_k\}\big) := \prod_{k=1}^{n} \frac{z_k - z}{1 - \overline{z}_k z}, \qquad \Delta_2(r, B_n) := 1 - \frac{1}{2\pi}\int_0^{2\pi} \big|B_n\big(re^{i\theta}, \{z_k\}\big)\big|^2\, d\theta. \tag{2}$$

In [5] the following formula for Δ_2 was derived:

$$\Delta_2(r, B_n) = \sum_{k=1}^{n} \frac{(1-r^2)(1-|z_k|^2)}{1-r^2|z_k|^2} \frac{\prod_{j\ne k} \frac{z_j - r^2 z_k}{1-\overline{z}_j r^2 z_k}}{\prod_{j \ne k} \frac{z_j - z_k}{1 - \overline{z}_j z_k}}. \tag{3}$$

We use the following notations for a Blaschke product in which there is one repeated zero, ρ, of multiplicity n:

$$B_n^*(z) = B_n\big(z, \{\rho, \rho, \ldots, \rho\}\big) = \left(\frac{\rho - z}{1 - \rho z}\right)^n \tag{4}$$

for the Blaschke product in which there is a repeated zero $\rho > 0$ of multiplicity n, and

$$B_n^{**}(z) = B_n\big(z, \big\{\rho e^{i(2\pi k)/n} : k = 1, \ldots, n\big\}\big), \tag{5}$$

for the Blaschke product when there are n zeroes evenly distributed around the circle of radius ρ.

By Theorems 2 and 3 we have

$$\Delta_2(r, B_n^*) = \sum_{k_1+k_2+k_3=0,\ k_i\geq 0}^{n-1} (-1)^{k_3} \binom{n}{k_1}\binom{n}{k_2}\binom{n+k_3-1}{k_3} \cdot \left[\frac{(1-r^2)^{n-k_1}(1-\rho^2)^{n-k_2}}{(1-r^2\rho^2)^{n+k_3}}\right] \cdot \rho^{2(k_2+k_3)} r^{2(k_1+k_3)} \quad \text{and} \tag{6}$$

$$\Delta_2(r, B_n^{**}) = \frac{(1-r^{2n})(1-\rho^{2n})}{1-r^{2n}\rho^{2n}}.$$

For the case $n = 2$ we have: (Note: In the following the computer algebra system Maple 14 was used to simplify complicated algebraic expressions).

Theorem 7 *For B_2^{**} and B_2^* as above, $\Delta_2(r, B_2^{**}) > \Delta_2(r, B_2^*)$ for $0 < r < 1$.*

Proof From (7) and (6) one has:

$$\Delta_2(r, B_2{}^{**}) = \frac{(1-r^4)(1-\rho^4)}{1-r^4\rho^4} \tag{7}$$

and the case of a repeated zero at ρ, i.e., $z_1 = z_2 = \rho$

$$\Delta_2(r, B_2{}^{*}) = \frac{(1-\rho^2)(1-r^2)(1+\rho^2+r^2+r^4\rho^2-6r^2\rho^2+r^2\rho^4+r^4\rho^4)}{(1-r^2\rho^2)^3}. \tag{8}$$

Thus

$$\Delta_2(r, B_2{}^{**}) - \Delta_2(r, B_2{}^{*}) = \frac{4r^2\rho^2(1-\rho^2)^2(1-r^2)^2}{(1-r^2\rho^2)^3(1+r^2\rho^2)} > 0. \tag{9}$$

□

Remark 5 In this result and in Theorem 9 we have an *explicit expression* for $\Delta_2(r, B_n{}^{**}) - \Delta_2(r, B_n{}^{*})$. A stronger result is obtained for a Blaschke product when the two zeroes are located anywhere on the circle of radius ρ (without loss of generality they may be conjugate roots).

Theorem 8 *For $0 < r < 1$, $\Delta_2(r, B_2(z, \{\rho e^{i\phi}, \rho e^{-i\phi}\}))$ is an increasing function of ϕ with*

$$\Delta_2(r, B_2{}^{*}) \leq \Delta_2\big(r, B_2\big(z, \{\rho e^{i\phi}, \rho e^{-i\phi}\}\big)\big) \leq \Delta_2(r, B_2{}^{**}), \quad 0 \leq \phi \leq \frac{\pi}{2}.$$

Proof From (1) with $z_1 = \rho e^{i\phi}, z_2 = \rho e^{-i\phi}$ one has

$$\begin{aligned} &\Delta_2\big(r, B_2\big(z, \{\rho e^{i\phi}, \rho e^{-i\phi}\}\big)\big) \\ &\quad = \frac{(1 - 8r^2\rho^2\cos^2(\phi) + r^2 + r^4\rho^2 + 2r^2\rho^2 + \rho^2 + r^2\rho^4 + r^4\rho^4)}{(1 - 4r^2\rho^2\cos^2(\phi) + 2r^2\rho^2 + r^4\rho^4)(1 - r^2\rho^2)} \\ &\quad\quad \times \big(1 - r^2 - \rho^2 + r^2\rho^2\big). \end{aligned} \tag{10}$$

Note that

$$\begin{aligned} \Delta_2\big(r, B_2\big(z, \{\rho e^{i0}, \rho e^{-i0}\}\big)\big) &= \Delta_2\big(r, B_2{}^*\big), \\ \Delta_2\big(r, B_2\big(z, \{\rho e^{i\frac{\pi}{2}}, \rho e^{-i\frac{\pi}{2}}\}\big)\big) &= \Delta_2\big(r, B_2^{**}\big). \end{aligned}$$

The derivative of (10) with respect to ϕ yields:

$$\begin{aligned} &\frac{d}{d\phi}\Delta_2\big(r, B_2\big(z, \{\rho e^{i\phi}, \rho e^{-i\phi}\}\big)\big) \\ &\quad = \frac{8r^2\rho^2\sin(\phi)\cos(\phi)(1-\rho^2)^2(1-r^2)^2(1+r^2\rho^2)}{(1-r^2\rho^2)(4r^2\rho^2\cos^2(\phi) - (1+r^2\rho^2)^2)} \end{aligned} \tag{11}$$

and hence $\Delta_2(r, B_2(z, \{\rho e^{i\phi}, \rho e^{-i\phi}\}))$ has endpoint minima at $\phi = 0, \pi$ and a maximum at $\phi = \pi/2$. This gives the desired result. □

The analogous result for a Blaschke product with three zeroes are contained in Theorems 9 and 10. The following are the expressions for a zero of multiplicity three:

$$\Delta_2\big(r, B_3^*\big) = \frac{(1-r^2)(1-\rho^2)\mu(r,\rho)}{(1-r^2\rho^2)^5}, \tag{12}$$

where

$$\begin{aligned} \mu(r,\rho) := {} & \rho^8 r^4 + r^8\rho^8 + \rho^8 r^6 - 4r^6\rho^6 - 13\rho^6 r^4 \\ & + 4\rho^6 r^2 + \rho^6 r^8 - 13\rho^4 r^6 + 42\rho^4 r^4 \\ & - 13\rho^4 r^2 + \rho^4 + r^8\rho^4 + 4r^6\rho^2 - 13\rho^2 r^4 \\ & - 4\rho^2 r^2 + \rho^2 + r^4 + r^2 + 1, \end{aligned}$$

and the expression for three equally separated zeroes

$$\Delta_2\big(r, B_3^{**}\big) = \frac{(1-r^6)(1-\rho^6)}{(1-r^6\rho^6)}. \tag{13}$$

Theorem 9 *For* $0 < r < 1$, $\Delta_2(r, B_3^{**}) > \Delta_2(r, B_3^*)$.

Proof A straightforward, but tedious calculation, gives

$$\begin{aligned}
&\Delta_2\big(r, B_3^{**}\big) - \Delta_2\big(r, B_3^{*}\big)\\
&\quad = 9\frac{r^2\rho^2(1-r^2)^2(1-\rho^2)^2(r^6\rho^4+r^4\rho^6-2r^4\rho^4-2r^2\rho^2+r^2+\rho^2)}{((1+r^2\rho^2)^2-r^2\rho^2)(1-r^2\rho^2)^5} > 0.
\end{aligned}$$

Note that $r^6\rho^4+r^4\rho^6-2r^4\rho^4-2r^2\rho^2+r^2+\rho^2$ can be shown to be positive by elementary calculus. We omit the details. □

Let $z_1=\rho$, $z_2=\rho e^{i\phi}$, $z_3=\overline{z_2}=\rho e^{-i\phi}$. Then one can show, using (13),

$$\Delta_2\big(r, B_3\big(z, \{z_1,\ z_2,\ z_3\}\big)\big) = \frac{(1-r^2)(1-\rho^2)p(r,\rho,\phi)}{(1-r^2\rho^2)q(r,\rho,\phi)}, \tag{14}$$

where $p(r,\rho,\phi)$ and $q(r,\rho,\phi)$ are polynomials of degree three in $\cos(\phi)$, namely

$$\begin{aligned}
&p(r,\rho,\phi)\\
&\quad = 1+r^2+\rho^2+2r^2\rho^2+r^4+\rho^6r^4+\rho^8r^4+\rho^8r^6+\rho^8r^8+2\rho^6r^2+r^8\rho^6\\
&\qquad +r^8\rho^4+\rho^4r^2+2r^4\rho^4+r^4\rho^2+r^6\rho^4+2r^6\rho^2+\rho^4+2r^6\rho^6\\
&\qquad +\big(-6\rho^4r^2-6r^4\rho^2-6\rho^6r^4+2\rho^6r^2-2r^2\rho^2+2r^6\rho^2-6r^6\rho^4-2r^6\rho^6\big)\\
&\qquad \times\cos(\phi)\\
&\qquad +\big(-8r^6\rho^4+16r^4\rho^4-8r^4\rho^2-4r^6\rho^6-8\rho^4r^2-4r^2\rho^2-8\rho^6r^4\big)\big(\cos(\phi)\big)^2\\
&\qquad +24r^4\rho^4\big(\cos(\phi)\big)^3,
\end{aligned}$$

and

$$\begin{aligned}
q(r,\rho,\phi) &= \rho^8r^8+2r^6\rho^6+2r^4\rho^4+2r^2\rho^2+1\big(-2r^6\rho^6-2r^2\rho^2-4r^4\rho^4\big)\cos(\phi)\\
&\quad +\big(-4r^2\rho^2-4r^6\rho^6\big)\big(\cos(\phi)\big)^2+8r^4\rho^4\big(\cos(\phi)\big)^3.
\end{aligned}$$

Again we note in the following theorem Maple 14 was used to perform a detailed analysis of the complicated expressions involved.

Theorem 10 *For $0<r<1$ and B_3^*, B_3^{**}, $B_3(z,\{z_1,z_2,z_3\})$ as above we have $\Delta_2(r, B_3(z,\{z_1,\ z_2,\ z_3\}))$ is an increasing function of ϕ with $\Delta_2(r, B_3^*) \le \Delta_2(r, B_3(z,\{z_1,\ z_2,\ z_3\})) \le \Delta_2(r, B_3^{**})$.*

Proof For $0<r<1$, $0<\rho<1$, $0\le\phi\le 2\pi/3$, $\Delta_2(r, B_3(z,\{z_1,\ z_2,\ z_3\}))$ has a global maximum at $\phi=2\pi/3$ and endpoint local minimum at $\phi=0$. We find the numerator of $\frac{d}{d\phi}\Delta_2(r, B_3(z,\{z_1,\ z_2,\ z_3\}))$ with respect to ϕ to be

$$-4\rho^2r^2\big(1-r^2\big)^2\big(1-\rho^2\big)^2\big(1+2\cos(\phi)\big)\sin(\phi)\,\mu(r,\rho,\phi), \tag{15}$$

where

$$\begin{aligned}
&\mu(r,\rho,\phi)\\
&\quad= \left(8\,r^4\rho^4 + 8\,\rho^6 r^6\right)\left(\cos(\phi)\right)^3 + \left(4\,\rho^6 r^6 - 4\,r^6\rho^4 + 4\,r^4\rho^4 - 4\,\rho^6 r^4\right)\left(\cos(\phi)\right)^2\\
&\qquad+ \left(-4\,\rho^6 r^4 - 4\,r^6\rho^4 - 10\,\rho^6 r^6 - 2\,\rho^8 r^6 - 6\,r^2\rho^2 - 6\,r^8\rho^8 - 2\,\rho^4 r^2\right.\\
&\qquad\left.- 2\,r^8\rho^6 - 2\,\rho^2 r^4 - 10\,r^4\rho^4\right)\cos(\phi) + r^{10}\rho^8 + \rho^{10} r^8 + r^8\rho^8 + 3\,r^8\rho^6\\
&\qquad+ 3\,\rho^8 r^6 + 3\,\rho^6 r^6 + 4\,r^6\rho^4 + 4\,\rho^6 r^4 + 3\,r^4\rho^4 + 3\,\rho^2 r^4 + 3\,\rho^4 r^2\\
&\qquad+ r^2\rho^2 + r^2 + \rho^2.
\end{aligned}$$

Then we have that $\mu(r,\rho,\phi)$ is a polynomial of degree three in $\cos(\phi)$, which vanishes only when $r = 0,\ \rho = 0$. The result follows. □

Conjecture 1 *Given the results of this section, it is reasonable to offer the following conjecture: For all* $n \geq 2$, $\Delta_2(r, B_n(z, \{z_1,\ z_2, \ldots, z_n\}))$ *is an increasing function of* ϕ *with* $\Delta_2(r, B_n^*) \leq \Delta_2(r, B_n(z, \{z_1,\ z_2, \ldots, z_n\})) \leq \Delta_2(r, B_n^{**})$ *where the* $z_k's$ *all lie on the circle of radius* ρ, $0 < \rho < 1$.

References

1. Bhatnagar, G.: A short proof of an identity of Sylvester. Int. J. Math. Math. Sci. **22**(2), 431–435 (1999)
2. Duren, P.: Theory of H^p Spaces. Academic Press, New York (1970)
3. Gluchoff, A.: The mean modulus of Blaschke products with zeroes in a non-tangential region. Complex Var. **I**, 311–326 (1983)
4. Gluchoff, A.: On inner functions with derivative in Bergman spaces. Ill. J. Math. **31**(1), 518–528 (1987)
5. Gluchoff, A., Gorkin, P., Mortini, R.: On some representation formulas involving the moduli of Blaschke products. Analysis **27**, 261–272 (2007)
6. Pau, J., Peláez, J.A.: Embedding theorems and integration operators on Bergman spaces with rapidly decreasing weights. J. Funct. Anal. **259**, 2727–2756 (2010)

Hyperbolic Derivatives Determine a Function Uniquely

Line Baribeau

Abstract The notion of hyperbolic derivative for functions from the unit disc to itself is well known. Recently, Rivard has proposed a definition for higher-order derivatives. We prove that the sequence of hyperbolic derivatives of order n ($n = 0, 1, 2, \ldots$) of a function f determines this function uniquely.

Keywords Hyperbolic derivative · Divided differences

Mathematics Subject Classification Primary 30E05 · 30F45 · Secondary 30C80 · 30B70

1 Preliminaries

Let $\mathbb{D}$ denote the unit disc, and denote by S the Schur class, i.e. the set of holomorphic functions from $\mathbb{D}$ to $\overline{\mathbb{D}}$. For $f \in S$, its *hyperbolic derivative* is defined as

$$f^h(z) := \frac{(1 - |z|^2) f'(z)}{1 - |f(z)|^2}.$$

This notion is well known. In a recent paper [7], P. Rivard has proposed a definition for higher-order hyperbolic derivatives. In this note, we prove that the sequence of the hyperbolic derivatives of all orders of a function f at a point z uniquely determines that function. Hence the hyperbolic derivatives of a function $f \in S$ can be seen as a sort of hyperbolic version of Taylor coefficients.

To understand the definition of higher-order hyperbolic derivatives, we need to recall the notion of hyperbolic divided differences. Let $f \in S$, and fix $z_1 \in \mathbb{D}$. The *hyperbolic divided difference with parameter* z_1 *of* f, which we denote $\Delta_{z_1} f$, is

L. Baribeau (✉)
Département de mathématiques et de statistique, Université Laval, 1045 av. de la Médecine, Québec, G1V 0A6, Canada
e-mail: Line.Baribeau@mat.ulaval.ca

J. Mashreghi, E. Fricain (eds.), *Blaschke Products and Their Applications*, Fields Institute Communications 65, DOI 10.1007/978-1-4614-5341-3_10,

defined as the function

$$\Delta_{z_1} f(z) := \begin{cases} \frac{[f(z), f(z_1)]}{[z, z_1]} & \text{if } f : \mathbb{D} \to \mathbb{D} \\ \eta & \text{if } f \equiv \eta \text{ for some } \eta \in \partial\mathbb{D} \end{cases} \tag{1}$$

where we have used the notation

$$[z, w] := \frac{w - z}{1 - \overline{w} z}.$$

It follows easily from the invariant form of the Schwarz lemma that the function $\Delta_{z_1} f$ also belongs to the class S. Therefore, we can iterate the process and construct, for distinct parameters $z_1, z_2, \ldots, z_n$, the *n-th order hyperbolic divided difference*:

$$\Delta^n_{z_1, z_2, \ldots, z_n} f(z) := \Delta_{z_n} \big(\Delta^{(n-1)}_{z_1, z_2, \ldots, z_{n-1}}(f)\big)(z).$$

These ideas first appeared in [2], where the authors made definition (1) and studied Δf using the hyperbolic metric. Higher-order hyperbolic divided differences were introduced in [1]. The operator Δ_{z_1} is a map of S onto itself. It can be used to characterize finite Blaschke products: f is a Blaschke product of degree $\leq n$ if and only if $\Delta^n f$ is a unimodular constant. Hyperbolic divided differences have nice applications to the Schwarz-Pick interpolation problem. Because $\Delta^n f$ takes its values in the unit disc (when it is not a constant), the distance between its values at different points can be measured using the hyperbolic metric. In conjunction with the Schwarz-Pick lemma, this can be used to formulate necessary and sufficient conditions on the data for the Schwarz-Pick problem to have a solution. Whereas the standard criterion is in terms of a matrix being positive semidefinite [4], here one only has to construct a table of hyperbolic divided differences and check that all entries are <1. It is then possible to construct the solutions using that table by a simple algorithm. It turns out that this construction corresponds to Schur's algorithm. Hence this point of view, using the language of hyperbolic divided differences, shows that Schur's algorithm is a hyperbolic version of classical Newton interpolation for polynomials. Further applications can be found in [3].

Hyperbolic derivatives can now be defined from hyperbolic divided differences, using a limiting process. More precisely, we have:

Definition 1 For $f \in S$ and $z, \zeta \in \mathbb{D}$, set

$$\Delta^n_z f(\zeta) := \lim_{z_n \to z} \lim_{z_{n-1} \to z} \cdots \lim_{z_1 \to z} \Delta^n_{z_1, \ldots, z_n} f(\zeta).$$

It is shown in [7] that this definition makes sense, and that the operator Δ^n_z so defined maps S into itself. If $\Delta^n_z f = \eta$ for a unimodular constant η, it follows from (1) that all subsequent hyperbolic derivatives are equal to η. This happens when f is a finite Blaschke product of degree less or equal to n.

Definition 2 For $f \in S$ and $z \in \mathbb{D}$, the *n-th order hyperbolic derivative of f at z* is the number

$$H^n f(z) := \Delta_z^n f(z).$$

For instance, for the first-order hyperbolic derivative, we find

$$H^1 f(z) = \lim_{\zeta \to z} \frac{\frac{f(z)-f(\zeta)}{1-\overline{f(z)}f(\zeta)}}{\frac{z-\zeta}{1-\bar{z}\zeta}} = f^h(z)$$

and we recover the standard hyperbolic derivative. Note that the hyperbolic derivative $H^1 f(z)$ is not a holomorphic function, and that $H^n f(z)$ *is not* the hyperbolic derivative of $H^{n-1} f(z)$.

Hyperbolic derivatives have the following nice invariance property:

$$\left|H^n(\psi \circ f \circ \phi)(z)\right| = \left|H^n f\big(\phi(z)\big)\right|$$

whenever ϕ and ψ are automorphisms of $\mathbb{D}$ [7]. In this respect, they are similar to the classical invariants of Peschl [6], which are the numbers $D^n f(z)$ defined as the Taylor coefficients of the function g defined by the formula

$$g(\zeta) := -\big[f\big([-\zeta, z]\big), f(z)\big] = \frac{f(\frac{z+\zeta}{1+\bar{z}\zeta}) - f(z)}{1 - \overline{f(z)}f(\frac{z+\zeta}{1+\bar{z}\zeta})} =: \sum_{n=1}^{\infty} \frac{D^n f(z)}{n!}\zeta^n. \qquad (2)$$

We saw above that $H^1 f$ coincides with the usual hyperbolic derivative, which also coincides with $D^1 f(z)$. However, for higher order, hyperbolic derivatives are different from Peschl invariants. The next values are

$$H^2 f(z) = \frac{D^2 f(z)}{2(1 - |H^1 f(z)|^2)},$$

for f not a Blaschke product of degree ≤ 1, and

$$H^3 f(z) = \frac{3\overline{H^1 f(z)}H^2 f(z) D^2 f(z) + D^3 f(z)}{6 - 6|H^1 f(z)|^2 - 3\overline{H^2 f(z)}D^2 f(z)},$$

for f not a Blaschke product of degree ≤ 2. (We note here that there is an unfortunate misprint in the formula for H^3 in [8]. The calculations which lead to the above formula are carried out in [9].)

The above formulas show that it is not easy to calculate hyperbolic derivatives. There exists a nice formula for calculating the Peschl invariants using Bell polynomials [5], but we do not know of a similar formula for hyperbolic derivatives. However, Rivard has obtained the following inductive formula [8], which we will use to prove our theorem.

Theorem 1 *Let $f \in S$, not a finite Blaschke product, and let $z \in \mathbb{D}$. We define*

$$\Phi_1(\zeta, z) := -g(-\zeta), \qquad \Psi_1(\zeta, z) := \zeta$$

and, for $k \geq 2$, we define recursively

$$\Phi_k(\zeta, z) := H^{k-1} f(z) \Psi_{k-1}(\zeta, z) - \Phi_{k-1}(\zeta, z); \tag{3}$$

$$\Psi_k(\zeta, z) := \zeta \left(\Psi_{k-1}(\zeta, z) - \overline{H^{k-1} f(z)} \Phi_{k-1}(\zeta, z)\right). \tag{4}$$

Then, for every $n \geq 2$,

$$H^n f(z) = \frac{H^{n-1} f(z) \Psi_{n-1}^{(n)}(0, z) - \Phi_{n-1}^{(n)}(0, z)}{n\left(\Psi_{n-1}^{(n-1)}(0, z) - \overline{H^{n-1} f(z)} \Phi_{n-1}^{(n-1)}(0, z)\right)}, \tag{5}$$

where $\Psi_{n-1}^{(\ell)}, \Phi_{n-1}^{(\ell)}$ are derivatives of order ℓ with respect to ζ, and $H^0 f(z) := f(z)$.

If f is a Blaschke product of degree n, then the above scheme can be carried out to calculate the hyperbolic derivatives up to order n, from which point they are all equal to the same unimodular constant.

2 The Main Theorem

We are now ready to state and prove our theorem.

Theorem 2 *For any fixed $z \in \mathbb{D}$, the numbers $z, H^0 f(z), H^1 f(z), H^2 f(z), \ldots$ uniquely determine the function f.*

The proof will be based on two lemmas. In the statement and the proof, we write $p(\cdot)$ to denote a polynomial expression in its variables and their conjugates, and for brevity we write H^k and D^k for the hyperbolic derivatives and Peschl invariants of f at z.

Lemma 1 *Suppose f is not a finite Blaschke product. Then for $n = 1, 2, \ldots,$ we have*

(i) *for $k = 1, 2, \ldots,$ the coefficient of ζ^k in $\Phi_n(\zeta)$ is of the form*

$$(-1)^{k+n} \frac{D^k}{k!} + p\left(z, H^0, H^1, \ldots, H^{n-1}, D^1, \ldots, D^{k-1}\right).$$

(ii) *for $k = 1, 2, \ldots,$ the coefficient of ζ^k in Ψ_n is*

$$p\left(z, H^0, H^1, \ldots, H^{n-1}, D^1, \ldots, D^{k-1}\right).$$

Proof This will be proved by induction on n.

Recall that, by definition, $\Phi_1(\zeta) = -g(-\zeta) = -\sum_{k=1}^{\infty}(-1)^k \frac{D^k}{k!}\zeta^k$. From this it follows that (i) holds for $n = 1$. Similarly, since $\Psi_1(\zeta) = \zeta$, (ii) holds for $n = 1$.

Now assume (i) and (ii) hold for n. Denote by $A_{j,k}$ and $B_{j,k}$ the coefficient of ζ^k in Φ_j and Ψ_j respectively. Then, from (3), (4), and the induction hypothesis, we have

$$\begin{aligned} A_{n+1,k} &= H^n B_{n,k} - A_{n,k} \\ &= H^n p\big(z, H^0, \dots, H^{n-1}, D^1, \dots, D^{k-1}\big) \\ &\quad - (-1)^{k+n}\frac{D^k}{k!} + p\big(z, H^0, \dots, H^{n-1}, D^1, \dots, D^{k-1}\big) \\ &= (-1)^{k+n+1}\frac{D^k}{k!} + p\big(z, H^0, \dots, H^{n-1}, H^n, D^1, \dots, D^{k-1}\big) \end{aligned}$$

and

$$\begin{aligned} B_{n+1,k} &= B_{n,k-1} - \overline{H^n} A_{n,k-1} \\ &= p\big(z, H^0, \dots, H^{n-1}, D^1, \dots, D^{k-2}\big) \\ &\quad - \overline{H^n}\Big[(-1)^{k+n-1}\frac{D^{k-1}}{(k-1)!} + p\big(z, H^0, \dots, H^{n-1}, D^1, \dots, D^{k-2}\big)\Big] \\ &= p\big(z, H^0, \dots, H^{n-1}, H^n, D^1, \dots, D^{k-1}\big), \end{aligned}$$

and (i) and (ii) follow by induction. □

Lemma 2 *Suppose f is not a finite Blaschke product.*

(i) *For $n = 1, 2, 3, \dots$, H^n can be written in the form*

$$\frac{D^n + p(z, H^0, H^1, \dots, H^{n-1}, D^1, \dots, D^{n-1})}{p(z, H^0, H^1, \dots, H^{n-1}, D^1, \dots, D^{n-1})}.$$

(ii) *For $n = 1, 2, 3, \dots$, the number D^n can be written as a function of the numbers $z, H^0, \dots, H^n$.*

Proof Since $H^1 = D^1$, assertion (i) is true for $n = 1$. For $n \geq 2$, it follows from formula (5) and Lemma 1.

Again, for $n = 1$, (ii) follows from $D^1 = H^1$. Assume it is true for $n \leq N$. Now, by (i), we have

$$D^{N+1} = H^{N+1} p\big(z, H^0, \dots, H^N, D^1, \dots, D^N\big) + p\big(z, H^0, \dots, H^N, D^1, \dots, D^N\big),$$

and, by the induction hypothesis, this can be rewritten as a function of the numbers $z, H^0, \dots, H^{N+1}$. By induction, (ii) holds for all n. □

Proof of Theorem 2 If the sequence of hyperbolic derivatives eventually stabilizes at some unimodular value, then f is a finite Blaschke product, and it is shown in [8] how we can recover f from its hyperbolic derivatives. So it is only necessary to consider the case where f is not a finite Blaschke product. For a fixed z, the numbers $z, f(z), D^1, D^2, \ldots$ entirely determine the function $g(\zeta)$. Hence, by Lemma 2(ii), the numbers $z, H^0, H^1, \ldots$ entirely determine g. Since, by formula (2), we have $g = \psi \circ f \circ \phi$ for some automorphisms ϕ and ψ depending only on z, f is also uniquely determined by $z, H^0, H^1, \ldots$. □

The interested reader is referred to [3, 7, 8] for applications of hyperbolic derivatives.

Acknowledgements This research was supported by NSERC (Canada) through the Discovery Grants program.

References

1. Baribeau, L., Rivard, P., Wegert, E.: On hyperbolic divided differences and the Nevanlinna-Pick problem. Comput. Methods Funct. Theory **9**, 391–405 (2009)
2. Beardon, A.F., Minda, D.: A multi-point Schwarz-Pick lemma. J. Anal. Math. **92**, 81–104 (2004)
3. Cho, K.Y., Kim, S.-A., Sugawa, T.: On a multi-point Schwarz-Pick lemma. arXiv:1102.0337v1 [math.CV]
4. Garnett, J.B.: Bounded Analytic Functions. Graduate Texts in Mathematics, vol. 236. Springer, New York (2007)
5. Kim, S.-A., Sugawa, T.: Invariant differential operators associated with a conformal metric. Mich. Math. J. **55**(2), 459–479 (2007)
6. Peschl, E.: Les invariants différentiels non holomorphes et leur rôle dans la théorie des fonctions. Rend. Semin. Mat. Messina **1**, 100–108 (1955)
7. Rivard, P.: A Schwarz-Pick theorem for higher-order hyperbolic derivatives. Proc. Am. Math. Soc. **139**, 209–217 (2011)
8. Rivard, P.: Some applications of higher-order hyperbolic derivatives. Complex Anal. Oper. Theory. doi:10.1007/s11785-011-0172-z
9. Rivard, P.: Sur la théorie des dérivées hyperboliques. Doctoral thesis, Université Laval (2011)

Hyperbolic Wavelets and Multiresolution in the Hardy Space of the Upper Half Plane

Hans G. Feichtinger and Margit Pap

Abstract A multiresolution analysis in the Hardy space of the unit disc was introduced recently (see Pap in J. Fourier Anal. Appl. 17(5):755–776, 2011). In this paper we will introduce an analogous construction in the Hardy space of the upper half plane. The levels of the multiresolution are generated by localized Cauchy kernels on a special hyperbolic lattice in the upper half plane. This multiresolution has the following new aspects: the lattice which generates the multiresolution is connected to the Blaschke group, the Cayley transform and the hyperbolic metric. The second: the nth level of the multiresolution has finite dimension (in classical affine multiresolution this is not the case) and still we have the density property, i.e. the closure in norm of the reunion of the multiresolution levels is equal to the Hardy space of the upper half plane. The projection operator to the nth resolution level is a rational interpolation operator on a finite subset of the lattice points. If we can measure the values of the function on the points of the lattice the discrete wavelet coefficients can be computed exactly. This makes our multiresolution approximation very useful from the point of view of the computational aspects.

Keywords Hyperbolic wavelets · Multiresolution on the upper half plane · Interpolation operator · Orthogonal rational wavelets

Mathematics Subject Classification Primary 30H10 · 33C47 · 41A20 · 42C40 · 43A32 · 43A65

H.G. Feichtinger · M. Pap (✉)
NuHAG, Faculty of Mathematics, University of Vienna, Alserbachstrasse 23, 1090 Wien, Austria
e-mail: margit.pap@univie.ac.at

H.G. Feichtinger
e-mail: hans.feichtinger@univie.ac.at

M. Pap
University of Pécs, Ifjúság útja 6, 7634 Pécs, Hungary
e-mail: papm@gamma.ttk.pte.hu

J. Mashreghi, E. Fricain (eds.), *Blaschke Products and Their Applications*,
Fields Institute Communications 65, DOI 10.1007/978-1-4614-5341-3_11,

1 Introduction

Analyzing continuous-time systems is of great significance, especially in association with high precision control applications, which are widely used today, for example in robotic surgery. The theory of wavelet constructions on the Hardy space of the unit disc (see [17]) can be associated with time frequency-domain description of discrete-time-invariant dynamical systems. In this paper we adapt this description to those in the half plane that is used in system theory to describe the spectral behavior of continuous-time-invariant systems.

The approximation and identification of transfer functions of a continuous-time-invariant system is an important part of system identification. Finding dense subsets in the Hardy space of the upper half plane is often very useful. However, unlike in the case of the Hardy space of the unit disc where the polynomials are dense, dense subsets in the Hardy space of the upper half plane are harder to find. Applying the Daubechies theory it can be shown that choosing as mother wavelet $\psi(y) = (1 + iy)^{-p}$ for $p \geq 2$ we can generate a frame for the Hardy space of the upper half plane. For $p = 3$ Ward and Partington in [29] described a rational wavelet decomposition of the Hardy–Sobolev class of the half plane. For $p = 1$ (the Cauchy kernel) does not fall under Daubechies theory since it does not have vanishing mean value, but Ward and Partington have shown that the system $\psi_{j,k} = 2^{j/2}\psi(2^j y - b_0 k)$, $j, k \in \mathbb{Z}$, does constitute a fundamental set for the upper half plan algebra.

1.1 Affine Multiresolution Analysis

Daubechies theory can be described in terms of the continuous affine wavelet transform, which is a voice transform generated by a representation of the affine group:

$$W_\psi f(a,b) = |a|^{-1/2} \int_{\mathbb{R}} f(t)\overline{\psi\left(a^{-1}t - b\right)}dt = \langle f, U_{(a,b)}\psi \rangle, \quad f, \psi \in L^2(\mathbb{R}),$$

where $U_{(a,b)}\psi(t) = |a|^{-1/2}\psi(a^{-1}t - b)$, $a > 0$, $b \in \mathbb{R}$ is a representation of the affine group on the $L^2(\mathbb{R})$ (see [10–13]). There is a rich bibliography of the affine wavelet theory (see for example [7, 9, 13, 15]). One important question is the construction of the discrete version, i.e., to find ψ so that the discrete translations and dilatations

$$\psi_{n,k} = 2^{-n/2}\psi\left(2^{-n}x - k\right) \quad n \in \mathbb{Z},\ k \in \mathbb{Z}$$

form a (orthonormal) basis or a frame in $L^2(\mathbb{R})$ which generates a multiresolution.

The general definition of the affine wavelet multiresolution analysis in $L^2(\mathbb{R})$ is the following.

Definition 1 Let V_j, $j \in \mathbb{Z}$ be a sequence of subspaces of $L^2(\mathbb{R})$. The collections of spaces $\{V_j,\ j \in \mathbb{Z}\}$ is called a multiresolution analysis with scaling function ϕ if the following conditions hold:

1. (nested) $V_j \subset V_{j+1}$,
2. (density) $\overline{\bigcup V_j} = L^2(R)$,
3. (separation) $\bigcap V_j = \{0\}$,
4. (basis) There exists a mother wavelet ϕ in V_0 such that the set $\{2^{n/2}\phi(2^n x - k),\ k \in \mathbb{Z}\}$ is a basis (orthonormal) in V_n.

The wavelet coefficients

$$\langle f, \psi_{n,k} \rangle = 2^{-n/2} W_\psi f\left(2^{-n}, k\right)$$

can be expressed by the values of the affine wavelet transform on the discrete lattice

$$\Lambda = \left\{\left(2^{-n}, -k\right) : n \in \mathbb{Z},\ k \in \mathbb{Z}\right\}.$$

The reconstruction of the function f if we know (measure) the wavelet coefficients is treated in the mentioned bibliographies.

Our aim is to introduce a multiresolution analysis in the Hardy space of the upper half plane using localized Cauchy kernels and to give a H^2 norm identification which is important in modern system theory. We take as starting point another special voice transform connected to the Blaschke group. We consider a discrete subset of the Blaschke group and then we will take the image of this set trough the Cayley transform. Using the localized Cauchy kernels for the upper half plane on this set we will construct the multiresolution in the Hardy space of the upper half plane.

1.2 The Blaschke Group

Let us denote by

$$B_a(z) := \epsilon \frac{z - b}{1 - \bar{b}z} \quad \left(z \in \mathbb{C}, a = (b, \epsilon) \in \mathbb{B} := \mathbb{D} \times \mathbb{T}\right), \tag{1}$$

the so called *Blaschke functions*, where

$$\mathbb{D} := \left\{z \in \mathbb{C} : |z| < 1\right\}, \qquad \mathbb{T} := \left\{z \in \mathbb{C} : |z| = 1\right\}. \tag{2}$$

If $a \in \mathbb{B}$, then B_a is an 1–1 map on $\mathbb{T}$ and $\mathbb{D}$, respectively. The restrictions of the Blaschke functions on the set $\mathbb{D}$ or on $\mathbb{T}$ with the operation $(B_{a_1} \circ B_{a_2})(z) := B_{a_1}(B_{a_2}(z))$ form a group. In the set of the parameters $\mathbb{B} := \mathbb{D} \times \mathbb{T}$ let us define the operation induced by the function composition in the following way: $B_{a_1} \circ B_{a_2} = B_{a_1 \circ a_2}$. The group $(\mathbb{B}, \circ)$ will be isomorphic with the group $(\{B_a, a \in \mathbb{B}\}, \circ)$. If we use the notations $a_j := (b_j, \epsilon_j)$, $j \in \{1, 2\}$ and $a := (b, \epsilon) =: a_1 \circ a_2$, then

$$\begin{aligned} b &= \frac{b_1\overline{\epsilon}_2 + b_2}{1 + b_1\overline{b}_2\overline{\epsilon}_2} = B_{(-b_2, 1)}(b_1\overline{\epsilon}_2), \\ \epsilon &= \epsilon_1 \frac{\epsilon_2 + b_1\overline{b}_2}{1 + \epsilon_2\overline{b}_1 b_2} = B_{(-b_1\overline{b}_2, \epsilon_1)}(\epsilon_2). \end{aligned} \tag{3}$$

The neutral element of the group $(\mathbb{B}, \circ)$ is $e := (0, 1) \in \mathbb{B}$ and the inverse element of $a = (b, \epsilon) \in \mathbb{B}$ is $a^{-1} = (-b\epsilon, \overline{\epsilon})$.

In [18, 19] we have studied the properties of the voice transform induced by the following representation of the Blaschke group on $H^2(\mathbb{T})$:

$$(U_{a^{-1}} f)(z) := \frac{\sqrt{e^{i\theta}(1-|b|^2)}}{(1-\overline{b}z)} f\left(\frac{e^{i\theta}(z-b)}{1-\overline{b}z}\right)$$
$$\left(z = e^{it} \in \mathbb{T}, a = \left(b, e^{i\theta}\right) \in \mathbb{B}\right). \tag{4}$$

The voice transform generated by U_a $(a \in \mathbb{B})$ is given by

$$(V_g f)\left(a^{-1}\right) := \langle f, U_{a^{-1}} g\rangle \quad \left(f, g \in H^2(\mathbb{T})\right). \tag{5}$$

1.3 The Hardy Space of the Upper Half Plane

Set $\mathbb{C}_+ = \{z \in \mathbb{C} : \mathrm{Im}(z) > 0\}$, $H(\mathbb{C}_+)$, $H(\mathbb{D})$ the set of holomorphic functions in $\mathbb{C}_+$, respectively, $\mathbb{D}$. We shall work with the Hardy spaces

$$H^2(\mathbb{C}_+) = \left\{h \in H(\mathbb{C}_+) : \sup\left\{\int_{\mathbb{R}} \left|h(x+iy)\right|^2 dx : y > 0\right\} < \infty\right\}, \tag{6}$$

$$H^2(\mathbb{D}) = \left\{h \in H(\mathbb{D}) : \sup\left\{\int_{-\infty}^{\infty} \left|h\left(re^{it}\right)\right|^2 dt : r \in (0,1)\right\} < \infty\right\}. \tag{7}$$

The basic properties of this spaces can be found for example in [6] and [16]. For each $f \in H^p(\mathbb{C}_+)$, $p \in [2, +\infty)$, there exists a non-tangential limit which belongs to $L^p(\mathbb{R})$. Likewise, for each $f \in H^p(\mathbb{D})$, $p \in [2, \infty)$ there exists a non-tangential limit which is in $L^p(\mathbb{T})$. For simplicity we shall use the same notation for a function in Hardy spaces as that for its non-tangential limits. $H^2(\mathbb{C}_+)$ and $H^2(\mathbb{D})$ are Hilbert spaces endowed with the following inner products:

$$\langle f, g\rangle_{H^2(\mathbb{C}_+)} := \int_{\mathbb{R}} f(t)\overline{g(t)}dt, \quad f, g \in H^2(\mathbb{C}_+), \tag{8}$$

$$\langle f, g\rangle_{H^2(\mathbb{D})} := \frac{1}{2\pi}\int_{-\pi}^{\pi} f\left(e^{it}\right)\overline{g\left(e^{it}\right)}dt, \quad f, g \in H^2(\mathbb{D}). \tag{9}$$

Although the unit disk $\mathbb{D}$ and the upper half-plane $\mathbb{C}_+$ can be mapped to one-another by means of Möbius transformations, they are not interchangeable as domains for Hardy spaces. Contributing to this difference is the fact that the unit circle has finite (one-dimensional) Lebesgue measure while the real line does not. However, these two spaces may still be connected through the Cayley transform which maps $\mathbb{C}_+$ to $\mathbb{D}$ and is defined by

$$K(\omega) = \frac{i-\omega}{i+\omega}, \quad \omega \in \mathbb{C}_+. \tag{10}$$

The correspondence between the boundaries is

$$e^{is} = K(t) = \frac{i-t}{i+t}, \quad t \in \mathbb{R}, \; s \in (-\pi, \pi),$$

which implies that $s = 2\arctan(t)$, $t \in \mathbb{R}$.

With the Cayley transform, the linear transformation from $H^2(\mathbb{D})$ to $H^2(\mathbb{C}_+)$ is defined for $f \in H^2(\mathbb{D})$ by

$$Tf := \frac{1}{\sqrt{\pi}} \frac{1}{\omega + i} (f \circ K) \tag{11}$$

and is an isomorphism between these spaces. Consequently the theory of the real line is a close analogy with what we have for the circle.

Suppose F is real-valued and $F \in L^2(\mathbb{R})$. Then the projection onto $\mathbb{H}^2(\mathbb{C}_+)$ is given by

$$F^+(z) = \frac{1}{2\pi i} \int_{-\infty}^{\infty} \frac{F(t)}{t-z} dt.$$

Denote the non-tangential limit of $F^+(z)$ by $F^+(t)$, then $F(t) = 2\,\mathrm{Re}\, F^+(t)$. It will suffice to decompose F^+.

The Cauchy formula for the upper half plane is the following: for any function $F \in \mathbb{H}^p(\mathbb{C}^+)$, $1 \le p < +\infty$ if $F(s)$ is its non-tangential boundary limit, then

$$F(z) = \frac{1}{2\pi i} \int_{-\infty}^{\infty} \frac{F(s)}{s-z} ds, \quad z \in \mathbb{C}_+. \tag{12}$$

The classical Fourier bases has been proved to be an efficient approach to represent a linear stationary signal. However, it is not efficient to represent a nonlinear and stationary signals (see [21]). For this purpose it is more efficient to use some special orthonormal basis of rational functions. In the case of the unit disc is used the well known Malmquist–Takenaka system. There is an analogue of this system for the upper half plane.

Let $\{\lambda_i\}_{i=0}^{\infty}$ an arbitrary sequence of complex numbers which lie in the upper half-plane $\mathbb{C}_+$, and let $\{\Phi_n\}_{n=0}^{\infty}$ defined by

$$\Phi_1(z) = \frac{\sqrt{\frac{\Im \lambda_1}{\pi}}}{z - \overline{\lambda_1}}, \qquad \Phi_n = \frac{\sqrt{\frac{\Im \lambda_n}{\pi}}}{z - \overline{\lambda_n}} \prod_{k=1}^{n-1} \frac{z - \lambda_k}{z - \overline{\lambda_k}} \quad (n = 2, 3, \ldots). \tag{13}$$

This is a system of rational functions associated with the set of poles $\{\overline{\lambda_i}\}_{i=0}^{\infty}$ lying in the lower half-plane.The linear-fractional transformation $z = i\frac{1-y}{1+y}$ changes this system into the Malmquist–Takenaka system over the unit circle.

The system of functions $\{\Phi_n\}_{n=0}^{\infty}$ is orthonormal on the entire axis $-\infty < x < +\infty$ in the following sense:

$$\int_{-\infty}^{+\infty} \Phi_n(x) \overline{\Phi_m(x)} dx = \delta_{mn}. \tag{14}$$

Moreover, if we have the following non Blaschke condition for the upper half plane:

$$\sum_{k=0}^{\infty} \frac{\Im \lambda_k}{1+|\lambda_k|} = \infty \tag{15}$$

then $\{\Phi_n\}_{n=0}^{\infty}$ is a complete orthonormal system for $\mathbb{H}^2(\mathbb{C}_+)$. In [8] Dzrbasjan proved the analogue of the Darboux–Christoffel formula for the upper half plane. For $n \geq 0$ let us consider the functions

$$B_n(z) = \prod_{n=0}^{n} \frac{z-\lambda_k}{z-\overline{\lambda_k}} \tau_k, \qquad \tau_k = \frac{|1+\lambda_k^2|}{1+\lambda_k^2}.$$

For arbitrary values of the variables $z \neq \xi$ and for any n, $0 \leq n < \infty$,

$$\sum_{k=0}^{n} \Phi_k(z)\overline{\Phi_k(\xi)} = \frac{1-\overline{B_n(\xi)}B_n(z)}{2i\pi(\overline{\xi}-z)}. \tag{16}$$

2 New Results

2.1 A Special Lattice in the Upper Half Plane

Let us denote

$$\mathbb{B}_1 = \left\{ (r_k, 1) : \ r_k = \frac{2^k - 2^{-k}}{2^k + 2^{-k}}, \ k \in \mathbb{Z} \right\}. \tag{17}$$

It can be proved that $(\mathbb{B}_1, \circ)$ is a subgroup of $(\mathbb{B}, \circ)$, and $(r_k, 1) \circ (r_n, 1) = (r_{k+n}, 1)$. The hyperbolic distance of the points r_k, r_n has the following property:

$$\rho(r_k, r_n) := \frac{|r_k - r_n|}{|1 - r_k \overline{r_n}|} = \left| \frac{\frac{2^k-2^{-k}}{2^k+2^{-k}} - \frac{2^n-2^{-n}}{2^n+2^{-n}}}{1 - \frac{2^k-2^{-k}}{2^k+2^{-k}} \frac{2^n-2^{-n}}{2^n+2^{-n}}} \right| = |r_{k-n}|. \tag{18}$$

Let us consider the set of points

$$A = \left\{ z_{k\ell} = r_k e^{i\frac{2\pi\ell}{2^{2k}}}, \ \ \ell = 0, 1, \ldots, 2^{2k}-1, \ \ k = 0, 1, 2, \ldots, \infty \right\}$$

and for a fixed $k \in \mathbb{N}$ let the level k be

$$A_k = \left\{ z_{k\ell} = r_k e^{i\frac{2\pi\ell}{2^{2k}}}, \ \ell \in \left\{0, 1, \ldots, 2^{2k}-1\right\} \right\}.$$

The inverse Cayley transform $K^{-1}(z) = i\frac{1-z}{1+z}$ takes the unit circle in the real axis and the unit disc in the upper half plane. Let us define

$$\begin{aligned} a_{k\ell} = K^{-1}(z_{k\ell}) &= \frac{2r_k \sin\frac{2\pi\ell}{2^{2k}}}{1 - 2r_k\cos\frac{2\pi\ell}{2^{2k}} + r_k^2} + i\frac{1-r_k^2}{1-2r_k\cos\frac{2\pi\ell}{2^{2k}} + r_k^2} \\ &= \alpha_{k\ell} + i\beta_{k\ell}, \end{aligned} \tag{19}$$

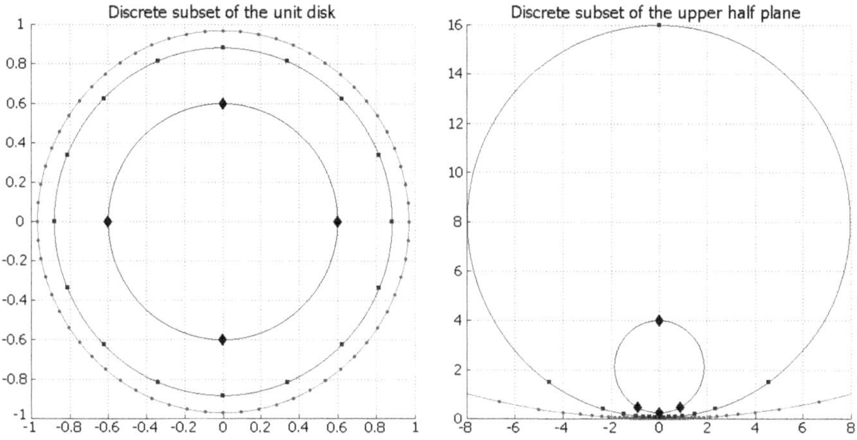

Fig. 1

$$B_k = \{a_{k\ell},\ \ell \in \{0, 1, \dots, 2^{2k} - 1\}\}, \tag{20}$$

$$B = \{a_{k\ell},\ \ \ell = 0, 1, \dots, 2^{2k} - 1,\ \ k = 0, 1, 2, \dots, \infty\}. \tag{21}$$

The points from B are in the upper half plane, and every point from B_k is on the circle with center $(0, \frac{1+r_k^2}{1-r_k^2})$ and radius $R_k = \frac{2r_k}{1-r_k^2}$, see Fig. 1. It is easy to show that the points from B do not satisfy the Blaschke condition for the upper half plane. Indeed,

$$\sum_{k=0}^{\infty} \sum_{\ell=0}^{2^{2k}-1} \frac{\beta_{k\ell}}{1 + |a_{k\ell}|^2} = \sum_{k=0}^{\infty} \sum_{\ell=0}^{2^{2k}-1} \frac{1 - r_k^2}{2(1 + r_k^2)} = \sum_{k=0}^{\infty} \frac{2^{2k}}{2^{2k} + 2^{-2k}} = \infty. \tag{22}$$

2.2 Multiresolution on the Upper Half Plane

In multiresolution analysis, one decomposes a function space in several resolution levels and the idea is to represent the functions from the function space by a low resolution approximation and adding to it the successive details that lift it to resolution levels of increasing detail.

Wavelet analysis couples the multiresolution idea with a special choice of basis for the different resolution spaces and for the wavelet spaces that represent the difference between successive resolution spaces. If V_n are the resolution spaces $V_0 \subset V_1 \subset \cdots \subset V_n \cdots$, then the wavelet spaces W_n are defined by the equality $W_n \oplus V_n = V_{n+1}$.

In the construction of affine wavelet multiresolutions the dilatation is used to obtain a higher level resolution ($f(x) \in V_n \Leftrightarrow f(2x) \in V_{n+1}$) and applying the translation we remain on the same level of resolution.

Using the lattice B we give a similar construction of the affine wavelet multiresolution in the space $H^2(\mathbb{C}_+)$. To show the analogy with the affine wavelet multiresolution we first represent the levels V_n by nonorthogonal bases and then we construct an orthonormal bases in V_n and we give also an orthogonal basis in W_n which is orthogonal to V_n. We will show that in the case of this discretization the corresponding Malmquist–Takenaka systems will span the resolution spaces and the density property will be fulfilled, i.e., $\overline{\bigcup_{k=1}^{\infty} V_k} = H^2(\mathbb{C}_+)$ in norm. In signal processing and system identification the rational orthogonal bases like the discrete Laguerre, Kautz and Malmquist–Takenaka systems are more efficient than the trigonometric system in the determination of the transfer functions. This field also has a rich bibliography (see [2–5, 14, 21–28] etc.).

We show that the projection $P_n f$ on the nth resolution level is an interpolation operator on the upper half plane until the nth level, which converges in $H^2(\mathbb{C}_+)$ norm to f. Let us introduce the analogue definition of multiresolution for the Hardy space of the upper half plane:

Definition 2 Let V_j, $j \in \mathbb{N}$, be a sequence of subspaces of $H^2(\mathbb{C}_+)$. The collections of spaces $\{V_j,\ j \in N\}$ is called a multiresolution if the following conditions hold:

1. (nested) $V_j \subset V_{j+1}$,
2. (density) $\overline{\bigcup V_j} = H^2(\mathbb{C}_+)$,
3. (analogue of dilatation) $(TU_{(r_1,1)^{-1}}T^{-1})V_n \subset V_{n+1}$,
4. (basis) There exists a $\psi_{n,\ell}$ (orthonormal) basis in V_n.

Let us consider the function $\varphi_{00} = \frac{1}{\sqrt{\pi}(z+i)}$, $V_0 = \{c\varphi_{00},\ c \in \mathbb{C}\}$ and let us consider the nonorthogonal hyperbolic wavelets at the first level

$$\varphi_{1,\ell}(z) = \sqrt{\frac{\beta_{1\ell}}{\pi}}\frac{1}{z - \overline{a_{1\ell}}} \qquad \ell = 0, 1, 2, 3.$$

Let us define the first resolution level as follows:

$$V_1 = \left\{ f : \mathbb{D} \to \mathbb{C},\ f(z) = c_{0,0}\varphi_{0,0} + \sum_{\ell=0}^{3} c_{1,\ell}\varphi_{1,\ell},\ c_{00}, c_{1,\ell} \in \mathbb{C},\ \ell = 0, 1, 2, 3 \right\}.$$

At the nth level the nonorthogonal wavelets are given by

$$\varphi_{n,\ell}(z) = \sqrt{\frac{\beta_{n\ell}}{\pi}}\frac{1}{z - \overline{a_{n\ell}}}, \qquad \ell = 0, 1, \ldots, 2^{2n} - 1, \tag{23}$$

and the nth resolution level is given by

$$V_n = \left\{ f : \mathbb{D} \to \mathbb{C},\ f(z) = \sum_{k=0}^{n} \sum_{\ell=0}^{2^{2k}-1} c_{k,\ell}\varphi_{k,\ell},\ c_{k,\ell} \in \mathbb{C} \right\}. \tag{24}$$

The closed subset V_n is spanned by

$$\{\varphi_{k,\ell},\ \ell=0,1,\ldots,2^{2k}-1,\ k=0,\ldots,n\}.$$

In this way we have obtained a sequence of closed, nested subspaces of $H^2(\mathbb{C}_+)$ for $z\in\mathbb{C}_+$,

$$V_0\subset V_1\subset V_2\subset\cdots V_n\subset\cdots H^2(\mathbb{C}_+).$$

The elements of B are different complex numbers consequently the corresponding finite subset of localized kernels

$$\left\{\frac{1}{z-\overline{a_{k\ell}}},\ \ell=0,1,\ldots,2^{2k}-1,\ k=0,1,\ldots,n\right\} \tag{25}$$

are linearly independent and they form a nonorthogonal basis in V_n. Applying the Gram–Schmidt orthogonalization for this set of analytic linearly independent functions we obtain the Malmquist–Takenaka system corresponding to the upper half plane and the set $\bigcup_{k=0}^{n} B_k$:

$$\psi_{m,\ell}(z)=\sqrt{\frac{\beta_{m\ell}}{\pi}}\frac{1}{z-\overline{a_{m\ell}}}\prod_{k=0}^{m-1}\prod_{j=0}^{2^{2k}-1}\frac{z-a_{kj}}{z-\overline{a_{kj}}}\prod_{j'=0}^{\ell-1}\frac{z-a_{mj'}}{z-\overline{a_{mj'}}}$$

$$\left(m=0,1,\ldots,n,\ \ell=0,1,\ldots,2^{2m}-1.\right) \tag{26}$$

From the Gram–Schmidt orthogonalization process it follows that

$$V_n=span\{\psi_{k,\ell},\ \ell=0,1,\ldots,2^{2k}-1,\ k=0,\ldots,n\}. \tag{27}$$

From (22) it follows that the Malmquist–Takenaka system corresponding to the set B is a complete orthonormal system of holomorphic functions in $H^2(\mathbb{C}_+)$.

From the completeness of the system $\{\psi_{k,\ell},\ \ell=0,1,\ldots,2^{2k}-1,\ \ k=\overline{0,\infty}\}$ in the Hilbert space $H^2(\mathbb{C}_+)$, it follows that this system is also a closed system, consequently the density property is valid in norm, i.e.:

$$\overline{\bigcup_{n\in\mathbb{N}} V_n}=H^2(\mathbb{C}_+). \tag{28}$$

The analogue of the dilatation and translation can be described in the following way: let $\phi=1$ and let

$$V_n'=span\{\phi_{k,\ell},\ \ell=0,1,\ldots,2^{2k}-1,\ k=0,\ldots,n\},$$

where

$$\phi_{n,\ell}(z)=(U_{(z_{n\ell},1)^{-1}}\phi)(z)=\frac{\sqrt{(1-r_n^2)}}{(1-\overline{z_{n\ell}}z)},\quad \ell=0,1,\ldots,2^{2n}-1.$$

These functions can be obtained from ϕ using the representation $U_{(r_n,1)^{-1}}$, and the translation

$$\phi_{n,\ell}(e^{it}) = (U_{(r_n,1)^{-1}}\phi)(e^{i(t-\frac{2\pi\ell}{2^{2n}})}).$$

We observe that taking the image of $\phi_{n,\ell}$ through the Cayley function, we have

$$\begin{aligned} T(\phi_{n,\ell})(\omega) &= \frac{1}{\sqrt{\pi}(i+\omega)} \frac{\sqrt{1-r_n^2}}{1-\overline{z_{n,\ell}}\frac{i-\omega}{i+\omega}} = \sqrt{\frac{\beta_{n,\ell}}{\pi}} \frac{\overline{i(i+a_{n,\ell})}}{\sqrt{2}|i+a_{n,\ell}|} \frac{1}{\omega-\overline{a_{n,\ell}}} \\ &= B_{k,\ell}\varphi_{n,\ell}(\omega), \end{aligned}$$

where $B_{k,\ell} = \frac{\overline{i(i+a_{n,\ell})}}{\sqrt{2}|i+a_{n,\ell}|}$ is a constant. From this we get $V_n = T(V_n')$. If a function $f \in V_n'$, then $U_{(r_1,1)^{-1}} f \in V_{n+1}'$. For this it is enough to show that

$$\begin{aligned} U_{(r_1,1)^{-1}}(\phi_{k,\ell})(e^{it}) &= U_{(r_1,1)^{-1}}\left[(U_{(r_k,1)^{-1}} p_0)\right](e^{i(t-\frac{2\pi\ell}{2^{2k}})}) \\ &= \left[(U_{(r_{k+1},1)^{-1}} p_0)\right](e^{i(t-\frac{2\pi 4\ell}{2^{2(k+1)}})}) \in V_{n+1}', \\ k = 1,\ldots,n,\ \ell &= 1,\ldots,2^{2k}-1. \end{aligned}$$

Consequently we have

$$T U_{(r_1,1)^{-1}} T^{-1} V_n \subset V_{n+1}. \tag{29}$$

The wavelet space W_n is the orthogonal complement of V_n in V_{n+1}. We will prove that

$$W_n = span\{\psi_{n+1,\ell},\ \ \ell = 0,1,\ldots,2^{2n+2}-1\}. \tag{30}$$

For an arbitrary $f(z) = \sum_{k=0}^{n}\sum_{\ell=0}^{2^{2k}-1} c_{k,\ell}\varphi_{k,\ell} \in V_n$ using the Cauchy formula we obtain

$$\begin{aligned} \langle \psi_{n+1j}, f\rangle &= \sum_{k=1}^{n}\sum_{\ell=0}^{2^{2k}-1} c_{k,\ell}\langle \psi_{n+1,j}, \varphi_{k,\ell}\rangle \\ &= \sum_{k=1}^{n}\sum_{\ell=0}^{2^{2k}-1} c_{k,\ell}\sqrt{\frac{\beta_{k\ell}}{\pi}} 2\pi i \psi_{n+1,\ell}(z_{k\ell}) = 0, \quad j = 0,1,\ldots,2^{2n+2}-1. \end{aligned}$$

Consequently,

$$\langle f, \psi_{n+1,j}\rangle = 0, \quad f \in V_n$$

which implies that

$$\psi_{n+1,j} \perp V_n \quad (j = 0,1,\ldots,2^{2n+2}-1). \tag{31}$$

From

$$V_{n+1} = V_n \oplus span\{\varphi_{n+1,j},\ j = 0, 1, \ldots, 2^{2n+2} - 1\} \tag{32}$$

it follows that W_n is an $2^{2(n+1)}$ dimensional space and

$$W_n = span\{\psi_{n+1,\ell},\ \ell = 0, 1, \ldots, 2^{2n+2} - 1\}. \tag{33}$$

2.3 The Projection Operator Corresponding to the nth Resolution Level

Let us consider the orthogonal projection operator of an arbitrary function $f \in H^2(\mathbb{C}_+)$ on the subspace V_n given by

$$P_n f(z) = \sum_{k=0}^{n} \sum_{\ell=0}^{2^{2k}-1} \langle f, \psi_{k,\ell}\rangle \psi_{k,\ell}(z). \tag{34}$$

This operator is called the projection of f at scale or resolution level n.

Theorem 1 *For $f \in H^2(\mathbb{C}_+)$ the projection operator $P_n f$ is an interpolation operator in the points*

$$a_{mj}\ \big(j = 0, \ldots, 2^{2m} - 1,\ \ m = 0, \ldots, n\big).$$

Proof Let us consider the kernel function of this projection operator

$$K_n(z, \xi) = \sum_{k=0}^{n} \sum_{\ell=0}^{2^{2k}-1} \overline{\psi_{k,\ell}(\xi)}\psi_{k,\ell}(z). \tag{35}$$

According to the result of Dzrbasjan (see [8])

$$K_n(z, \xi) = \frac{1}{2i\pi(\overline{\xi} - z)}\left(1 - \prod_{k=0}^{n}\prod_{\ell=0}^{2^{2k}-1} \frac{z - a_{k\ell}}{z - \overline{a_{k\ell}}}\tau_{k\ell} \overline{\prod_{k=0}^{n}\prod_{\ell=0}^{2^{2k}-1} \frac{\xi - a_{k\ell}}{\xi - \overline{z_{k\ell}}}\tau_{k\ell}}\right). \tag{36}$$

From this relation it follows that the values of the kernel function in the points a_{mj} $(j = 0, \ldots, 2^{2m} - 1,\ m = 0, \ldots, n)$ are equal to

$$K(a_{mj}, \xi) = \frac{1}{2i\pi(\overline{\xi} - a_{mj})}. \tag{37}$$

Using the Cauchy integral formula we get

$$P_n f(a_{mj}) = \frac{1}{2\pi i}\int_{-\infty}^{+\infty} \frac{f(t)}{t - a_{mj}} dt = f(a_{mj})$$
$$\big(j = 0, \ldots, 2^{2m} - 1,\ m = 0, \ldots, n\big). \tag{38}$$

Let us denote by $\hat{f}_n$ the solution of the minimal-norm interpolation problem

$$\|\hat{f}_n\|_{H^2(\mathbb{C}_+)} = \min_{f_n \in H^2(\mathbb{C}_+)} \|f_n\|_{H^2(\mathbb{C}_+)}, \tag{39}$$

satisfying the interpolation conditions

$$f_n(a_{mj}) = f(a_{mj}) \quad \big(j = 0, \ldots, 2^{2m} - 1, \ \ m = 0, \ldots, n\big). \qquad \square$$

Remark 1 The projection $P_n f$ is the solution of the minimal-norm interpolation problem (39),

$$\|f - P_n f\|_{H^2(\mathbb{C}_+)} \to 0, \quad n \to \infty.$$

Proof This result follows from Theorem 2.2, Theorem 5.3.1 of [20] and (28). Because $\{\psi_{k,\ell},\ k = \overline{0, \infty},\ \ell = 0, 1, \ldots, 2^{2k} - 1\}$ is a complete set in the Hilbert space $H^2(\mathbb{C}_+)$ is also closed set. This implies that $\|f - P_n f\|_{H^2(\mathbb{C}_+)} \to 0$ as $n \to \infty$. From Theorem 2.2 and Theorem 5.3.1 of [20] it follows that the best approximant is given by $\hat{f}_n(z) = P_n f(z)$. $\square$

It is a natural question to ask what we can say about of $H^p(\mathbb{C}_+)$ norm convergence of $P_n f$. In analogy to Theorem 5.1 of [1] (proved for the right half plane) it can be proved that for all $1 < p < \infty$ and $f \in H^p(\mathbb{C}_+)$

$$\|f - P_n f\|_{H^p(\mathbb{C}_+)} \to 0, \quad n \to \infty.$$

For the error term we have the following estimation. Let us denote

$$e_n(f, p) = \inf_{f_n \in V_n} \|f_n - f\|_{H^p(\mathbb{C}_+)},$$

the best $H^p(\mathbb{C}_+)$-norm approximation error of f in V_n. Then there exists a constant C_p depending only on p such that

$$\|f - P_n f\|_{H^p(\mathbb{C}_+)} \le C_p e_n(f, p).$$

In what follows we propose a computational scheme for the best approximant in the wavelet base $\{\psi_{k,\ell},\ \ell = 0, 1, \ldots, 2^{2k} - 1,\ k = 0, \ldots, n\}$.

2.4 Reconstruction Algorithm

The projection of $f \in H^2(\mathbb{C}_+)$ onto V_{n+1} can be written in the following way:

$$P_{n+1} f = P_n f + Q_n f, \tag{40}$$

where

$$Q_n f(z) := \sum_{\ell=0}^{2^{2(n+1)}-1} \langle f, \psi_{n+1,\ell} \rangle \psi_{n+1,\ell}(z). \tag{41}$$

This operator has the following properties:

$$Q_n f(z_{k\ell}) = 0, \quad k = 1, \ldots, n, \ \ell = 0, 1, \ldots, 2^{2n} - 1. \tag{42}$$

Consequently P_n contains information on low resolution, i.e., until the level B_n, and Q_n is the high resolution part. After n steps,

$$P_{n+1} f = P_1 f + \sum_{k=1}^{n} Q_n f. \tag{43}$$

Thus

$$V_{n+1} = V_0 \oplus W_0 \oplus W_1 \oplus \cdots \oplus W_n.$$

The set of coefficients of the best approximant $P_n f$

$$\left\{ b_{k\ell} = \langle f, \psi_{k,\ell} \rangle, \ \ell = 0.1, \ldots, 2^{2k} - 1 \ \ k = 0, 1, \ldots, n \right\} \tag{44}$$

is the (discrete) hyperbolic wavelet transform of the function f. Thus it is important to have an efficient algorithm for the computation of the coefficients.

The coefficients of the projection operator $P_n f$ can be computed if we know the values of the functions on $\bigcup_{k=0}^{n} B_k$. For this reason we express first the function $\psi_{k,\ell}$ using the bases $(\varphi_{k',\ell'} \ \ell' = 0, 1, \ldots, 2^{2k'} - 1, \ k' = 0, \ldots, k)$, i.e. we write the partial fraction decomposition of $\psi_{k\ell}$:

$$\psi_{k,\ell} = \sum_{k'=0}^{k-1} \sum_{\ell'=0}^{2^{2k'}-1} c_{k',\ell'} \frac{-1}{2\pi i (\xi - \overline{a_{k'\ell'}})} + \sum_{j=0}^{\ell} c_{k,j} \frac{-1}{2\pi i (\xi - \overline{a_{kj}})}. \tag{45}$$

Using the orthogonality of the functions $(\psi_{k',\ell'} \ \ell' = 0, 1, \ldots, 2^{2k'} - 1, \ k' = 0, \ldots, k)$ and the Cauchy formula we get

$$\begin{aligned} \delta_{kn}\delta_{\ell m} &= \langle \psi_{nm}, \psi_{k\ell} \rangle \\ &= \sum_{k'=0}^{k-1} \sum_{\ell'=0}^{2^{2k'}-1} \overline{c_{k',\ell'}} \psi_{n,m}(a_{k'\ell'}) + \sum_{j=0}^{\ell} \overline{c_{k,j}} \psi_{n,m}(a_{kj}) \end{aligned}$$

$$\left(m = 0, 1, \ldots, 2^{2n} - 1, \ n = 0, \ldots, k \right). \tag{46}$$

If we order these equalities so that we write first the relations (46) for $n = k$ and $m = \ell, \ell - 1, \ldots, 0$, respectively, then for $n = k - 1$ and $m = 2^{2(k-1)} - 1, \ 2^{2(k-1)} -$

$2, \ldots, 0$, etc., this is equivalent to

$$\begin{pmatrix} 1 \\ 0 \\ 0 \\ \cdot \\ \cdot \\ \cdot \\ 0 \end{pmatrix} = \begin{pmatrix} \psi_{k,\ell}(a_{k,\ell}) & 0 & 0 & \ldots & 0 \\ \psi_{k,\ell-1}(a_{k,\ell}) & \psi_{k,\ell-1}(a_{k,\ell-1}) & 0 & \ldots & 0 \\ \psi_{k,\ell-2}(a_{k,\ell}) & \psi_{k,\ell-2}(a_{k,\ell-1}) & \psi_{k,\ell-2}(a_{k,\ell-2}) & \ldots & 0 \\ \vdots & & & \vdots & \\ \psi_{00}(a_{k,\ell}) & \psi_{00}(a_{k,\ell-1}) & \psi_{00}(a_{k,\ell-2}) & \ldots & \psi_{00}(a_{00}) \end{pmatrix}$$

$$\times \begin{pmatrix} \overline{c_{k,\ell}} \\ \overline{c_{k,\ell-1}} \\ \overline{c_{k,\ell-2}} \\ \vdots \\ \overline{c_{00}} \end{pmatrix}. \tag{47}$$

This system has a unique solution $(\overline{c_{k,\ell}}, \overline{c_{k,\ell-1}}, \overline{c_{k,\ell-2}}, \ldots, \overline{c_{00}})^T$. If we determine this vector, then we can compute the exact value of $\langle f, \psi_{k,\ell}\rangle$ knowing the values of f on the set $\bigcup_{k=0}^n B_k$.

Indeed, using again the partial fraction decomposition of $\psi_{k,\ell}$ and the Cauchy integral formula we get

$$\langle f, \psi_{k,\ell}\rangle = \sum_{k'=0}^{k-1} \sum_{\ell'=0}^{2^{2k'}-1} \overline{c_{k',\ell'}} f(a_{k',\ell'}) + \sum_{j=0}^{\ell} \overline{c_{k,j}} f(a_{k,j}). \tag{48}$$

Summary In this chapter, we have generated a multiresolution in $H^2(\mathbb{C}_+)$ and we have constructed a rational orthogonal wavelet system which generates this multiresolution. Measuring the values of the function f in the points of the set $B = \bigcup_{k=0}^n B_k \subset \mathbb{C}_+$ we can write $(P_n f,\ n \in \mathbb{N})$, the projection operator on the nth resolution level, which is convergent in $H^2(\mathbb{C}_+)$ norm to f, is the best approximant interpolation operator on the set the $\bigcup_{k=0}^n B_k$ and $P_n f(z) \to f(z)$ uniformly on every compact subset of the upper half plane.

Acknowledgements This chapter was developed during a fruitful stay of the second author at NuHAG group at the University of Vienna as Marie Curie fellow FP7-People-IEF-2009.

References

1. Akay, H., Ninnes, B.: Orthonormal basis functions for continuous-time systems and L^p convergence. Math. Control Signals Syst. **12**, 295–305 (1999)
2. Bokor, J., Athans, M.: Frequency domain identification of the MIT interferometer tested in generalized orthogonal basis. In: Proceedings of the 11th IFAC Symposium on System Identification, Kiayushu, Japan, vol. 4, pp. 1735–1739 (1997)

3. Bokor, J., Schipp, F., Szabó, Z.: Identification of rational approximate models in H^∞ using generalized orthonormal basis. IEEE Trans. Autom. Control **44**(1), 153–158 (1999)
4. Bultheel, A., González-Vera, P.: Wavelets by orthogonal rational kernels. Contemp. Math. **236**, 101–126 (1999)
5. Bultheel, A., González-Vera, P., Hendriksen, E., Njastad, O.: Orthogonal Rational Functions. Cambridge Monographs on Applied and Computational Mathematics, vol. 5. Cambridge University Press, Cambridge (1999)
6. Cima, J., Ross, W.: The Backward Shift on the Hardy Space. Mathematical Surveys and Monographs, vol. 79. Am. Math. Soc., Providence (2000). pp. xii+199. MR1761913
7. Daubechies, I.: Orthonormal bases of compactly supported wavelets. Commun. Pure Appl. Math. **41**, 909–996 (1988)
8. Dzrbasjan, M.M.: Biorthogonal systems of rational functions and best approximant of the Cauchy kernel on the real axis. Math. USSR Sb. **24**(3), 409–433 (1974)
9. Evangelista, G., Cavaliere, S.: Discrete frequency-warped wavelets: theory and applications. IEEE Trans. Signal Process. **46**(4), 874–883 (1998)
10. Feichtinger, H.G., Gröchenig, K.: A unified approach to atomic decompositions trough integrable group representations. In: Cwinkel, M., et al. (eds.) Functions Spaces and Applications. Lecture Notes in Math., vol. 1302, pp. 307–340. Springer, Berlin (1989)
11. Feichtinger, H.G., Gröchenig, K.: Banach spaces related to integrable group representations and their atomic decomposition I. J. Funct. Anal. **86**(2), 307–340 (1989)
12. Goupilland, P., Grossman, A., Morlet, J.: Cycle-octave and related transforms in seismetic signal analysis. Geoexploration **25**, 85–102 (1984)
13. Heil, C.E., Walnut, D.F.: Continuous and discrete wavelet transforms. SIAM Rev. **31**(4), 628–666 (1989)
14. de Hoog, T.J.: Rational Orthonormal Basis and Related Transforms in Linear System Modeling. Ponsen and Looijn, Wageningen (2001)
15. Mallat, S.: Theory of multiresolution signal decomposition: the wavelet representation. IEEE Trans. Pattern Anal. Math. Intell. **11**(7), 674–693 (1989)
16. Mashreghi, J.: Representation Theorems in Hardy Spaces. Cambridge University Press, Cambridge (2009)
17. Pap, M.: Hyperbolic wavelets and multiresolution in $H^2(\mathbb{T})$. J. Fourier Anal. Appl. **17**(5), 755–776 (2011). doi:10.1007/s00041-011-9169-2
18. Pap, M., Schipp, F.: The voice transform on the Blaschke group I. PU.M.A. **17**(3–4), 387–395 (2006)
19. Pap, M., Schipp, F.: The voice transform on the Blaschke group II. Ann. Univ. Sci. (Budapest), Sect. Comput. **29**, 157–173 (2008)
20. Partington, J.: Interpolation, Identification and Sampling. London Mathematical Society Monographs, vol. 17. Oxford University Press, London (1997)
21. Qian, T.: Intrinsic mono-component decomposition of functions: an advance of Fourier theory. Math. Methods Appl. Sci. (2009). doi:10.1002/mma.1214. www.interscience.wiley.com.
22. Soumelidis, A., Bokor, J., Schipp, F.: Signal and system representations on hyperbolic groups: beyond rational orthogonal bases. In: ICC 2009 7th IEEE International Conference on Computational Cybernetics, Palma de Mallorca. ISBN: 978-1-4244-5311-5
23. Soumelidis, A., Bokor, J., Schipp, F.: Detection of changes on signals and systems based upon representations in orthogonal rational bases. In: Proc. of 5th IFAC Symposium on Fault Detection Supervision and Safety for Technical Processes, SAFEPROSS 2003, Washington D.C., USA (June 2003), on CD
24. Soumelidis, A., Bokor, J., Schipp, F.: Representation and approximation of signals and systems using generalized Kautz functions. In: Proc. of the 36th Conference on Decisions and Control, San Diego, CA, pp. 3793–3796 (1997), CDC'97
25. Soumelidis, A., Bokor, J., Schipp, F.: Frequency domain representation of signals in rational orthogonal bases. In: Proc. of the 10th Mediterranean Conference on Control and Automation, Lisbon, Portugal (2002), on CD. Med'(2002)

26. Soumelidis, A., Pap, M., Schipp, F., Bokor, J.: Frequency domain identification of partial fraction models. In: Proc. of the 15th IFAC World Congress, Barcelona, Spain, June, pp. 1–6 (2002)
27. Szabó, Z.: Interpolation and quadrature formula for rational systems on the unit circle. Ann. Univ. Sci. (Budapest), Sect. Comput. **21**, 41–56 (2002)
28. Ward, N.F.D., Partington, J.R.: Robust identification in the disc algebra using rational wavelets and orthonormal basis functions. Int. J. Control **64**, 409–423 (1996)
29. Ward, N.F.D., Partington, J.R.: A construction of rational wavelets and frames in Hardy-Sobolev spaces with applications to system modeling. SIAM J. Control Optim. **36**(2), 654–679 (1998)

Norms of Composition Operators Induced by Finite Blaschke Products on Möbius Invariant Spaces

María J. Martín and Dragan Vukotić

Abstract We obtain an asymptotic formula for the norms of composition operators induced by finite Blaschke products on analytic (quotient) Besov spaces in terms of their degree. We also compute the norms of such operators on the true Bloch and Dirichlet spaces.

Keywords Finite Blaschke product · Analytic Besov space · Bloch space · Dirichlet space · Composition operator · Operator norm

Mathematics Subject Classification Primary 47B33 · Secondary 30D45 · 31C25

1 Introduction

Let $\mathbb{D} = \{z \in \mathbb{C} : |z| < 1\}$ denote the unit disk in the complex plane. By a *self-map* of $\mathbb{D}$ we mean a function ϕ analytic in $\mathbb{D}$ such that $\phi(\mathbb{D}) \subset \mathbb{D}$. For any such ϕ, we can define C_ϕ, the *composition operator* with *symbol* ϕ, by $C_\phi f = f \circ \phi$. It is well known that such an operator maps boundedly any classical function space such as Hardy, Bergman, or Bloch into itself. However, this is not the case with the Dirichlet space or, more generally, with the conformally invariant analytic Besov spaces B^p. This was first studied in [2] in terms of Carleson measures. It is known (and easy to show) that composition operators whose symbol is a finite Blaschke product are bounded operators on these spaces as well.

At the international conference "Recent Advances in Operator Related Function Theory" held in Dublin in 2004, Professor J. Cima asked the second author whether one could compute or control the norm of a composition operator induced by a finite Blaschke product and acting on the Dirichlet space (or another conformally invariant space) in terms of the degree of the product. It turns out that the answer

M.J. Martín · D. Vukotić (✉)
Departamento de Matemáticas, Módulo 17, Ciencias, Universidad Autónoma de Madrid, 28049 Madrid, Spain
e-mail: dragan.vukotic@uam.es

M.J. Martín
e-mail: mariaj.martin@uam.es

J. Mashreghi, E. Fricain (eds.), *Blaschke Products and Their Applications*, Fields Institute Communications 65, DOI 10.1007/978-1-4614-5341-3_12,

can be deduced from two papers by Arazy, Fisher, and Peetre [2, 3]. In the present paper we review some details of their findings and combine it with other results to obtain an asymptotic formula for the norms of such operators, acting either on an analytic Besov space or on the Bloch space, in terms of the degree of the symbol. To be more precise, we show that if B is a Blaschke product of degree n, then the norms of the induced composition operators C_B on the quotient space $B^p/\mathbb{C}$, $1 \le p \le \infty$, are comparable to $n^{1/p}$, independently of the product chosen. We also compute the norms of C_B on the true Bloch and Dirichlet spaces (taking constant values into account). The answer in this case, of course, depends not only on n but also on the value $B(0)$.

2 Background

In this section we review all the necessary basic material and fix the notation.

2.1 *Finite Blaschke Products*

A *finite Blaschke product* is a function of the form

$$B(z) = \lambda \prod_{k=1}^{n} \frac{a_k - z}{1 - \overline{a}_k z}$$

whose zeros a_k, $1 \le k \le n$ all lie in $\mathbb{D}$, and $|\lambda| = 1$. The number n of factors (and also zeros) of B is called the *degree* of the Blaschke product. It follows readily from Rouché's theorem that for every such B and each $w \in \mathbb{D}$ there are exactly n values $z \in \mathbb{D}$, counting the multiplicities, for which $B(z) = w$.

Finite Blaschke products are obviously continuous in the closed disk $\overline{\mathbb{D}}$ and have modulus one on the unit circle. This turns out to be their defining property: every non-constant self-map of $\mathbb{D}$ which is continuous in the closed disk, has modulus one at every point on the unit circle, and has finitely many zeros in $\mathbb{D}$ must be a finite Blaschke product. This follows easily from the maximum modulus principle; see [5, § I.2].

Lemma 1 *If B_m and B_n are two Blaschke products of degrees m and n respectively, then $B_m \circ B_n$ is again a finite Blaschke product of degree mn.*

Proof Note that $B_m \circ B_n$ is again continuous in the closed disk and has modulus one on the unit circle. Also, since B_m has m zeros and each one of them has n pre-images in $\mathbb{D}$ under B_n (counting the multiplicities), it follows that $B_m \circ B_n$ has exactly mn zeros in $\mathbb{D}$ and is therefore a finite Blaschke product of degree mn. □

2.2 Disk Automorphisms. The Hyperbolic Metric

It is a well-known fact that every disk automorphism (i.e., a bijective self-map of $\mathbb{D}$) is a composition of a rotation (multiplication by a constant of modulus one) and an automorphism ϕ_α, where $\phi_\alpha(z) = (\alpha - z)/(1 - \overline{\alpha} z)$, $\alpha \in \mathbb{D}$. Note that ϕ_α is an involution, meaning that $\phi_\alpha(\phi_\alpha(z)) = z$ for all $z \in \mathbb{D}$.

The *hyperbolic distance* between two points z and w in $\mathbb{D}$ is defined as

$$\varrho(z, w) = \frac{1}{2} \log \frac{1 + |\phi_w(z)|}{1 - |\phi_w(z)|}.$$

The metric defined in this way is complete and conformally invariant, meaning that $\varrho(\phi(z), \phi(w)) = \varrho(z, w)$ for all z, $w \in \mathbb{D}$ and every disk automorphism ϕ.

2.3 Hyperbolic Derivative

The *hyperbolic derivative* of a self-map ϕ of $\mathbb{D}$ is the quantity

$$\phi^*(z) = \frac{(1 - |z|^2)\phi'(z)}{1 - |\phi(z)|^2}.$$

The Schwarz-Pick lemma tells us that $|\phi^*(z)| \le 1$ for every self-map ϕ of $\mathbb{D}$, and equality holds at some (and therefore at every point) of $\mathbb{D}$ if and only if ϕ is a disk automorphism.

The following result due to M. Heins [6] will be fundamental. We only state one part of it which is relevant for our purpose.

Theorem 1 (Heins Theorem) *An analytic function ϕ from $\mathbb{D}$ into itself is a finite Blaschke product if and only if*

$$\lim_{|z| \to 1^-} \left|\phi^*(z)\right| = 1.$$

2.4 The Bloch Space

The *Bloch space* $\mathcal{B}$ is defined as the set of all analytic functions in the unit disk $\mathbb{D}$ whose *invariant derivative* $(1 - |z|^2) f'(z)$ is bounded. It is a Banach space when equipped with the norm

$$\|f\|_{\mathcal{B}} = \left|f(0)\right| + s(f) = \left|f(0)\right| + \sup_{z \in \mathbb{D}} (1 - |z|^2) \left|f'(z)\right|.$$

Obviously, the quantity $s(f)$ defined above is a seminorm in $\mathcal{B}$ which becomes a true norm in the quotient space $\mathcal{B}/\mathbb{C}$, the Bloch space modulo constants.

Any function in $\mathcal{B}$ satisfies the following growth condition:

$$\left|f(\alpha) - f(0)\right| \leq \varrho(0,\alpha)\, \|f\|_{\mathcal{B}}, \quad \alpha \in \mathbb{D}. \tag{1}$$

This is easily deduced by integrating the inequality $|f'(z)| \leq \|f\|_{\mathcal{B}}(1-|z|^2)^{-1}$ along the line segment from 0 to z.

2.5 Analytic Besov Spaces

Denote by dA the normalized Lebesgue area measure in $\mathbb{D}$:

$$dA(z) = \frac{1}{\pi}\, dx\, dy = \frac{1}{\pi}\, r\, dr\, d\theta, \quad z = x + yi = r\, e^{i\theta} \in \mathbb{D}.$$

The *Dirichlet space* $\mathcal{D}$ is the Hilbert space of analytic functions in $\mathbb{D}$ with a square integrable derivative, with the natural norm given by

$$\|f\|_{\mathcal{D}}^2 = \left|f(0)\right|^2 + \int_{\mathbb{D}} \left|f'(z)\right|^2 dA(z).$$

More generally, the *analytic Besov space* B^p is defined for $1 < p < \infty$ as the set of all analytic functions in the disk such that

$$\|f\|_{B^p}^p = \left|f(0)\right|^p + (p-1)\int_{\mathbb{D}} \left|f'(z)\right|^p \left(1-|z|^2\right)^{p-2} dA(z) < \infty.$$

The second summand on the right is only a seminorm in B^p but is, of course, a true norm in the quotient space $B^p/\mathbb{C}$, the space B^p modulo constants. These quotient spaces $B^p/\mathbb{C}$ are *conformally invariant*: $\|f \circ \phi\|_{B^p} = \|f\|_{B^p}$ for all disk automorphisms ϕ.

The space B^1 cannot be defined as above but there is a naturally related way of defining it as the space of functions analytic in $\mathbb{D}$ whose second derivative is area-integrable, equipped with the following norm:

$$\|f\|_{B^1} = \left|f(0)\right| + \left|f'(0)\right| + \int_{\mathbb{D}} \left|f''(z)\right| dA(z).$$

This norm is not conformally invariant but the space itself is. It is well-known that the corresponding quotient space $B^1/\mathbb{C}$ (often also denoted by $\mathcal{M}$) is minimal in the sense of inclusion among all "reasonable" conformally invariant spaces; see [2] or [11]. An alternative definition of this minimal space can be given in terms of an "atomic decomposition": it is the space of all analytic functions in $\mathbb{D}$ which can be represented as

$$f = \sum_{k=1}^{\infty} c_k \phi_{\alpha_k} \tag{2}$$

for some absolutely summable sequence (c_k) and some sequence (α_k) in $\mathbb{D}$, where ϕ_{α_k} are the corresponding disk automorphisms as defined earlier. The norm on $\mathcal{M} = B^1/\mathbb{C}$ is given by

$$\|f\|_{B^1/\mathbb{C}} = \inf \sum_{k=1}^{\infty} |c_k|,$$

the infimum being taken over all possible representations as in (2).

The Bloch space $\mathcal{B}$ can be understood as the limit case of B^p as $p \to \infty$ (and $\mathcal{B}/\mathbb{C}$ as the limit case of $B^p/\mathbb{C}$). We refer the reader to [2] or [11].

2.6 Change of Variables

We will need a change of variables formula from measure theory which is standard in the theory of composition operators (see [2] or [4, Theorem 2.32]): if g and W are two non-negative functions defined on $\mathbb{D}$, measurable with respect to dA and ϕ is a self-map of $\mathbb{D}$, then

$$\int_{\mathbb{D}} g(\phi(z)) |\phi'(z)|^2 W(z) dA(z) = \int_{\phi(\mathbb{D})} g(w) \left(\sum_{j \geq 1} W(z_j(w)) \right) dA(w),$$

where $z_j(w)$ are the zeros of $\phi(z) - w$ repeated according to multiplicity. We will actually only need the case when $\phi = B$, a finite Blaschke product of degree n, and

$$g = |f'|^p, \quad f \in B^p, \qquad W(z) = \left((1 - |z|^2) |B'(z)| \right)^{p-2}.$$

We formulate the statement in this special case as a lemma.

Lemma 2 *If B is a Blaschke product of degree n and $f \in B^p$ then*

$$\int_{\mathbb{D}} |(f \circ B)'(z)|^p (1 - |z|^2)^{p-2} dA(z)$$
$$= \int_{\mathbb{D}} |f'(w)|^p \sum_{j=1}^{n} \left((1 - |z_j(w)|^2) |B'(z_j(w))| \right)^{p-2} dA(w),$$

where for any $w \in \mathbb{D}$ we denote by $z_j(w)$, $1 \leq j \leq n$, the n pre-images of w by B.

2.7 Composition Operators on Conformally Invariant Spaces

It is an easy consequence of the Schwarz-Pick lemma that, for every self-map ϕ of $\mathbb{D}$, the induced composition operator C_ϕ maps $\mathcal{B}$ boundedly into itself. However, it need not map any B^p into itself. There are, of course, symbols ϕ for which this is always true. For example, every finitely valent symbol and, in particular, every finite Blaschke product induces a bounded composition operator on every B^p, $1 \le p \le \infty$.

It should be observed that since the composition operators fix constant functions, every composition operator C_ϕ defined on B^p induces a well-defined composition operator on $B^p/\mathbb{C}$, denoted again by C_ϕ in order not to burden the notation. The same holds for the quotient Bloch space $\mathcal{B}/\mathbb{C}$.

3 Norms of Composition Operators on Quotient Besov Spaces

The norms of arbitrary composition operators on the quotient Bloch space $\mathcal{B}/\mathbb{C}$ have been computed precisely by Montes-Rodríguez [9]. Combining his result with Heins' Theorem 1, we get the following.

Theorem 2 (Heins and Montes-Rodríguez Theorem) *For any holomorphic self-map ϕ of the disk, the norm of the composition operator C_ϕ on the quotient Bloch space $\mathcal{B}/\mathbb{C}$ is given by*

$$\|C_\phi\|_{\mathcal{B}/\mathbb{C}\to\mathcal{B}/\mathbb{C}} = \sup_{z\in\mathbb{D}} |\phi^*(z)|.$$

In particular, for every finite Blaschke product B we have

$$\|C_B\|_{\mathcal{B}/\mathbb{C}\to\mathcal{B}/\mathbb{C}} = 1.$$

Using interpolation, among other techniques, Arazy, Fisher, and Peetre [3] proved the following result about the norms of finite Blaschke products in quotient Besov spaces. (Of course, they were formulated in terms of the seminorm in B^p, which is completely equivalent.)

Theorem 3 (Arazy, Fisher, and Peetre Theorem) *If $1 \le p \le \infty$, then there exist absolute positive constants m_p and M_p such that*

$$m_p n^{1/p} \le \|B\|_{B^p/\mathbb{C}} \le M_p n^{1/p},$$

for any finite Blaschke product B of degree n.

In other words, the constants above are independent of the degree and location of the zeros of B.

The key point which will do the hard work for us is another theorem from the influential paper of Arazy, Fisher, and Peetre mentioned [2, Theorem 13]. Recall that, by definition, μ is a (B^p, p)*-Carleson measure* if there exists a universal constant M such that

$$\int_{\mathbb{D}} |f'|^p \, d\mu \le M \int_{\mathbb{D}} |f'(z)|^p (1-|z|^2)^{p-2} \, dA(z)$$

for all $f \in B^p$. It is customary in the theory of Carleson measures to consider the so-called *Carleson windows*:

$$S(h,\theta) = \{re^{it} \in \mathbb{D} : |re^{it} - e^{i\theta}| < h\}, \quad \theta \in [0,2\pi),\ h > 0.$$

It was shown in [2, Theorem 13] that the following conditions are equivalent whenever $1 < p < \infty$:

(i) μ is a (B^p, p)-Carleson measure;
(ii) there is a constant a such that $\mu(S(h,\theta)) \le ah^p$ for all $h \in (0,1)$ and $\theta \in [0,2\pi)$;
(iii) there is a constant b such that $\int_{\mathbb{D}} |\phi'_\alpha|^p d\mu \le b$ for all $\alpha \in \mathbb{D}$;
(iv) there is a constant c such that $(\int_{\mathbb{D}} |f'|^p d\mu)^{1/p} \le c\|f\|_1$ for all $f \in B^1$.

In particular, this statement allows us to check that finite Blaschke products induce bounded composition operators; see [2]. However, we will need this result stated in a little more precise form. A careful inspection of the proof of the above theorem reveals the following. If (i) holds, that is, if μ is a (B^p, p)-Carleson measure then (iii) holds. Writing

$$b = \sup_{\alpha\in\mathbb{D}} \int_{\mathbb{D}} |\phi'_\alpha|^p d\mu,$$

their proof of (iii) $\Rightarrow$ (ii) [2, p. 129] shows that $\mu(S(h,\theta)) \le 5^p b \cdot h^p$; that is,

$$\mu\big(S(h,\theta)\big) \le \left(5^p \sup_{\alpha\in\mathbb{D}} \int_{\mathbb{D}} |\phi'_\alpha|^p d\mu\right) \cdot h^p.$$

Also, the proof of the implication (ii) $\Rightarrow$ (i) in [2, pp. 129–130] shows that

$$\int_{\mathbb{D}} |f'|^p \, d\mu \le 9 \cdot 2^p \cdot a \cdot \|f\|^p_{B^p/\mathbb{C}},$$

where a is the constant from the previous bound: $a = 5^p \sup_{\alpha\in\mathbb{D}} \int_{\mathbb{D}} |\phi'_\alpha|^p d\mu$. We summarize this in one single inequality as a theorem.

Theorem 4 (Arazy, Fisher, and Peetre Theorem) *Let* $1 < p < \infty$. *Whenever* f *is analytic in* $\mathbb{D}$ *and* $\|f\|^p_{B^p/\mathbb{C}} = 1$, *we have*

$$\int_{\mathbb{D}} |f'|^p \, d\mu \le 9 \cdot 10^p \sup_{\alpha\in\mathbb{D}} \int_{\mathbb{D}} |\phi'_\alpha|^p d\mu.$$

Using this key result, we can now easily prove our theorem on the norm of the composition operator C_B on the quotient Besov spaces.

Theorem 5 *Given a finite Blaschke product* B *of degree* n, *the norm of the composition operator* C_B *induced by it on the quotient analytic Besov space* $B^p/\mathbb{C}$, $1 \le p \le \infty$, *satisfies the inequality*

$$m_p n^{1/p} \le \|C_B\|_{B^p/\mathbb{C} \to B^p/\mathbb{C}} \le M_p n^{1/p},$$

where the constants m_p and M_p (not necessarily the same as in Theorem 3) depend only on p but not on B.

Proof Denote by Id the identity map: $Id(z) = z$. The lower inequality follows directly from the facts that $C_B(Id) = B$, $\|Id\|_{B^p/\mathbb{C}} = 1$ and from Theorem 3:

$$\|C_B\| \geq \left\|C_B(Id)\right\|_{B^p/\mathbb{C}} = \|B\|_{B^p/\mathbb{C}} \geq m_p n^{1/p}.$$

The upper inequality has to be discussed case by case. Understanding B^∞ as $\mathcal{B}$, the result in this case is the combination of the theorems of Montes-Rodríguez and Heins mentioned earlier: $\|C_B\|_{\mathcal{B}/\mathbb{C}\to\mathcal{B}/\mathbb{C}} = 1$; that is, $m_\infty = M_\infty = 1$.

The upper inequality when $2 \leq p < \infty$ can be deduced in various manners. The simplest proof follows directly from the Schwarz-Pick lemma:

$$\left(1-|z|^2\right)\left|B'(z)\right| \leq 1 - \left|B(z)\right|^2,$$

the change of variable $B(z) = w$ and the n-fold covering property of B:

$$\begin{aligned}
\|C_B f\|^p_{B^p/\mathbb{C}} &= (p-1)\int_{\mathbb{D}} \left|f'\left(B(z)\right)\right|^p \left|B'(z)\right|^p \left(1-|z|^2\right)^{p-2} dA(z) \\
&\leq (p-1)\int_{\mathbb{D}} \left|f'\left(B(z)\right)\right|^p \left|B'(z)\right|^2 \left(1-\left|B(z)\right|^2\right)^{p-2} dA(z) \\
&= n(p-1)\int_{\mathbb{D}} \left|f'(w)\right|^p \left(1-|w|^2\right)^{p-2} dA(w) \\
&= n\|f\|^p_{B^p/\mathbb{C}}.
\end{aligned}$$

The most delicate case of the upper inequality is $1 < p < 2$. This is where an analysis like in Theorem 4 of Arazy, Fisher, and Peetre is needed (of course, it also works when $p \geq 2$ but we have preferred a simpler proof as above). First of all, we need to produce an appropriate Carleson measure. Applying the formula from Lemma 2, we can define

$$d\mu(w) = \sum_{j=1}^{n} \left(\left(1-\left|z_j(w)\right|^2\right)\left|B'\left(z_j(w)\right)\right|\right)^{p-2} dA(w),$$

so that

$$\|f \circ B\|^p_{B^p/\mathbb{C}} = \int_{\mathbb{D}} |f'|^p \, d\mu$$

by Lemma 2. We know that μ defined in this way is a Carleson measure because it was already shown in [2] that finite Blaschke products induce bounded composition operators on B^p. In view of this, by Theorem 4 we have

$$\int_{\mathbb{D}} |f'|^p \, d\mu \leq 9 \cdot 10^p \sup_{\alpha \in \mathbb{D}} \int_{\mathbb{D}} |\phi'_\alpha|^p \, d\mu$$

for all $f \in B^p$ of unit norm. Since $\phi_\alpha \circ B$ is again a Blaschke product of degree n by Lemma 1, it follows from the change of variable (applied twice), the above inequality, and Theorem 3 that

$$\|f \circ B\|^p_{B^p/\mathbb{C}} = \int_{\mathbb{D}} |f'|^p \, d\mu \le 9 \cdot 10^p \sup_{\alpha \in \mathbb{D}} \int_{\mathbb{D}} |\phi'_\alpha|^p d\mu = 9 \cdot 10^p \sup_{\alpha \in \mathbb{D}} \|\phi_\alpha \circ B\|^p_{B^p/\mathbb{C}}$$
$$\le 9 \cdot 10^p K\, n = M_p^p\, n,$$

for every function f of norm one in $B^p/\mathbb{C}$. This shows that

$$\|C_B\|_{B^p/\mathbb{C} \to B^p/\mathbb{C}} \le M_p n^{1/p}.$$

Finally, in the case $p = 1$ we use the atomic decomposition. Writing f as in (2), we get

$$f \circ B = \sum_{k=1}^{\infty} c_k (\phi_{a_k} \circ B) = \sum_{k=1}^{\infty} c_k B_k,$$

where each B_k is again a Blaschke product of degree n by Lemma 1. The generalized triangle inequality for the norms and Theorem 3 yield

$$\|f \circ B\|_{B^1/\mathbb{C}} \le \sum_{k=1}^{\infty} |c_k| \cdot \|B_k\|_{B^1/\mathbb{C}} \le \sum_{k=1}^{\infty} |c_k| \cdot Mn.$$

Taking the infimum over all possible representations of f, we get

$$\|f \circ B\|_{B^1/\mathbb{C}} \le Mn \|f\|_{B^1/\mathbb{C}}.$$

This completes the proof. □

4 Exact Norm Computations on the True Bloch and Dirichlet Spaces

The problem of norm computation becomes more delicate when one considers the true analytic Besov spaces, taking the constants into account. In this section we compute the norms of C_B on two distinguished spaces in the scale of B^p spaces: on the Bloch space $\mathcal{B}$ and on the Dirichlet space $\mathcal{D}$.

4.1 Norms on the True Bloch Space

We first compute the norms of C_B as an operator acting on the Bloch space $\mathcal{B}$ for any finite Blaschke product B. For some further norm estimates of more general composition operators on $\mathcal{B}$ in terms of the hyperbolic derivative of the symbol, the reader is referred to [10], [7, Chap. 2], and [1].

Note that the result below states that to the expression already known from Theorem 6 for the quotient space one should also add a value involving $B(0)$ and the hyperbolic distance as a basic measure of growth of Bloch functions. We use the notation defined in Sects. 2.2 and 2.4.

Theorem 6 *Let B be a finite Blaschke product. Then, as operators acting on the true Bloch space, $\|C_B\|_{\mathcal{B}\to\mathcal{B}} = \varrho(0, B(0)) + 1$, for any finite Blaschke product B, no matter what its degree is.*

Proof Let f be a function of unit norm in $\mathcal{B}$ and $\alpha = B(0)$. By the Schwarz-Pick lemma and the basic growth estimate (1), we have

$$\begin{aligned}
\|f \circ B\|_{\mathcal{B}} &= \big|f(\alpha)\big| + \sup_{z\in\mathbb{D}}\big(1-|z|^2\big)\big|B'(z)\big|\big|f'\big(B(z)\big)\big| \\
&\le \big|f(\alpha)\big| + \sup_{z\in\mathbb{D}}\big(1-\big|B(z)\big|^2\big)\big|f'\big(B(z)\big)\big| \\
&\le \big|f(\alpha)-f(0)\big| + \big|f(0)\big| + s(f) \\
&\le \big|f(\alpha)-f(0)\big| + 1 \\
&\le \varrho(0,\alpha)+1.
\end{aligned}$$

The inequality $\|C_B\|_{\mathcal{B}\to\mathcal{B}} \le \varrho(0,\alpha)+1$ follows.

For the reverse inequality $\|C_B\|_{\mathcal{B}\to\mathcal{B}} \ge \varrho(0,\alpha)+1$, it is convenient to distinguish between two cases. When $\alpha = 0$, taking into account that composition operators fix the constant function one, it follows immediately that $\|C_B\|_{\mathcal{B}\to\mathcal{B}} \ge 1$. This gives the desired equality in this special case.

For all other values $\alpha \in \mathbb{D}\setminus\{0\}$, we can consider the function

$$f(z) = \frac{1}{2}\log\frac{1+\lambda z}{1-\lambda z}, \qquad \lambda = \overline{\alpha}/|\alpha|.$$

Since B is a finite Blaschke product, we know that $B(\mathbb{D}) = \mathbb{D}$. Hence for an arbitrary increasing sequence $(r_n)_{n=1}^{\infty}$ of positive numbers such that $\lim_{n\to\infty} r_n = 1$ we can find points z_n in $\mathbb{D}$ for which $B(z_n) = \overline{\lambda}\sqrt{r_n}$, hence $\lambda^2 B^2(z_n) = |B(z_n)|^2$. By the Chain Rule, the definition of the Bloch norm, and the choice of our function f, we have

$$\begin{aligned}
\|f \circ B\|_{\mathcal{B}} &= \varrho(0,\alpha) + \sup_{z\in\mathbb{D}} \frac{(1-|z|^2)|B'(z)|}{|1-\lambda^2 B^2(z)|} \\
&\ge \varrho(0,\alpha) + \limsup_{n\to\infty} \frac{(1-|z_n|^2)|B'(z_n)|}{|1-\lambda^2 B^2(z_n)|}
\end{aligned}$$

$$= \varrho(0, \alpha) + \limsup_{n \to \infty} \frac{(1 - |z_n|^2)|B'(z_n)|}{1 - |B(z_n)|^2}$$

$$= \varrho(0, \alpha) + 1,$$

where the last inequality follows from Heins' Theorem 1. □

4.2 Norms on the True Dirichlet Space

Following an idea from our earlier paper [8], we can also compute the norms of C_B on the Dirichlet space for an arbitrary finite Blaschke product B. As one would expect, the norm is comparable to the square root of the degree of B and also takes into account the value $B(0)$. The quantity L defined below appears because Dirichlet functions grow at most like the square root of the logarithm.

Theorem 7 *Let B be a finite Blaschke product of degree n and write*

$$L = \log \frac{1}{1 - |B(0)|^2}.$$

Then we have the formula

$$\|C_B\|_{\mathcal{D} \to \mathcal{D}} = \sqrt{\frac{n + L + 1 + \sqrt{(n + L - 1)^2 + 4L}}{2}}.$$

The norm is attained for the following function F of norm one:

$$F(z) = \sqrt{1 - K^2 L} + K \log \frac{1}{1 - \overline{B(0)} z},$$

where the constant K is chosen so that

$$K^2 = \frac{1}{2L} + \frac{1}{2L} \frac{n + L - 1}{\sqrt{(n + L - 1)^2 + 4L}}.$$

Proof Let us write again $\alpha = B(0)$. Given $f \in \mathcal{D}$, we can use the change of variable $w = B(z)$ and Lemma 2 as before to get:

$$\|f \circ B\|_{\mathcal{D}}^2 = |f(\alpha)|^2 + n \int_{\mathbb{D}} |f'|^2 dA = \left(|f(\alpha)|^2 - n|f(0)|^2\right) + n\|f\|_{\mathcal{D}}^2.$$

Restricting to the functions of unit norm, we obtain

$$\|C_B\|_{\mathcal{D} \to \mathcal{D}}^2 = \sup\left\{n + |f(\alpha)|^2 - n|f(0)|^2 : \|f\|_{\mathcal{D}} = 1\right\}.$$

Let $f(z)=\sum_{k=0}^{\infty} c_k z^k$ be the Taylor series expansion of f in $\mathbb{D}$. Clearly, $c_0 = f(0)$. Recalling our assumptions that

$$B(0)=\alpha, \qquad \|f\|_{\mathcal{D}}^2 = |c_0|^2 + \sum_{k=1}^{\infty} k|c_k|^2 = 1, \tag{3}$$

and using the triangle and Cauchy-Schwarz inequalities, we get

$$\begin{aligned} & n + \left|f(\alpha)\right|^2 - n\left|f(0)\right|^2 \\ & = n - n|c_0|^2 + \left|c_0 + \sum_{k=1}^{\infty} c_k \alpha^k\right|^2 \\ & \le n - n|c_0|^2 + \left(|c_0| + \sum_{k=1}^{\infty} \sqrt{k}|c_k| \cdot \frac{|\alpha|^k}{\sqrt{k}}\right)^2 & (4) \\ & \le n - n|c_0|^2 + \left(|c_0| + \sqrt{\sum_{k=1}^{\infty} k|c_k|^2} \cdot \sqrt{\log \frac{1}{1-|\alpha|^2}}\right)^2 . & (5) \end{aligned}$$

Next, we give a precise upper bound on the quantity

$$n - n|c_0|^2 + \left(|c_0| + \sqrt{\sum_{k=1}^{\infty} k|c_k|^2} \cdot \sqrt{\log \frac{1}{1-|\alpha|^2}}\right)^2$$

subject to the conditions (3). To simplify the notation, let us write

$$x = \sum_{k=1}^{\infty} k|c_k|^2 \quad \text{and} \quad L = \log \frac{1}{1-|\alpha|^2}.$$

Keeping in mind that $|c_0|^2 = 1 - x$, the problem reduces to maximizing the function of one variable:

$$\Phi(x) = n - n(1-x) + \left(|c_0| + \sqrt{L}\sqrt{x}\right)^2 = 1 + (n+L-1)x + 2\sqrt{L}\sqrt{x-x^2}.$$

Since

$$\Phi'(x) = n + L - 1 + \sqrt{L}\frac{1-2x}{\sqrt{x-x^2}}, \qquad \Phi''(x) = \frac{-\sqrt{L}}{2(x-x^2)^{3/2}} < 0 \quad (0<x<1),$$

we know by elementary calculus that the function Φ achieves its maximum value at its only critical point:

$$x_0 = \frac{1}{2} + \frac{1}{2}\frac{n+L-1}{\sqrt{(n+L-1)^2+4L}},$$

hence

$$\Phi(x) \le \Phi(x_0) = \frac{n+L+1+\sqrt{(n+L-1)^2+4L}}{2}, \quad \text{for all } x \in [0,1]. \tag{6}$$

This yields the desired upper bound:

$$\|C_B\|^2_{\mathcal{D}\to\mathcal{D}} \le \frac{n+L+1+\sqrt{(n+L-1)^2+4L}}{2}.$$

It is now only left to verify that equality can actually hold throughout the chain of inequalities (4), (5), and (6) obtained in the process. Equality is obtained in (4) and (5) by choosing the Taylor coefficients of f to be

$$c_k = K\frac{\overline{\alpha}^k}{k}, \quad k = 1, 2, 3, \ldots,$$

with $K > 0$ and $c_0 > 0$, so that also $c_k\alpha^k > 0$ for all $k \ge 0$. This yields the function

$$F_\alpha(z) = c_0 + \sum_{k=1}^{\infty} \frac{(\overline{\alpha}z)^k}{k} = c_0 + K\log\frac{1}{1-\overline{\alpha}z}.$$

The exact value of c_0 ought to be chosen as follows:

$$c_0 = F_\alpha(0) = \sqrt{1-K^2L} = \sqrt{1-K^2\log\frac{1}{1-|\alpha|^2}},$$

and we still have the freedom of choosing K so as to get

$$K^2 = \frac{1}{2L} + \frac{1}{2L}\frac{n+L-1}{\sqrt{(n+L-1)^2+4L}}.$$

This guarantees that the maximum of $\Phi(x)$, as in (6), will be achieved at the value x_0 indicated above. The proof is now complete. □

Acknowledgements The authors thankfully acknowledges partial support from MICINN grant MTM2009-14694-C02-01, Spain. The second author was also partially supported by the European ESF Network HCAA ("Harmonic and Complex Analysis and Its Applications").

References

1. Allen, R.F., Colonna, F.: On the isometric composition operators on the Bloch space in $\mathbb{C}^n$. J. Math. Anal. Appl. **355**(2), 675–688 (2009)
2. Arazy, J., Fisher, S.D., Peetre, J.: Möbius invariant function spaces. J. Reine Angew. Math. **363**, 110–145 (1985)
3. Arazy, J., Fisher, S.D., Peetre, J.: Besov Norms of Rational Functions. Lecture Notes Math., vol. 1302, pp. 125–129. Springer, Berlin (1988)

4. Cowen, C., MacCluer, B.: Composition Operators on Spaces of Analytic Functions. Studies in Advanced Mathematics. CRC Press, Boca Raton (1995)
5. Garnett, J.B.: Bounded Analytic Functions. Academic Press, San Diego (1981). Reprinted by Springer, New York (2010)
6. Heins, M.: Some characterizations of finite Blaschke products of positive degree. J. Anal. Math. **46**, 162–166 (1986)
7. Martín, M.J.: Composition operators and geometric function theory (Spanish), 138 pp. Doctoral Thesis, Universidad Autónoma de Madrid (December 2005)
8. Martín, M.J., Vukotić, D.: Norms and spectral radii of composition operators acting on the Dirichlet space. J. Math. Anal. Appl. **304**, 22–32 (2005)
9. Montes-Rodríguez, A.: The Pick-Schwarz lemma and composition operators on Bloch spaces. Rend. Circ. Mat. Palermo (2) Suppl. **56**, 167–170 (1998). International Workshop on Operator Theory (Cefalù, 1997)
10. Xiong, C.: Norm of composition operators on the Bloch space. Bull. Aust. Math. Soc. **70**(2), 293–299 (2004)
11. Zhu, K.: Analytic Besov spaces. J. Math. Anal. Appl. **157**, 318–336 (1991)

On the Computable Theory of Bounded Analytic Functions

Timothy H. McNicholl

Abstract The theory of bounded analytic functions is reexamined from the viewpoint of computability theory.

Keywords Computability theory · Computable analysis · Complex analysis · Blaschke products

Mathematics Subject Classification Primary 03F60 · Secondary 30J05 · 30J10

1 Introduction

The first electronic computers were produced in the mid-20th century. At the time of this writing, it is likely that the reader has in his pocket a computing device that dwarfs these early machines in computational power. Despite the astonishing progress and dazzling possibilities for the future, it is worth taking a step back and noting that the mathematical limitations of such computing devices were established by A.M. Turing in 1936 [25]. The resulting mathematical theory is known as *computability theory* and is a key component of the *theory of computation*. Since Turing's seminal work, computability theory has also experienced a pullulating development.

One exciting application of computability is to reexamine a mathematical theory from the viewpoint of computability. For example, Boone's demonstration that the word problem is incomputable [2]. Such investigations not only yield novel results which lead us to reconsider the nature of a theory, but also illuminate fundamental limitations and possibilities for its practical applications.

Here, we survey a selection of results on the computable theory of bounded analytic functions; in particular, with regards to Blaschke products and inner functions. In doing so, we will be led on a tour of many of the fundamental and beautiful results of computability theory and *computable analysis*. That is, the theory of computation with continuous data.

T.H. McNicholl (✉)
Department of Mathematics, Iowa State University, Ames, IA 50011, USA
e-mail: mcnichol@iastate.edu

J. Mashreghi, E. Fricain (eds.), *Blaschke Products and Their Applications*,
Fields Institute Communications 65, DOI 10.1007/978-1-4614-5341-3_13,

The paper is organized as follows. In Sect. 2, we give a brief introduction to classical computability theory. In Sect. 3, we introduce the fundamental notions of computable analysis specialized to the unit disk. Sections 4 and 5 contain the results on Blaschke products. Inner functions are treated in Sects. 7 and 8. Proofs will be presented only for the sake of elucidating some broader point or to improve earlier work.

Before continuing, let us settle a few matters of notation. Let $\mathbb{N}$ denote the set of natural numbers by which we mean the set of all non-negative integers. We write $f :\subseteq A \to B$ when f is a function whose domain is contained in A and whose range is contained in B. If $\mathrm{dom}(f) \subset A$, then we say f is a *partial function from A into B*. We denote a countably infinite sequence $a_0, a_1, a_2, \ldots$ by $\{a_n\}_{n\in\mathbb{N}}$.

2 Basic Computability Theory

We summarize here just those elements of computability theory necessary for our exposition on bounded analytic functions. A more expansive development can be found in [4].

Broadly speaking, computability theory is a mathematical theory that delineates the limitations and potential of discrete computing devices such as the modern digital computer. When it is determined that some given problem can be solved by such a device, the investigation usually turns to limitations on efficiency. That is, what are the bounds on how quickly a computer can solve the problem or how much memory must be used. See, e.g. [1]. But, if it is determined that no such device can solve the problem, research may turn to ranking its unsolvability relative to other such problems. Such ranking methods lead to development of *degree theory* [13, 22].

Although physical computers are never far out of sight, the fundamental notion of computability theory is that of an algorithm (or 'program' for computation). Loosely speaking, an algorithm is a procedure that consists of a sequence of steps that can be followed without thinking. The viewpoint of computability theory is that computers are simply devices for implementing algorithms. Thus, the primary objects of study of computability theory are algorithms not computers.

2.1 Computable Functions and Sets

The fundamental definition of computability theory then is that of a computable function which we render as follows.

Definition 1 A function $f :\subseteq \mathbb{N}^k \to \mathbb{N}$ is *computable* if there is an algorithm that, given any $(n_1, \ldots, n_k) \in \mathbb{N}^k$ as input, produces $f(n_1, \ldots, n_k)$ as output if $(n_1, \ldots, n_k) \in \mathrm{dom}(f)$ and does not halt if $(n_1, \ldots, n_k) \notin \mathrm{dom}(f)$.

Fig. 1

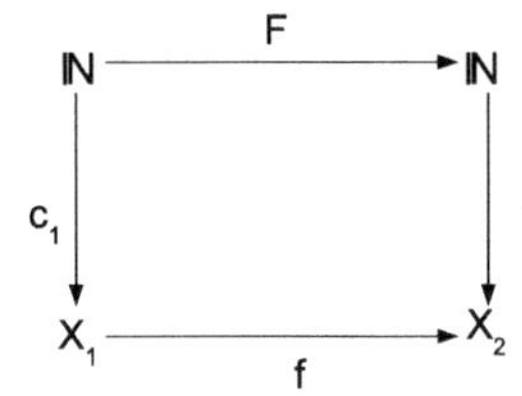

Examples of computable functions are addition, multiplication, and division (where $a \div b$ is declared to be undefined if b does not divide a). For example, the following algorithm computes the division function; (a, b) denotes the pair given as input.

Step 1: Set $j = 1$.
Step 2: If $b \times j = a$, then stop and output j; otherwise set $j = j + 1$ and repeat this step.

In fact, it is likely that any function the reader can think of is computable. Nevertheless, most functions from $\mathbb{N}^k$ into $\mathbb{N}$ are incomputable. A more explicit statement and demonstration of this point will be given later.

The non-halting condition in Definition 1 might raise some eyebrows as well since failure to halt is usually seen as programmer error. For example, in the division algorithm just presented, the search could be terminated when the counter has exceeded the first input value. Nevertheless, there are some functions whose computation requires the use of a search procedure for which no such test can be added so as to avoid searches that do not terminate. This point will be made more explicit and demonstrated in the next subsection.

Of course, the purview of algorithms ranges far beyond the natural numbers. For example, we also want to compute with the rational numbers. We transfer computability concepts from $\mathbb{N}$ to other domains by means of *codings*. Formally, a coding of a set X is a surjection $c :\subseteq \mathbb{N} \to X$. When $c(n) = x$, we refer to n as a *code* of x (with respect to c). When X_1 and X_2 are sets for which we have established codings c_1 and c_2 respectively, and when $f :\subseteq X_1 \to X_2$, then we say that f is *computable with respect to* c_1 *and* c_2 if there is a computable $F :\subseteq \mathbb{N} \to \mathbb{N}$ such that the diagram in Fig. 1 commutes. More precisely, $fc_1(n) = c_2F(n)$ whenever $c_1(n) \in \text{dom}(f)$, and $n \notin \text{dom}(F)$ whenever $c_1(n) \notin \text{dom}(f)$.

Let $\langle , \rangle : \mathbb{N}^2 \to \mathbb{N}$ denote Cantor's bijection of $\mathbb{N}^2$ with $\mathbb{N}$. That is,

$$\langle m, n \rangle = \frac{1}{2}\big((m+n)^2 + 3m + n\big).$$

Let $()_0 : \mathbb{N} \to \mathbb{N}$ and $()_1 : \mathbb{N} \to \mathbb{N}$ denote the inverse functions of $\langle , \rangle$. That is, $\langle (n)_0, (n)_1 \rangle = n$ for all n. By elaborating on Cantor's bijection of $\mathbb{N}$ onto $\mathbb{N} \times \mathbb{N}$, we can establish natural codings of the integers, the rational numbers, the set of all isomorphism equivalence classes of finite graphs, etc. We can thereby transfer computability to all these domains. There are, of course, numerous codings of each of these sets. However, all "reasonable" codings yield the same class of computable functions. Thus, henceforth we shall not mention them. Furthermore, we often identify objects with their codes.

In addition, by means of codings, we can reduce the consideration of multivariable functions to single variable functions. Namely, if we begin by fixing a coding c of $\mathbb{N}^k$, obtained by iterating Cantor's coding, and if for each $j = 1, \ldots, k$ we have established a coding c_j of a set X_j, then, by composition, we obtain a coding of $X_1 \times \cdots \times X_k$.

Before proceeding further, let us work through an example. Define a coding of $\mathbb{Z}$, $c_{\mathbb{Z}}$, by

$$c_{\mathbb{Z}}(n) = (n)_0 - (n)_1.$$

Then, a coding of $\mathbb{Z}^2$, $c_{\mathbb{Z}^2}$, is yielded by setting

$$c_{\mathbb{Z}^2}(n) = \big(c_{\mathbb{Z}}\big((n)_0\big), c_{\mathbb{Z}}\big((n)_1\big)\big).$$

To show that addition is a computable operation on $\mathbb{Z}$ with respect to this coding, set

$$F(n) = \big\langle \big((n)_0\big)_0 + \big((n)_1\big)_0, \big((n)_0\big)_1 + \big((n)_1\big)_1 \big\rangle.$$

Clearly, F is computable. Also, $c_{\mathbb{Z}}(F(n)) = c_{\mathbb{Z}}((n)_0) + c_{\mathbb{Z}}((n)_1)$. Thus, with respect to this coding, addition is a computable operation on $\mathbb{Z}$.

At this point, the reader may be disconcerted by the fact that no precise definition of "algorithm" has been proffered. Doubtless, we all have a good intuitive idea of what an algorithm is, but such is not a sound foundation for a mathematical theory. There have been numerous attempts to mathematically formalize the notion of 'algorithm'; or, at least, to mathematically formalize the class of computable functions. For example, Turing machines, partial recursive functions, flowchart computability, and unlimited register machines. The first three of these are described in [18], and the last is described in [5]. The unlimited register machine forms the theoretical basis for the modern computer.

What is miraculous is that all of these notions yield the same class of computable functions. For this reason, the notion of "computable function" is regarded as very stable, and most of the discipline of computability theory can be discussed without reference to any such formalization.

2.2 *Computable vs. Computably Enumerable Sets*

A useful distinction, and a theme which runs through many applications of computability theory, is that between the notion of a *computable* set and that of a *computably enumerable* (*c.e.*) set. The definitions are as follows.

Definition 2 A set $A \subseteq \mathbb{N}$ is *computable* if its characteristic function is computable.

Definition 3 A set $A \subseteq \mathbb{N}$ is *computably enumerable* if it is empty or if there is a sequence $\{a_n\}_{n=0}^{\infty}$ such that $A = \{a_0, a_1, \ldots\}$ and such that $n \mapsto a_n$ is a computable function. Such a sequence is called a *computable enumeration* of A.

A few immediate consequences of Definitions 2 and 3 are worth noting here. The first is that every finite set $A \subseteq \mathbb{N}$ is computable. This is trivial if A is empty (just output 0 on any input!). Suppose $A = \{b_1, \ldots, b_k\}$. Then, on input n, we search through all $j \in \{1, \ldots, k\}$ until either we exhaust the elements of this set or we find one such that $n = b_j$. In the latter case, we output 1 and in the former case we output 0.

Another consequence is that every computable set $A \subseteq \mathbb{N}$ is computably enumerable. For, if A is finite and non-empty, say $A = \{b_0, \ldots, b_k\}$, then we may write A as $A = \{a_0, a_1, \ldots\}$ where

$$a_n = \begin{cases} b_n & \text{if } n \leq k, \\ b_k & \text{if } n > k \end{cases}$$

and clearly $n \mapsto a_n$ is computable. If A is infinite, then we can define a_n to be the $(n+1)$-st element of A when its elements are listed in increasing order. Again, it follows that $n \mapsto a_n$ is computable.

However, we have the following.

Theorem 1 *There is a* c.e. *set that is not computable.*

We will discuss the proof of Theorem 1 in Sect. 2.3. Right now, we will use it to settle one pending issue. Namely, we can now demonstrate the necessity of the non-halting condition in Definition 1. For, let A be a *c.e.* and incomputable set. Clearly, A is not empty. So, let $\{a_n\}_{n=0}^{\infty}$ be a computable enumeration of A. Then, define a function $f :\subseteq \mathbb{N} \to \mathbb{N}$ by defining $f(n)$ to be 1 if $n \in A$ and declaring $f(n)$ to be undefined if $n \notin A$. It follows that f is computable: on input n, search for $k \in \mathbb{N}$ such that $n = a_k$ and output 1 as soon as one is found (and if there is no such k, then the algorithm will not halt!).

2.3 From Whence All This Comes: The Fundamental Theorem of Computability Theory

The following theorem underlies many of the claims we have made about computable functions and sets.

Theorem 2 (Fundamental Theorem of Computability Theory) *There is a computable function $U :\subseteq \mathbb{N} \times \mathbb{N} \to \mathbb{N}$ with the property that if $f :\subseteq \mathbb{N} \to \mathbb{N}$ is any computable function, then there is a number $e \in \mathbb{N}$ such that $f(n) = U(e, n)$ for all $n \in \mathbb{N}$.*

A precise proof of Theorem 2 requires the sort of framework alluded to at the end of Sect. 2.1 and which we wish to avoid. However, for those familiar with programming in a language like BASIC, C, or JAVA, it is possible to give a fairly convincing and concise proof sketch. To begin, we start with the premise that every algorithm can be coded in such a language. A program in such a language is a sequence of symbols. If we fix the language, then it is possible to produce a coding of all such sequences. The function U can then be viewed as an interpreter for this language.

It is now possible to give a proof of Theorem 1. Namely, let

$$K = \{e \in \mathbb{N} : (e, e) \in \mathrm{dom}(U)\}.$$

The set K is known as *Turing's Halting Set*. It is the domain of the computable function $n \mapsto U(n, n)$. It is not difficult to show that the domain of a computable function $g :\subseteq \mathbb{N} \to \mathbb{N}$ is a *c.e.* set. (Namely, assuming dom(g) is infinite, begin running the computations of $g(0), g(1), \ldots$ in parallel and let a_n be the $(n + 1)$st computation to halt.) Hence, K is *c.e.* If K were computable, then the function g that is defined to be 1 on all $n \in \mathbb{N} - K$ and is declared to be undefined on all other numbers, would be computable. But, if e is such that $g(n) = U(e, n)$ for all n, then we obtain

$$e \in K \quad \Leftrightarrow \quad e \in \mathrm{dom}(g) \quad \Leftrightarrow \quad e \notin K,$$

which is a contradiction.

Another important consequence of Theorem 2 is that there are countably many computable functions and sets. This makes precise the assertion that most partial functions from $\mathbb{N}$ into $\mathbb{N}$ and most subsets of $\mathbb{N}$ are not computable.

2.4 Uniform vs. Non-uniform Computability

Another theme which pervades much of computability theory and its applications is the distinction between uniform and non-uniform computability. While these notions have no overarching and precise definitions, they are exemplified by the following.

Definition 4 A family of sets $\{A_n\}_{n \in \mathbb{N}}$ is *uniformly computable* if there is an algorithm that given any $n, k \in \mathbb{N}$ as input determines if $k \in A_n$.

It is quite easy to construct a family of *computable* sets that is not uniformly computable. For, let $A \subseteq \mathbb{N}$ be any incomputable set. For each $n \in \mathbb{N}$, let

$$A_n = \begin{cases} \{1\} & n \in A, \\ \{0\} & n \notin A. \end{cases}$$

Since every finite set is computable, each A_n is. But, if the family $\{A_n\}_{n \in \mathbb{N}}$ were computable, then A would be as well.

The distinction between uniform and non-uniform computability generally arises when considering hypothetical claims of computability. That is, claims of the form

"If $A, B, C, \ldots$ are computable then so is X."

Such a claim is said to hold *uniformly* if there is an algorithm that, given as input algorithms for computing $A, B, C, \ldots$, produces an algorithm for computing X. In order to consider algorithms as input or output of other algorithms, one must assume they are represented in some formalism such as flowcharts or Turing machines as discussed in Sect. 2.1.

3 Computable Analysis on the Unit Disk

Roughly speaking, computable analysis is the theory of computation with *continuous* data, e.g. Euclidean space and its hyperspaces. The classical computability theory, which is what we have outlined in Sect. 2, only deals with discrete data. However, much scientific computation involves models in which the underlying data are assumed to be continuous. We can not use codings to bridge the gap since, for example, it follows from Cantor's Non-Denumerability Theorem that there is no coding of $\mathbb{R}$. However, if we are careful, we can bridge the gap by means of approximation of continuous data by discrete data. Mathematically speaking, this means we use the discipline of topology to bridge the gap. That is, a topology is viewed as a notion of approximation.

Despite the simple nature of this idea, there are several approaches to the foundations of computable analysis. See e.g. [3, 8, 10, 11, 19, 27]. However, when restricted to the unit disk, they all yield the same theory; that is, they produce the same classes of computable points, sequences, functions, etc. We will proceed via definitions which would be acceptable to all of these schools of thought.

To begin, let $D_r(z)$ denote the open disk with radius r and center z. If r is a rational number, and if z is a rational point (that is, a point whose coordinates are rational numbers), then $D_r(z)$ shall be referred to as an *open rational disk* and $\overline{D_r(z)}$ shall be referred to as a *closed rational disk*. Again, by elaborating on Cantor's coding of $\mathbb{N} \times \mathbb{N}$, we can produce a reasonable coding of the set of all open rational disks. We can now define what it means for a point in the plane to be computable.

Definition 5 We say that a point $z \in \mathbb{C}$ is *computable* if there is an algorithm that, given any $k \in \mathbb{N}$ as input, produces an open rational disk D that contains z and whose diameter is at most 2^{-k}.

In other words, this algorithm never tells us z exactly, but rather gives us a set of possible positions for z. With larger k as input, a narrower set of possibilities is obtained.

There are many examples of computable points. For instance, every rational point q is computable: on input k, simply output the open disk with center q and radius

$2^{-(k+1)}$. The irrational numbers π, $\sqrt{2}$, e are also computable. This is a consequence of the fact that there are numerous algorithms for computing their decimal expansions. By imitating the proofs that addition, multiplication, and division are continuous, it can be shown that the set of computable points is a subfield of $\mathbb{C}$. In fact, most points the reader is likely to think of are computable. But, again, there are only countably many computable points in the plane.

By extending this definition a little, we get the notion of a computable sequence of complex numbers.

Definition 6 A sequence of complex numbers $\{z_n\}_{n\in\mathbb{N}}$ is *computable* if there is an algorithm that, given any $n, k \in \mathbb{N}$ as input, produces an open rational disk D that contains z_n and whose radius is at most 2^{-k}.

It follows that if $\{z_n\}_{n\in\mathbb{N}}$ is computable, then each z_n is computable. The converse does not hold however. For, let $A \subseteq \mathbb{N}$ be any incomputable set. Let

$$z_n = \begin{cases} 1 & n \in A, \\ 0 & n \notin A. \end{cases}$$

Since each rational number is computable, each z_n is computable. However, if the sequence $\{z_n\}_{n\in\mathbb{N}}$ were computable, then there would be an algorithm that, given any $n \in N$ as input, computes an open rational disk of radius $1/2$ that contains z_n. Such a disk must contain exactly one of 0, 1. This yields an algorithm that computes the characteristic function of A—a contradiction.

Much has been written about the definition of 'computable function' on continuous spaces such as $\mathbb{C}$. We give a definition for functions on the unit disk which is equivalent to what would be obtained from almost any of the more general notions in the literature and which also agrees with the picture painted by practice.

Definition 7 A function $f : \mathbb{D} \to \mathbb{C}$ is *computable* if there is an algorithm P that has the following three properties.

- *Approximation*: Given as input an open rational disk $D_1 \subseteq \mathbb{D}$ such that $\overline{D_1} \subseteq \mathbb{D}$, either P does not halt, or it produces an open rational disk D_2 as output.
- *Correctness*: If P halts on D_1, and if D_2 is produced as output, then $f[D_1] \subseteq D_2$.
- *Convergence*: If U is a neighborhood of $f(z)$, then there is an open rational disk D_1 that contains z and such that when it is provided as input to P, P produces a disk D_2 that is contained in U.

Loosely speaking, the approximation property states that P maps approximations (to points) to approximations. Namely, the center of D_1 should be regarded as an approximation to a point $z \in \mathbb{D}$, and its radius should be viewed as an upper bound on the error of this approximation. Similarly, the center of D_2 should be regarded as an approximation of $f(z)$ and its radius should be viewed as an upper bound on the error. The correctness property states that if the input is an approximation to $z \in \mathbb{D}$, then the output must approximate $f(z)$. The third says that we must be able to obtain arbitrarily good approximations to $f(z)$ by providing sufficiently good approximations to z as input.

An easy, but possibly disconcerting, consequence of this definition is that every computable function is continuous. This may seem problematic. For example, the function

$$f(z) = \begin{cases} 0 & z = 0, \\ 1 & z \neq 0 \end{cases}$$

is discontinuous, but the simplicity of its definition may lead one to believe at first glance that it is computable. In fact, it would seem that in many languages a code fragment like

$$\text{if } (z = 0) \text{ output } 0 \text{ else output } 1$$

would compute f. But, this code must contend with approximations to z; presumably decimal expansions. Any reasonable theory of computation with continuous data must have a convergence criterion: the code should obtain arbitrarily good approximations to $f(z)$ from sufficiently good approximations to z. Furthermore, there ought to be some sort of *error control*: we should be able to compute a bound on the error in the output approximation, and these bounds should tend to zero as the input approximations become more accurate. Everything now turns on how the test in this code is evaluated. If z is approximated by $0.0\ldots 0 + 0.0\ldots 0i$, and if the number of decimal places is sufficiently large,then the code must produce an output approximation with error no larger than $1/2$. Presumably, this output is zero. But, better approximation to z, e.g. as would be obtained by more accurate observation of a physical quantity z represents might reveal that a further decimal place of the real part of z is 9, and 1 would be precluded as an output.

So, in fact, the continuity consequence of Definition 7 is no deficiency at all but a sharp clarification.

Another elementary consequence of the definitions presented thus far is that if $f : \mathbb{D} \to \mathbb{C}$ is computable, and if z is a computable point in $\mathbb{D}$, then $f(z)$ is computable.

As for examples, again it is likely that almost any function the reader will think of is computable. But again, most continuous functions on the disk are incomputable.

We will also need a definition of computability for functions from [0, 1] into $\mathbb{C}$. This is obtained by suitable modification of Definition 7 as follows. First, define an interval to be rational if its endpoints are rational numbers.

Definition 8 A function $f : [0, 1] \to \mathbb{C}$ is *computable* if there is an algorithm P that has the following three properties.

- *Approximation*: Given as input an open rational interval I_1, either P does not halt, or it produces an open rational disk D_2 as output.
- *Correctness*: If P halts on I_1, then $f[I_1 \cap [0, 1]] \subseteq D_2$.
- *Convergence*: If U is a neighborhood of $f(x)$, then there is an open rational interval I_1 that contains x and such that when it is provided as input to P, P produces a disk D_2 that is contained in U.

4 Some Basic Computability Results on Blaschke Products

Whenever $a \in \mathbb{D} - \{0\}$, we let

$$b_a(z) = \frac{|a|}{a} \frac{a - z}{1 - \overline{a}z}.$$

And, when $A = \{a_n\}_{n=0}^{\infty}$ is a sequence of non-zero points in $\mathbb{D}$, we let:

$$B_A = \prod_{n=0}^{\infty} b_{a_n}$$

$$\Sigma_A = \sum_{n=0}^{\infty} 1 - |a_n|.$$

The function B_A is called a *Blaschke product*. The sequence A is said to be a *Blaschke sequence* if $\Sigma_A < \infty$. We say that B_A *converges at* z if $B_A(z) \neq 0$. It is well-known that B_A converges at all points of the disk except the terms of A if and only if A is a Blaschke sequence and that otherwise B_A converges nowhere on the disk. Other properties of Blaschke products can be found in [7] and [21].

The first question to consider is "If A is a computable Blaschke sequence, does it follow that B_A is a computable function?" The following theorem, which first appeared in [14], shows that it does not.

Theorem 3 *There is a computable Blaschke sequence A such that B_A is incomputable.*

The proof of Theorem 3 turns on the following result of E. Specker [23].

Theorem 4 *There is a computable and decreasing sequence of positive rational numbers $1 > r_0 > r_1 > \cdots > 0$ whose limit is incomputable.*

Proof Let $A \subseteq \mathbb{N}$ be *c.e.* and incomputable, and let $\{a_n\}_{n=0}^{\infty}$ be a computable enumeration of A. Since A is infinite, it is possible to choose this enumeration so that it contains no repetitions. Let

$$r_k = 1 - \sum_{j=0}^{k} 2^{-(a_k+1)}.$$

Thus, $1 > r_0 > r_1 > \cdots > 0$. Let $r = \lim_{k \to \infty} r_k$. If r were computable, then $1 - r$ would also be computable. Hence, it would be possible to compute its base 2 expansion. However, the n-th place to the right of the '.' in this expansion is 1 if and only if $(n - 1) \in A$. Hence, it would also follow that A is computable—a contradiction. Thus, r is incomputable. □

Proof of Theorem 3 Let $1 > r_0 > r_1 \ldots$ be as in Theorem 4. Let

$$a_0 = r_0$$
$$a_{n+1} = \frac{r_{n+1}}{r_n}$$
$$A = \{a_n\}_{n=0}^{\infty}.$$

Thus, A is computable. But,

$$B_A(0) = \lim_{k\to\infty} \prod_{n=0}^{k} b_{a_n}(0)$$
$$= \lim_{k\to\infty} r_{k+1}.$$

Thus, $B_A(0)$ is incomputable. Hence, B_A is incomputable. Note also that, since $B_A(0)$ is incomputable, it is non-zero and so $\Sigma_A < \infty$. □

Our next goal is to show that if B is a computable and not identically zero Blaschke product, then there is a computable Blaschke sequence A such that $B = B_A$ [14]. The proof of this result will lead us on a tour of several fundamental results and techniques of computable analysis. We begin with the following definition.

Definition 9 A closed set $X \subseteq \mathbb{C}$ is said to be *computably closed* if the set of all open rational disks that contain a point of X is computably enumerable.

Almost any natural closed subset of the plane D is computably closed. However, if z is an incomputable point in $\mathbb{C}$, then $\{z\}$ is not computably closed.

Our first preliminary goal is to show the following.

Theorem 5 *The zero set of a non-constant, computable, and analytic function* $f : \mathbb{D} \to \mathbb{C}$ *is computably closed.*

In general, it does not follow that the zero set of a computable function is computably closed. In fact, E. Specker showed that there is a computable function $f : [0, 1] \to \mathbb{R}$ whose zero set is not computably closed and even has positive Lebesgue measure [24].

The essence of the problem is to list just those open rational disks that contain a zero of f. When f is an analytic, there are many zero-finding methods available to us. Perhaps the simplest of these is to use the Argument Principle. Namely, the number of zeros of f is an open disk D is

$$\mathrm{Arg}(f; D) =_{df} \frac{1}{2\pi i} \int_{\partial D} \frac{f'}{f}$$

provided there is no zero of f on the boundary of D.

Upon considering the application of the Argument Principle, we are led to two key principles of computable analysis: integration is a computable operator but differentiation is not. Namely, we have the following two theorems. The first is relatively well-known, and the second is due to J. Myhill [16].

Theorem 6 *If $f : [0, 1] \to \mathbb{R}$ is computable, then $\int_0^1 f$ is a computable real.*

Theorem 7 *There is a computable and continuously differentiable $f : [0, 1] \to \mathbb{R}$ whose derivative is not computable.*

A highly useful feature of Theorem 6 is that it is *uniformly* true. That is, there is an algorithm that, given as input an algorithm that computes a function $f : [0, 1] \to \mathbb{R}$, produces an algorithm that computes $\int_0^1 f$.

Cauchy's Formula allows us to express differentiation of analytic functions in terms of integration. If we put these observations together, we obtain the following which was previously observed by P. Hertling [9].

Proposition 1 *The derivative of a computable analytic function on the disk is computable.*

Furthermore, Proposition 1 holds *uniformly* in the sense that there is an algorithm, that given as input an algorithm for computing an analytic function on the disk, produces an algorithm that computes its derivative.

The next difficulty we must contend with regards to the application of the Argument Principle is that there is no algorithm which will tell us whether the boundary of a rational disk contains no zero of f. More precisely, we have the following.

Proposition 2 *There is a computable Blaschke product B such that*

$$\left\{D : D \text{ is an open rational disk and } \partial D \cap B^{-1}\left[\{0\}\right] = \emptyset\right\}$$

is incomputable.

The proof of Proposition 2 illustrates the technique of coding a *c.e.* incomputable set into a problem. However, we delay the proof until we cover some supporting material in Sect. 5.

Fortunately, with a little care, the obstacle illustrated by Proposition 2 is easy to surmount. Namely, we begin by enumerating all closed rational disks $D \subseteq \mathbb{D}$ while simultaneously estimating the value of $\mathrm{Arg}(f; D)$. That is, by simultaneously running an algorithm for estimating the value of this integral, which may be undefined. If at some stage in this process, this algorithm tells us that the value of $\mathrm{Arg}(f; D)$ lies in an interval that contains only one positive integer, then we do not list D as a disk that contains a zero of f, but we *do* list every open rational disk D' such that $D \subseteq D' \subseteq \mathbb{D}$ as one that contains a zero of f. For, if ∂D is zero-free, then it must be the case that D and hence D' contains a zero of f. But, if ∂D contains a zero of f, it is still the case that D' contains a zero of f. Hence, every open rational disk listed by this process contains a zero of f. Furthermore, since the zeros of f are isolated, it follows that every rational disk that contains a zero of f is eventually listed by this process. We have thus proven Theorem 5.

The computability properties of the complement of the zero set of f will be useful. So, we introduce the notion of a computably open set.

Definition 10 An open set $U \subseteq \mathbb{C}$ is *computably open* if the set of all closed rational disks $D \subseteq U$ is computably enumerable.

Proposition 3 *If f is a computable function on $\mathbb{D}$, then $\mathbb{D} - f^{-1}[\{0\}]$ is computably open.*

Proof Fix an algorithm for computing f. Begin running this algorithm on all open rational disks in parallel. Whenever it halts on an open rational disk D_1 and produces as output a disk D_2 that does not contain 0, we can list D_1 as an *open* rational disk that contains no zero of f. We then list a closed rational disk D as containing no zero of f whenever we discover (from the first list) open rational disks $D_{2,1}, \ldots, D_{2,n}$ that cover D and that contain no zero of f. □

Note that the proof of Proposition 3 is uniform.

We now turn to the following definitions from [15].

Definition 11 Suppose $f : \mathbb{D} \to \mathbb{C}$ is an analytic function with infinitely many zeros.

1. A *zero sequence* of f is a sequence $\{a_n\}_{n\in\mathbb{N}}$ whose terms are precisely the zeros of f and such that the number of times a zero of f appears in this sequence is its multiplicity as a zero of f.
2. A *truncated zero sequence* of f is a sequence $\{a_n\}_{n\in\mathbb{N}}$ whose terms are precisely the zeros of f and such that each zero of f appears exactly once in this sequence.

The heart of the matter then is to show that a computable and not identically zero Blaschke product has a computable zero sequence. As an intermediate step, we show the following. The proof illustrates the technique of constructing a sequence through approximations.

Theorem 8 *If f is a non-constant, computable, and analytic function on $\mathbb{D}$ with infinitely many zeros, then f has a computable truncated zero sequence.*

Proof By Theorem 5, there is a computable enumeration $\{D_n\}_{n\in\mathbb{N}}$ of all open rational disks $D \subseteq \mathbb{D}$ that contain a zero of f. Our goal now is to construct a computable array $\{D_{m,n}\}_{m,n\in\mathbb{N}}$ of rational disks such that $\bigcap_n D_{m,n}$ consists of a single point for each m and such that this point is a zero of f. We additionally ensure that for each zero of f, z_0, there is exactly one m such that $z_0 \in \bigcap_n D_{m,n}$.

We construct $\{D_{m,n}\}_{m,n}$ inductively and by stages. At stage 0, we set $D_{0,0} = D_0$. At stage $t+1$, we do the following. First of all, let $m_0, n_0, \ldots, n_{m_0}$ denote the numbers such that at the end of stage t we have defined $D_{m,n}$ precisely when $m \leq m_0$ and $n \leq n_m$. Let $r = (t+1)_0$. If

$$D_r \cap \left(\bigcap_{n \leq n_m} D_{m,n} \right) = \emptyset$$

whenever $m \leq m_0$, then we let

$$D_{m_0+1,0} = D_r.$$

If there is an $m \leq m_0$ such that $D_r \subseteq D_{m,n_m}$, and if the diameter of D_r is no larger than one half of the diameter of D_{m,n_m}, then choose the least such m and set $D_{m,n_m+1} = D_r$. If neither of these cases holds, then we do not define any more terms of $\{D_{m,n}\}_{m,n}$ at stage $t+1$. This completes the construction of $\{D_{m,n}\}_{m,n}$.

We now show this construction builds a double sequence with the required properties. First of all, we must show it defines $D_{m,n}$ for every m, n. To this end, we first show that it defines $D_{m,0}$ for every m. For, suppose otherwise. Let m_1 be the least number such that $D_{m_1,0}$ is never defined. Hence, $m_1 > 0$. Also, $D_{m_1+1,0}$, $D_{m_1+2,0}, \ldots$ are never defined. $D_{0,0}, D_{1,0}, \ldots, D_{m_1-1,0}$ contain finitely many zeros of f. So, let z_0 be a zero of f that does not belong to any of these disks. Then, there is a rational disk D that contains z_0 and such that $D \cap D_{j,0} = \emptyset$ whenever $j \leq m_1$. At the same time, $D = D_r$ for some r. There are infinitely many t such that $(t+1)_0 = r$. So, there is such a value of t for which it is also true that $D_{0,0}, \ldots, D_{m_1-1,0}$ have been defined by the end of stage $t+1$. Let $m_0, n_0, \ldots, n_{m_0}$ be as in the description of stage $t+1$. Hence, $m_0 = m_1 - 1$. The construction ensures that $D_{m,n} \supseteq D_{m,n+1}$ whenever $D_{m,n}$ and $D_{m,n+1}$ are defined. Thus, $D_{m,n} \cap D_r = \emptyset$ whenever $m \leq m_0$ and $n \leq n_m$. But, this implies that $D_{m_1,0} = D_r$, contrary to what we assumed. Thus, $D_{m,0}$ is defined for all m.

We now claim that for each m, $D_{m,n}$ is defined for all n. For, suppose m, n are such that $D_{m,n}$ is never defined. Let n_1 be the largest number such that D_{m,n_1} is defined. By construction, D_{m,n_1} contains a zero of f. So, there is a rational disk $D \subseteq D_{m,n_1}$ that contains a zero of f and whose diameter is not larger than one half of the diameter of D_{m,n_1}. Then, there is a t such that $D_{(t+1)_0} = D$ and such that D_{m,n_1} is defined at stage t. Let $m_0, n_0, \ldots, n_{m_0}$ be as in the description of stage $t+1$. By working through the cases in the description of stage $t+1$, we see that $D_{m',n'}$ must be set $= D$ at stage $t+1$ for some m', n' and that $m' \leq m$. However, the construction ensures that $D_{m',n} \cap D = \emptyset$ whenever $m' < m$ and $n \leq n_{m'}$. So, it must be that $m' = m$ and this is a contradiction.

Thus, $D_{m,n}$ is defined for all m, n. Furthermore, our construction of the array $\{D_{m,n}\}_{m,n}$ provides an algorithm for computing $D_{m,n}$ from m, n. It also follows that $\bigcap_n D_{m,n}$ consists of a single point for each m and that this point is a zero of f. For each m, we define a_m to be the unique point in $\bigcap_n D_{m,n}$. We thusly obtain a computable sequence $\{a_m\}_{m \in \mathbb{N}}$. It follows from the construction of $\{D_{m,n}\}_{m,n}$ that this sequence contains no repetitions. It only remains to show that every zero of f appears in this sequence. By way of contradiction, suppose z_0 is a zero of f that does not appear in this sequence. It follows from the construction that z_0 is a limit point of $\{a_0, a_1, \ldots\}$. But, this is impossible since f is analytic and non-constant. The proof is complete. □

We have thus shown that we can algorithmically discover the zeros of f and arrange them into a sequence with no repetitions. Our final step is to show we can algorithmically augment this sequence so that each zero is repeated according to its multiplicity. We will need the following lemma.

Lemma 1 *Suppose $f : \mathbb{D} \to \mathbb{C}$ is computable, non-constant, and analytic. Suppose also that $a \in \mathbb{D}$ is a zero of f. Let*

$$f_1(z) = \lim_{z' \to z} \frac{f(z')}{z' - a}$$

for all $z \in \mathbb{D}$. Then, f_1 is computable.

Proof Since the zeros of f are isolated, it follows from Theorem 5 that a is a computable point. Fix an algorithm for computing f.

The following algorithm computes f_1. Suppose we are given as input an open rational disk D_1 such that $\overline{D_1} \subseteq \mathbb{D}$. We begin by computing, for each $k \in \mathbb{N}$, an open rational disk S_k that contains a and whose diameter is smaller than 2^{-k}. We continue to do so until we find S_{k_0} that is contained in $\mathbb{C} - \partial D_1$. It may be that no such S_{k_0} is found (i.e. in the case when a belongs to the boundary of D_1). In this case, the algorithm will never halt.

Suppose S_{k_0} is contained in the complement of $\overline{D_1}$. We then know that $a \notin D_1$. So, we proceed by following the steps in the algorithm for computing f with D_1 as input. If no output is thereby yielded, then this algorithm does not halt either. So, suppose D_2 is yielded as output. Compute the least $k_1 \in \mathbb{N}$ so that 2^{-k_1} is smaller than the diameter of D_2. From D_2 and S_{k_1} we can compute an open rational disk D_3 that contains $f_1[D_1]$. Furthermore, we can do this in such a way that the diameter of D_3 tends to zero as the diameter of D_2 approaches zero.

On the other hand, suppose S_{k_0} is contained in D_1; hence, $a \in D_1$. Let D_2 denote the disk that is concentric with D_1 and whose radius is twice the radius of D_1. By the Cauchy Integral Formula,

$$f_1(z) = \frac{1}{2\pi i} \int_{\partial D_2} \frac{f_1(\zeta)}{\zeta - z} d\zeta$$

for each $z \in D_1$. It follows that we can compute a rational disk D_3 that contains $f_1[D_1]$. Furthermore, we can do so in such a way that the radius of D_3 tends to zero as the radius of D_1 tends to zero. We output D_3. This completes the description of our algorithm.

We now verify that this algorithm satisfies the conditions set forth in Definition 7. It is clear that the approximation and correctness conditions are met. In order to verify convergence, suppose $z \in \mathbb{D}$ and that U is a neighborhood of $f(z)$. We first consider the case $z \neq a$. In this case, it suffices to show that the algorithm for computing f satisfies a stronger convergence criterion. Namely, if U_1 is any neighborhood of $f(z)$, then z belongs to an open rational disk D_1 such that $a \in \mathbb{D} - \overline{D_1}$ and such that when provided as input, the algorithm yields an open rational disk contained in U_1. We prove this by contradiction. Fix $\epsilon_0 > 0$ such that $D_{\epsilon_0}(f(z)) \subseteq U_1$. Then, for each positive $\epsilon < \epsilon_0$, there is an open rational disk $D_{1,\epsilon}$ that contains z and such that the algorithm for computing f not only halts when provided $D_{1,\epsilon}$ as input but also produces a disk D_2 that is contained in $D_\epsilon(f(z))$. Each $\overline{D_{1,\epsilon}}$ contains the line segment from z to a. But, this entails that $f(\zeta) = f(z)$ whenever ζ belongs to the line segment from z to a. Since f is analytic, we are lead to the conclusion that f is a constant—a contradiction.

We now consider the case $z = a$. In this case, the key point is to show that the algorithm for computing f must halt on arbitrarily small rational disks that contain a. Suppose this is not the case. It then follows from the convergence criterion of Definition 7 that f is constant on a closed disk that contains a. But, since f is analytic, this again entails that f is constant. This completes the proof. □

We note that the proof of Lemma 1 is uniform in that it provides an algorithm that transforms algorithms for computing f and a into an algorithm for computing f_1. This point is crucial for the proof of the following.

Theorem 9 *If $f : \mathbb{D} \to \mathbb{C}$ is a non-constant, computable, and analytic function with infinitely many zeros, then f has a computable zero sequence.*

Proof By Theorem 8, there is a computable truncated zero sequence for f, $\{a_n\}_{n \in \mathbb{N}}$. We build a zero sequence of f, $\{a'_n\}_{n \in \mathbb{N}}$, by stages as follows. To begin, set $n_0 = 0$. It follows, as in the proof of Proposition 3, that there is a computable enumeration of all closed rational disks $D \subseteq \mathbb{D}$ whose boundary contains no zero of f. (Search for coverings of the boundary instead of coverings of the closed disk.) Thus, we can compute a rational number $0 < r_0 < 1$ such that $\partial D_{r_0}(a_0)$ contains no zero of f but $\mathrm{Arg}(f, D_{r_0}(a_0)) = 1$. Thus, a_0 is the only zero of f in $\overline{D_{r_0}(a_0)}$. We define $a'_0 = a_0$ and set

$$f_1(z) = \lim_{z' \to z} \frac{f(z')}{z' - a_0}.$$

Suppose we have defined n_t, a'_t, f_t, and r_t. By way of induction, $\partial D_{r_t}(a_t)$ contains no zero of f_t and $D_{r_t}(a_t)$ contains no zero of f_t except possibly a_{n_t}. So, we test if $\mathrm{Arg}(f_t, D_{r_t}(a_t)) \neq 0$. If so, then we set:

$$n_{t+1} = n_t$$

$$f_{t+1}(z) = \lim_{z' \to z} \frac{f_t(z')}{z' - a_{n_{t+1}}}$$

$$a'_{t+1} = a_{n_t}$$
$$r_{t+1} = r_t.$$

Otherwise, we compute a rational number $0 < r_{t+1} < 1$ such that $\partial D_{r_{t+1}}(a_{n_t+1})$ contains no zero of f and such that $\mathrm{Arg}(f, D_{r_{t+1}}(a_{n_{t+1}})) = 1$. (So that a_{n_t+1} is the only zero of f_t in $D_{r_{t+1}}(a_{n_t+1})$.) We then set

$$n_{t+1} = n_t + 1$$
$$f_{t+1}(z) = \lim_{z' \to z} \frac{f_t(z')}{z' - a_{n_{t+1}}}$$
$$a'_{t+1} = a_{n_{t+1}}.$$

This completes our description of the construction of $\{a'_n\}_{n\in\mathbb{N}}$. The idea of the construction is that we "sit on" a zero and repeatedly divide by the corresponding linear factor, adding to $\{a'_n\}_{n\in\mathbb{N}}$ each time, until it is exhausted as a sero. It follows that $\{a'_n\}_{n\in\mathbb{N}}$ is a zero sequence of f and that its construction also provides an algorithm for its computation. □

Corollary 1 *If B is a computable and not identically zero Blaschke product, then there is a computable Blaschke sequence A such that $B = B_A$.*

5 The Missing Parameter

One interpretation of Theorem 3 is that a Blaschke sequence does not provide enough information to compute the corresponding Blaschke product. This leads to the question as to what additional information is required. It turns out the Blaschke sum provides the exact amount of necessary information. The following is Theorem 4.6 of [14]. We give a more direct proof here.

Theorem 10 *If A is a computable Blaschke sequence whose Blaschke sum is a computable real, then B_A is computable.*

Proof Let $A = \{a_n\}_{n\in\mathbb{N}}$. Thus, $B_k =_{df} \prod_{n=0}^{k} b_{a_n}$ is computable uniformly in k. We now give an algorithm that computes B_A. Suppose we are given as input an open rational disk D_1 such that $\overline{D_1} \subseteq \mathbb{D}$. We can then compute the least $k \in \mathbb{N}$ such that $\overline{D_1}$ is contained in the open disk of center 0 and radius $1 - \frac{1}{2^{k+1}}$; denote this disk by $\mathbb{D}_k$. Compute the least positive integer m such that $1/m$ is smaller then the radius of D_1. (The idea is that as D_1 converges to a point, the value of m will be pushed towards infinity.) Since Σ_A is computable, we can now compute N_0 so that

$$\sum_{n=N_0}^{\infty} 1 - |a_n| \leq \frac{1}{2^m(2^{k+2} - 1)}.$$

Since A is computable, by direct computation we can compute a rational disk D_2' such that

$$B_{N_0}[D_1] \subseteq D_2'.$$

And, we can perform this computation in such a way that the diameter of D_2' tends to zero as the diameter of D_1 tends to zero. We output the disk that is concentric with D_2' and whose radius is the sum of $\frac{1}{2^{m-2}}$ and the radius of D_2'.

We now verify correctness. Suppose D_1 is given as input and that $\overline{D_1} \subseteq \mathbb{D}$. Suppose D_2 is output, and let N_0, m, k be as in the description of the algorithm. By means of the identity

$$1 - b_{a_n}(z) = \frac{(1-|a_n|)(a_n + |a_n|z)}{a_n(1-\overline{a_n}z)}$$

it follows that

$$\begin{aligned} \left|1 - b_{a_n}(z)\right| &\leq \frac{1+|z|}{1-|z|}(1-|a_n|) \\ &\leq \left(2^{k+2}-1\right)\left(1-|a_n|\right) \end{aligned}$$

for all $z \in \overline{\mathbb{D}_k}$. The other key inequalities are that

$$e^x - 1 \leq 4x \tag{1}$$

when $0 \leq x \leq 1$ and that

$$\left|\prod_{n=1}^{N}(1+u_n) - 1\right| \leq \exp\left(|u_1| + \cdots + |u_N|\right) - 1 \tag{2}$$

whenever $u_1, \ldots, u_N \in \mathbb{C}$. The first follows from elementary calculus. The second follows from Lemma 15.3 of [21]. To apply (2), choose $z \in \mathbb{D}_k$ and set $u_n = b_{a_n}(z) - 1$. We obtain that when $M > N_0$,

$$\begin{aligned} \left|B_M(z) - B_{N_0}(z)\right| &= \left|B_{N_0}(z)\right|\left|\frac{B_M(z)}{B_{N_0}(z)} - 1\right| \\ &\leq \left|B_{N_0}(z)\right|\left(\exp\left(|u_{N_0+1}| + \cdots + \left|u_M(z)\right|\right) - 1\right) \\ &\leq \exp\left(\frac{1}{2^m}\right) - 1 \\ &\leq \frac{1}{2^{m-2}}. \end{aligned}$$

Therefore, for all $z \in \mathbb{D}_k$, $|B_A(z) - B_{N_0}(z)| \leq \frac{1}{2^{m-2}}$. It follows that $B_A[D_1] \subseteq D_2$. Thus, correctness.

We now demonstrate convergence. Suppose $z \in \mathbb{D}$ and that U is a neighborhood of $B_A(z)$. Let k be the smallest positive integer such that $z \in \mathbb{D}_k$. Let D_1 be an open rational disk such that $z \in D_1$ and $\overline{D_1} \subseteq \mathbb{D}_k$. It follows that the algorithm halts on input D_1; let D_2 be the disk output. Since k is fixed, it follows that the radius of D_2 tends to zero as the radius of D_1 approaches zero. Thus, convergence. □

Theorem 10 holds uniformly in that its proof provides an algorithm that transforms an algorithm for computing A and an algorithm for computing Σ_A into an algorithm for computing B_A.

Theorem 10 thus tells us that Σ_A provides, when combined with A, *sufficient* information for the computation of B_A. The question now arises as to whether Σ_A provides the exact amount of additional information required or whether some weaker parameter suffices. It turns out that it does as can be demonstrated by showing that Σ_A can be computed from A and B_A. However, an even stronger result holds. Namely, the following theorem which proceeds from the results in [15].

Theorem 11 *If A is a computable Blaschke sequence, and if $B_A(0)$ is computable, then Σ_A is computable.*

Proof sketch Let $A = \{a_n\}_{n\in\mathbb{N}}$. Note that $B_A(0) = \prod_{n=0}^{\infty} |a_n|$. To estimate the error in approximating Σ_A by $\sum_{n=0}^{k} 1 - |a_n|$, we use the inequality

$$1 - x \leq e^{-x}$$

from which we obtain

$$\frac{1}{\prod_{n=k+1}^{\infty} |a_n|} \geq \exp\left(\sum_{n=k+1}^{\infty} 1 - |a_n|\right).$$

Thus, the error in this approximation is bounded above by

$$-\ln\left(\prod_{n=k+1}^{\infty} |a_n|\right)$$

which tends to 0 as k approaches infinity and which can be computed uniformly from k (in the sense that there is an algorithm that from k, m computes a rational number whose distance from this quantity is at most 2^{-m}). □

Corollary 2 *If A is a computable Blaschke sequence, and if $B_A(0)$ is computable, then B_A is computable.*

Again, Theorem 11 and Corollary 2 hold uniformly.

It is worth contrasting Corollary 2 with the following result of Caldwell and Pour-El [20].

Theorem 12 *There is a computable sequence of complex numbers $\{a_n\}_{n\in\mathbb{N}}$ such that the power series $\sum_{n=0}^{\infty} a_n z^n$ converges everywhere and defines a function that is computable on every open rational disk but is not computable on $\mathbb{C}$.*

The computability of a function on an open rational disk or $\mathbb{C}$ is defined by appropriate elaboration on Definition 7.

Finally, we are now in position to prove Proposition 2.

Proof of Proposition 2 Let $C \subseteq \mathbb{N}$ be *c.e.* and incomputable. Let $\{c_n\}_{n\in\mathbb{N}}$ be a computable enumeration of C. For all t, set

$$C_t = \{c_n : n < t\}.$$

For all n, t, let $a_{n,t} = 1 - 2^{-(n+1)}$ if $n \notin C_t$. But, if $n \in C_t$, and if s is the smallest integer such that $n \in C_s$, then set $a_{n,t} = 1 - (2^{-(n+1)} + 2^{-(s+n+1)})$. And, let $a_n = \lim_{t\to\infty} a_{n,t}$. It follows that $|a_{n,t} - a_n| \leq 2^{-(t+n+1)}$. Thus, $A =_{df} \{a_n\}_{n\in\mathbb{N}}$ is computable. Also,

$$\sum_{n=k}^{\infty} 1 - |a_n| \leq \sum_{n=k}^{\infty} 2^{-n} = 2^{-k+1}.$$

And,

$$\left| \sum_{n=0}^{k} a_{n,t} - \sum_{n=0}^{k} a_n \right| \leq \sum_{n=0}^{k} 2^{-(n+t+1)} \leq 2^{-t}.$$

Hence,

$$\left| \sum_{n=0}^{t} a_{n,t} - \Sigma_A \right| \leq 2^{-t+3}.$$

It follows that Σ_A and hence B_A are computable.

By way of contradiction, suppose there is an algorithm that, given as input a rational disk D such that $\overline{D} \subseteq \mathbb{D}$, determines if ∂D contains a zero of B_A. For all n, $n \notin C$ if and only if the boundary of $D_{\epsilon_n}(0)$ contains a zero of B_A where $\epsilon_n = 1 - 2^{-(n+1)}$. So, it follows that C is computable—a contradiction. □

6 Interpolating Sequences and Naftalévich's Theorem

We return again to Theorem 3. It is natural to ask whether geometric conditions on a Blaschke sequence can influence the computability of the corresponding Blaschke product. Since, when A is computable, the computability of B_A depends only on that of Σ_A, the consideration of interpolating sequences leads to an examination of Naftalévich's Theorem [17]:

Theorem 13 *If $A = \{a_n\}_{n\in\mathbb{N}}$ is a Blaschke sequence, then there is an interpolating sequence $A' = \{a'_n\}_{n\in\mathbb{N}}$ such that $|a'_n| = |a_n|$ for all n.*

In [26], V. Andreev and T. McNicholl prove the following (although it is buried in the middle of a computability proof).

Theorem 14 *If $A = \{a_n\}_{n\in\mathbb{N}}$ is a Blaschke sequence, and if for each n we let*

$$\theta_n = \sum_{k=0}^{n} 1 - |a_k|$$

$$a'_n = |a_n|e^{i\theta_n},$$

then $A' =_{df} \{a'_n\}_{n\in\mathbb{N}}$ is an interpolating sequence.

The proof of Theorem 14 is contained in the proofs of Lemmas 5.2 through 5.6 of [26].

From Theorem 14 and the other results we have discussed, we obtain the following.

Corollary 3 *There is a computable interpolating sequence A such that B_A is incomputable.*

Moreover, Theorem 14 shows that there is a very simple procedure for producing interpolating sequences from Blaschke sequences.

7 Inner Functions—Frostman's Theorem

Let $M_a(z) = \frac{z-a}{1-\bar{a}z}$. We begin with the following statement of Frostman's Theorem [6, 7].

Theorem 15 *If $u : \mathbb{D} \to \mathbb{D}$ is an inner function with infinitely many zeros, then for all $\alpha \in \mathbb{D}$, except in a set of logarithmic capacity zero, the function $M_\alpha \circ u$ can be expressed in the form*

$$M_\alpha \circ u(z) = \lambda z^k B_A(z) \tag{3}$$

where $|\lambda| = 1$ and A is a Blaschke sequence.

An immediate consequence of Theorem 15 is the following result on estimation.

Corollary 4 *If $u : \mathbb{D} \to \mathbb{D}$ is an inner function with infinitely many zeros, and if $\epsilon > 0$, then there is a Blaschke sequence A, a natural number k, and a point $\lambda \in \partial\mathbb{D}$ such that $|u(z) - \lambda z^k B_A(z)| < \epsilon$ for all $z \in \mathbb{D}$.*

Computable versions of Theorem 15 and Corollary 4 are proven in [15]. Namely, we have the following.

Theorem 16 *If $u : \mathbb{D} \to \mathbb{D}$ is a computable inner function with infinitely many zeros, and if $N \in \mathbb{N}$, then there is a computable $\alpha \in \mathbb{D}$ such that $M_\alpha \circ u$ is expressible in the form* (3) *and $|\alpha| < 2^{-n}$.*

Corollary 5 *If $u : \mathbb{D} \to \mathbb{D}$ is a computable inner function with infinitely many zeros, and if $n \in \mathbb{N}$, then there is a computable Blaschke product B_A, a computable $\lambda \in \partial\mathbb{D}$, and a $k \in \mathbb{N}$, such that $|u(z) - \lambda z^k B_A(z)| < 2^{-n}$ for all $z \in \mathbb{D}$.*

Furthermore, Theorem 16 and Corollary 5 hold uniformly. We sketch the proof of Theorem 16.

When $u : \mathbb{D} \to \mathbb{C}$, let

$$m_u(r) = \frac{1}{2\pi} \int_0^{2\pi} \log\left|u\left(re^{i\theta}\right)\right| d\theta.$$

The key complex analysis results in the proof of Theorem 16 are the following.

Theorem 17 *If u is an inner function, then the following are equivalent.*

1. *u can be expressed in the form*

$$u(z) = \lambda z^k B_A(z)$$

for some Blaschke sequence A, $\lambda \in \partial\mathbb{D}$, and $k \in \mathbb{N}$.

2. $\lim_{r \to 1} m_u(r) = 0$.

Lemma 2 *If u is an inner function, then m_u is increasing.*

The key principle from computable analysis that is used in the proof of Theorem 16 is that maximum-finding is a computable operation on compact sets. For instance, we have the following.

Proposition 4 *If $f : \mathbb{D} \to \mathbb{R}$ is computable, and if $D \subseteq \mathbb{D}$ is a closed rational disk, then*

$$\max\left\{f(z) : z \in \mathbb{D}\right\}$$

is a computable real.

And again, this result is uniform. A generalization of Proposition 4, which handles arbitrary compact sets, appears in Chap. 5 of [27]. As an aside, we mention the following theorem of E. Specker [24].

Theorem 18 *If* $f : [0, 1] \to \mathbb{R}$ *is computable, then* $\max\{f(x) : x \in [0, 1]\}$ *is a computable real number. However, there is a computable* $f : [0, 1] \to \mathbb{R}$ *that does not attain its maximum value at any computable real.*

In other words, maximum-finding is a computable operation, but finding *where* a maximum occurs is not!

With these elements, we can now sketch a proof of Theorem 16. To begin, abbreviate $M_\alpha \circ u$ by u_α and set $m(u, \alpha, r) = m_{u_\alpha}(r)$. Note that $m(u, \alpha, r)$ is continuous as a function of α. Moreover, it is computable as a function of α if u, r are computable. We then search for a rational number $r \in (0, 1)$ and a closed rational disk $D_1 \subseteq D_{2^{-(n+2)}}(0)$ whose diameter is smaller than $1/8$ and such that

$$\min\{m(u, \alpha, r_1) : \alpha \in D_1\} > -1/2.$$

It follows from Frostman's Theorem, Theorem 17, and Lemma 2, that this search must succeed. We then search for a rational number $r_2 \in (0, 1)$ and a closed rational disk $D_2 \subseteq D_1$ whose diameter is smaller than $1/16$ and such that

$$\min\{m(u, \alpha, r_2) : \alpha \in D_2\} > -1/4.$$

Again, this search must succeed; otherwise, by Theorem 17 and Lemma 2, Frostman's Theorem would be contradicted.

Let $D_3, D_4, \ldots$ and $r_3, r_4, \ldots$ by obtained by continuing in this fashion. Let α be the unique point in $\bigcap_j D_j$. It follows from Theorem 17 and Lemma 2 that $M_\alpha \circ u$ has the form (3). Moreover, it follows that α is computable since the sequence of disks $\{D_n\}_{n \in \mathbb{N}}$ is computable.

8 Inner Functions—Factorization

We begin with the following statement of the Factorization Theorem for inner functions.

Theorem 19 *Suppose u is an inner function with infinitely many zeros. Then, u can be written in the form*

$$u(z) = \lambda z^k \sigma(z) B(z) \tag{4}$$

where $|\lambda| = 1$, $k \in \mathbb{N}$, σ *is an inner function, and* B *is a Blaschke product. Furthermore,* λ, k, σ, *and* B *are unique.*

Accordingly, when u has been written in the form (4), let $\lambda_u = \lambda$, $k_u = k$, $\sigma_u = \sigma$, and $B_u = B$.

It is natural to ask whether factorization of inner functions is a computable operation. For example, we might ask "If u is a computable inner function, is k_u computable?" Unfortunately, the answer to this question is a trivial "yes": k_u, being a rational number, is automatically computable even when u is incomputable! Clearly, the question does not capture what we wanted to ask. A better approach is to ask about uniform computation. That is, is there an algorithm that given as input an algorithm for computing an inner function u, computes k_u? The answer is that there is not, and the demonstration of this fact showcases one of the gems of classical computability theory: the Recursion Theorem.

Recall from Sect. 2.3, that there is a computable $U :\subseteq \mathbb{N} \times \mathbb{N} \to \mathbb{N}$ such that whenever $f :\subseteq \mathbb{N} \to \mathbb{N}$ is computable, there is a number $e \in \mathbb{N}$ such that $f(x) = U(e, x)$ for all x. Accordingly, let $\phi_e(x) = U(e, x)$.

Theorem 20 (The Recursion Theorem) *Suppose $f : \mathbb{N} \to \mathbb{N}$ is computable. Then, there is a number $e_0 \in \mathbb{N}$ such that $\phi_{e_0} = \phi_{f(e_0)}$.*

The Recursion Theorem was proven by Kleene in 1938 [12]. An elegant proof appears in [18]. It is the chief weapon in defeating claims of uniformity. The Recursion Theorem is sometimes referred to as the Fixed Point Theorem. However, it is important to note that $f(e_0)$ may not equal e_0; these numbers merely index the same computable function.

We now prove the following.

Theorem 21 *There is no algorithm that, given as input an algorithm that computes an inner function u, computes k_u.*

Proof By way of contradiction, suppose otherwise. Then, there is a computable function $h :\subseteq \mathbb{N} \to \mathbb{N}$ such that if ϕ_e computes an inner function u, then $h(e) = k_u$. Fix an algorithm for computing h.

For each $e, t \in \mathbb{N}$, define a rational number $a_{e,t}$ as follows. First, run the computation of $h(e)$ for t steps. If no number is output, or if a number besides 1 is output, set $a_{e,t} = 2^{-t}$. But, if 1 is output, then set $a_{e,t} = 2^{-s}$ where $s \leq t$ is the exact number of steps required for the computation of $h(e)$. Let $a_e = \lim_{t \to \infty} a_{e,t}$. Hence, $\{a_{e,t}\}_{e \in \mathbb{N}}$ is computable.

Let $B = \prod_{n=0}^{\infty} b_{a_n}$ where $a_n = 1 - \frac{1}{n^2}$.

It is now possible to define a computable function $f : \mathbb{N} \to \mathbb{N}$ with the property that for each $e \in \mathbb{N}$, $\phi_{f(e)}$ computes $M_{a_e} \cdot B$. So, by the Recursion Theorem, there is a number $e_0 \in \mathbb{N}$ such that $\phi_{f(e_0)} = \phi_{e_0}$. Thus, ϕ_{e_0} computes $u =_{df} M_{a_{e_0}} \cdot B$. Thus, $h(e_0)$ must be defined and $h(e_0) = k_u$. But, by construction, if $h(e_0) = 1$, then $a_e \neq 0$, and so $M_{a_{e_0}}(0)B(0) \neq 0$. So, it must be that $h(e_0) \neq 1$. But the construction of f then yields that $M_{a_e} \cdot B$ has a zero of order 1 at 0; again a contradiction! Since both of the only possible two cases yield a contradiction, we must conclude the existence of the function h is impossible. □

Since even the most elementary component of the factorization of u can not be computed from u alone, we now turn to the search for sufficient additional parameters. It turns out that the Blaschke sum provides the right amount of information.

Namely, when u is an inner function with infinitely many zeros, let

$$\Sigma_u = \sum_{n=0}^{\infty} 1 - |a_n|$$

where $\{a_n\}_{n=0}^{\infty}$ is a zero sequence of B_u. The following is essentially Lemma 5.4 of [15].

Lemma 3 *There is an algorithm that, given as input an algorithm that computes an inner function u with infinitely many zeros and an algorithm that computes Σ_u, computes k_u.*

The zero sequence of B_u can be computed from u as in the proof of Theorem 9. We now easily obtain the following which is essentially Theorem 5.5.1 of [15].

Proposition 5 *There is an algorithm that, given as input an algorithm for computing an inner function u with infinitely many zeros and an algorithm for computing Σ_u, produces k_u and algorithms for computing λ_u, σ_u and B_u.*

It then follows from Theorem 11 that Σ_u provides the right amount of additional information necessary for computing the factorization of an inner function with infinitely many zeros.

Acknowledgements I first of all want to thank my departed colleague Alec Matheson for introducing me to the beautiful world of bounded analytic functions and for setting in motion the research direction described in this paper. Many thanks to the referee and to Barry Cooper for their helpful comments. I also thank Javad Mashreghi for his encouragement. Finally, I thank my wife Susan for her support.

References

1. Arora, S., Barak, B.: Computational Complexity: A Modern Approach. Cambridge University Press, Cambridge (2009)
2. Boone, W.W.: The word problem. Proc. Natl. Acad. Sci. USA **44**, 1061–1065 (1958)
3. Braverman, M., Cook, S.: Computing over the reals: foundations for scientific computing. Not. Am. Math. Soc. **53**(3), 318–329 (2006)
4. Barry Cooper, S.: Computability Theory. Chapman & Hall/CRC, Boca Raton (2004)
5. Cutland, N.: Computability: An Introduction to Recursive Function Theory. Cambridge University Press, Cambridge (1980)
6. Frostman, O.: Potential d'equilibre et capacité des ensembles avec quelques applications à la théorie des fonctions. Meddelanden Lunds Univ. Mat. Sem. **3**, 1–118 (1935)
7. Garnett, J.B.: Bounded Analytic Functions. Academic Press, New York (1981)
8. Grzegorczyk, A.: On the definitions of computable real continuous functions. Fundam. Math. **44**, 61–71 (1957)
9. Hertling, P.: An effective Riemann mapping theorem. Theor. Comput. Sci. **219**, 225–265 (1999)

10. Kalantari, I., Welch, L.: Point-free topological spaces, functions and recursive points; filter foundation for recursive analysis. I. Ann. Pure Appl. Log. **93**(1–3), 125–151 (1998)
11. Kalantari, I., Welch, L.: Recursive and nonextendible functions over the reals; filter foundation for recursive analysis. II. Ann. Pure Appl. Log. **98**(1–3), 87–110 (1999)
12. Kleene, S.: On notation for ordinal numbers. J. Symb. Log. **3**(4), 150–155 (1938)
13. Lerman, M.: Degrees of Unsolvability: Local and Global Theory. Perspectives in Mathematical Logic. Springer, Berlin (1983)
14. Matheson, A., McNicholl, T.H.: Computable analysis and Blaschke products. Proc. Am. Math. Soc. **136**(1), 321–332 (2008)
15. McNicholl, T.H.: Uniformly computable aspects of inner functions: estimation and factorization. Math. Log. Q. **54**(5), 508–518 (2008)
16. Myhill, J.: A recursive function defined on a compact interval and having a continuous derivative that is not recursive. Michigan J. Math. **18**, 97–98 (1971)
17. Naftalevič, A.G.: On interpolation by functions of bounded characteristic. Vilniaus Valst. Univ. Moksl. Darb. Mat. Fiz. Chem. Moksl. Ser. **5**, 5–27 (1956)
18. Odifreddi, P.G.: Classical Recursion Theory. The Theory of Functions and Sets of Natural Numbers, 1st edn. North-Holland, Amsterdam (1989)
19. Pour-El, M.B., Richards, J.I.: Computability in Analysis and Physics. Perspectives in Mathematical Logic. Springer, Berlin (1989)
20. Pour-El, M.B., Caldwell, J.: On a simple definition of computable function of a real variable—with applications to functions of a complex variable. Z. Math. Log. Grundl. Math. **21**, 1–19 (1975)
21. Rudin, W.: Real and Complex Analysis, 3rd edn. McGraw-Hill, New York (1987)
22. Soare, R.I.: Recursively Enumerable Sets and Degrees. Springer, Berlin (1987)
23. Specker, E.: Nicht konstruktiv beweisbare Sätze der Analysis. J. Symb. Log. **14**, 145–158 (1949)
24. Specker, E.: Der Satz vom Maximum in der rekursiven Analysis. In: Heyting, A. (ed.) Constructivity in Mathematics: Proceedings of the Colloquium Held at Amsterdam, 1957. Studies in Logic and the Foundations of Mathematics, pp. 254–265. North-Holland, Amsterdam (1959)
25. Turing, A.M.: On computable numbers, with an application to the entscheidungsproblem. Proc. Lond. Math. Soc., Ser. 2 **42**, 220–265 (1936)
26. Andreev, V.V., McNicholl, T.H.: Computing interpolating sequences. Theory Comput. Syst. **46**(2), 340–350 (2010)
27. Weihrauch, K.: Computable Analysis. Texts in Theoretical Computer Science. An EATCS Series. Springer, Berlin (2000)

Polynomials Versus Finite Blaschke Products

Tuen Wai Ng and Chiu Yin Tsang

Abstract The aim of this chapter is to compare polynomials of one complex variable and finite Blaschke products and demonstrate that they share many similar properties. In fact, we collect many known results as well as some very recent results for finite Blaschke products here to establish a dictionary between polynomials and finite Blaschke products.

Keywords Polynomials · Finite Blaschke products · Ritt's theorems · Chebyshev polynomials · Approximation

Mathematics Subject Classification Primary 30J10 · Secondary 30C10 · 30E10 · 30D05 · 39B12

1 Introduction

The main goal of this paper is to compare polynomials of one complex variable and finite Blaschke products and show that they share many similar properties and hence one can establish a dictionary between polynomials and finite Blaschke products. The underlying common feature for polynomials and finite Blaschke products is the fact that both classes of functions are finite self mappings. Therefore, we shall consider general finite mappings first.

A continuous mapping between two locally compact spaces is called *proper* if the preimage of every compact set is compact. A holomorphic mapping $f : X \to Y$ between Riemann surfaces X and Y is said to be *finite* if f is nonconstant and proper. The concept of finite holomorphic maps was introduced by T. Radó in [39]. T. Radó proved that if a surjective mapping $f : X \to Y$ is finite, then there exists a natural number n such that f takes every value $c \in Y$, counting with multiplicities,

T.W. Ng (✉) · C.Y. Tsang
Department of Mathematics, The University of Hong Kong, Pokfulam, Hong Kong
e-mail: ntw@maths.hku.hk

C.Y. Tsang
e-mail: ssuperhk@yahoo.com.hk

J. Mashreghi, E. Fricain (eds.), *Blaschke Products and Their Applications*, Fields Institute Communications 65, DOI 10.1007/978-1-4614-5341-3_14,

n times (see [19, pp. 28–30] for a proof). This number n is called the *degree* of f and is denoted by $\deg f$.

For the case when X and Y are the complex plane $\mathbb{C}$, we can see easily that such a surjective finite map must be a *polynomial* of degree n, that is, a function of the form

$$c_n z^n + c_{n-1} z^{n-1} + \cdots + c_1 z + c_0,$$

where $c_i \in \mathbb{C}$ with $c_n \neq 0$.

Now let us consider the case when both X and Y are the open unit disk $\mathbb{D}$. Then such a surjective finite map must be a *finite Blaschke product* of degree n, that is, a rational function B of the form

$$B(z) = e^{i\theta} \frac{z - z_1}{1 - \overline{z_1} z} \cdot \frac{z - z_2}{1 - \overline{z_2} z} \cdot \cdots \cdot \frac{z - z_n}{1 - \overline{z_n} z},$$

where $z_i \in \mathbb{D}$ and $\theta \in \mathbb{R}$. This was proved by P. Fatou in [18] and we refer the reader to the proof and the historical note of Fatou and Radó's results on finite mappings in [40, pp. 211–217].

It was J. Walsh [48] who first suggested that finite Blaschke products should be considered as "polynomials" in the unit disk or hyperbolic plane $\mathbb{D}$ and he has proven a version of Gauss-Lucas Theorem for finite Blaschke products. This point of view was also propagated by A. Beardon and D. Minda [5], as well as D.A. Singer [44]. In this paper, we will confirm this point of view by establishing a dictionary between polynomials and finite Blaschke products. In fact we divide each of following sections of the paper into Part One and Part Two. Part One will focus on results for polynomials and Part Two will consider the corresponding results for finite Blaschke products. At the end of each section, a table is given to indicate the correspondence between them.

This paper is organized as follows. Section 2 includes some basic results of polynomials and finite Blaschke products in order to show the analogy between them. In Sect. 3, some elementary results about Chebyshev polynomials are reviewed. This section also studies Chebyshev-Blaschke products, which were first introduced in [49] and then also considered and studied in [32] and [50]. Some analogous results about Chebyshev-Blaschke products are also given and the details of their proofs can be found in [30]. Section 4 discusses the problems about two finite Blaschke products sharing a set. Such problems for polynomials were considered by many people (for example, C.C. Yang [51], T.C. Dinh [12, 13], F.B. Pakovich [34–36]). The related results for polynomials are stated first. Some analogous results for finite Blaschke products are also given and their proof can be found in [31].

2 Elementary Results

In this section, we review some classical results on polynomials and finite Blaschke products. Reference will be given to these results (except those results which can be proved easily).

2.1 Part One: Polynomials

2.1.1 Some Simple Properties

Let $n_f(w; K)$ denote the number of solutions in $K \subset \mathbb{C}$ for the equation $f(z) = w$, counting with multiplicity.

Theorem 1 *Let f be analytic in $\mathbb{C}$ with $f(\mathbb{C}) = \mathbb{C}$. Then f is a polynomial of degree k if and only if $n_f(w; \mathbb{C}) \equiv k$ for all $w \in \mathbb{C}$.*

Theorem 2 (Uniqueness Theorem) *If P_1 and P_2 are polynomials of degree not exceeding n and if the equation*

$$P_1(z) = P_2(z) \tag{1}$$

is satisfied at $n + 1$ distinct points in $\mathbb{C}$, then $P_1 \equiv P_2$.

Remark 1 If P_1 and P_2 are monic polynomials of degree n and if (1) is satisfied at n distinct points in $\mathbb{C}$, then $P_1 \equiv P_2$.

(A polynomial is said to be *monic* if its leading coefficient is 1.)

Theorem 3 *Let f be analytic in $\mathbb{C}$ and suppose that*

$$\lim_{|z|\to\infty} \left|f(z)\right| = +\infty.$$

Then f is a polynomial.

For any $f : \mathbb{C} \to \mathbb{C}$, the difference quotient of f is given by

$$f^{\#}(z, w) = \frac{f(z) - f(w)}{z - w}.$$

Theorem 4 *For a given w, $P(z)$ is a polynomial of degree k if and only if $P^{\#}(z, w)$ is a polynomial of degree $k - 1$ in the variable z.*

Theorem 5 *Let f be an entire function on $\mathbb{C}$. Then there is a sequence $\{P_k\}$ of polynomials that converges to f pointwise on $\mathbb{C}$.*

2.1.2 Critical Points and Critical Values

Let P be a polynomial of degree n. The point $z \in \mathbb{C}$ is called a *critical point* of P if $P'(z) = 0$. And w is called a *critical value* of P if $w = P(z)$ for some critical point z of P. Let $\{w_1, w_2, \ldots, w_k\}$ be the set of the critical values of P in $\mathbb{C}$. For each w_j, the inverse image of $P^{-1}(w_j)$ has less than n points, say $n - \delta_P(w_j)$ points. We call $\delta_P(w_j)$ the *deficiency* of P at w_j. It is well known that $\sum_{j=1}^{k} \delta_P(w_j) = n - 1$ (for example, see [4, p. 352]). Conversely, we have the following result.

Theorem 6 (For Instance, [4, Theorem 6.2]) *Let $w_1, w_2, \ldots, w_k$ be distinct points of $\mathbb{C}$ and let $\delta(w_1), \delta(w_2), \ldots, \delta(w_k) \in \mathbb{N}$ such that $\sum_{j=1}^{k} \delta(w_j) = n - 1$. Then there is a polynomial P of degree n with $\{w_1, w_2, \ldots, w_k\}$ as its set of the critical values such that $\delta_P(w_j) = \delta(w_j)$ for all $j = 1, \ldots, k$.*

The following result shows how the set of critical points can characterize a polynomial.

Theorem 7 *Two polynomials f, g have the same critical points, counted with multiplicity, if and only if $f = l \circ g$ for some linear polynomial l.*

S. Smale [45, p. 33] proved the following inequality which relates the critical points and critical values of a polynomial.

Theorem 8 ([45]) *Let P be a non-linear polynomial with critical points ζ_j. If z is not a critical point of P, then*

$$\min_j \left| \frac{P(z) - P(\zeta_j)}{z - \zeta_j} \right| \le 4 \left| P'(z) \right|. \tag{2}$$

Smale then asked whether one can replace the factor 4 in the upper bound in (2) by 1, or even possibly by $(d-1)/d$. This problem remains open. It is easy (see [6]) to show that Smale's conjecture is equivalent to the following conjecture.

Conjecture 1 (Normalized) *Let P be a monic polynomial of degree $d \ge 2$ such that $P(0) = 0$ and $P'(0) \ne 0$. Let $\{\zeta_1, \ldots, \zeta_{d-1}\}$ be its critical points. Then*

$$\min_j \left| \frac{P(\zeta_j)}{\zeta_j} \right| \le N \left| P'(0) \right|$$

holds for $N = 1$ (or even $(d-1)/d$).

The estimate of the constant N has been considered by many people and we refer the reader to [10, 11, 20, 23, 29] and [6] for more details.

The following theorem concerns the relative geometric locations between the zeros and the critical points of a polynomial.

Theorem 9 (Gauss-Lucas) *Let P be a polynomial of degree n with zeros $z_1, \ldots, z_n$. The critical points of P lie in the convex hull of the set $\{z_1, \ldots, z_n\}$.*

2.1.3 Factorizations of Polynomials

Let P be a non-linear polynomial in one complex variable. We say that P is *prime* if and only if there do not exist two complex polynomials P_1 and P_2 both with degree greater than one such that $P(z) = P_1(P_2(z))$. Otherwise, P is called *composite* or *factorized*.

Clearly, for a given polynomial P, one can always factorize it as a composition of prime polynomials only and this factorization will be called a *prime factorization*. The number of prime polynomials in a prime factorization is called the *length* of this prime factorization.

In 1922, J.F. Ritt [41] proved the following three fundamental results on the factorizations of complex polynomials.

Theorem R1 ([41]) *A non-linear polynomial P is composite if and only if the monodromy group of F is imprimitive* (definition of the monodromy groups will be given in Sect. 3).

Theorem R2 ([41]) *The length of a non-linear polynomial P is independent of its prime factorizations.*

Theorem R3 ([41]) *Given two prime factorizations of a non-linear polynomial P, one can pass from one to the other by repeatedly use of the following operations:*

1. $\tilde{P} \circ \hat{P} = (\tilde{P} \circ L) \circ (L^{-1} \circ \hat{P})$, *with polynomials $\tilde{P}$, $\hat{P}$ and a linear polynomial L;*
2. $T_m \circ T_n = T_n \circ T_m$, *where T_k is the Chebyshev polynomial of degree k;*
3. $z^r[P_0(z)]^k \circ z^k = z^k \circ [z^r P_0(z^k)]$, *with integers r, k and a polynomial P_0.*

2.2 Part Two: Finite Blaschke Products

2.2.1 Some Simple Properties

Recall that $n_f(w; K)$ denote the number of solutions in $K \subset \mathbb{C}$ for the equation $f(z) = w$, counting with multiplicity.

Theorem 10 *Let f be a finite Blaschke product of degree k. Then for all $w \in \mathbb{D}$,*

$$n_f(w; \mathbb{D}) \equiv k.$$

The converse is also true.

Theorem 11 (Fatou [15–17], Radó [39]) *Let f be analytic in $\mathbb{D}$ with $f(\mathbb{D}) = \mathbb{D}$. Suppose $n_f(w; \mathbb{D}) \equiv k$ for all $w \in \mathbb{D}$. Then f is a finite Blaschke product of degree k.*

Theorem 12 (Uniqueness Theorem) *If B_1 and B_2 are finite Blaschke products of degree not exceeding n and if the equation*

$$B_1(z) = B_2(z) \tag{3}$$

is satisfied at $n + 1$ distinct points in $\mathbb{D}$, then $B_1 \equiv B_2$.

Remark 2 ([24]) If B_1 and B_2 are monic Blaschke products of degree n and if (3) is satisfied at n distinct points in $\mathbb{D}$, then $B_1 \equiv B_2$.

(A finite Blaschke product is said to be *monic* if its normalizing constant $e^{i\theta}$ is 1.)

Theorem 13 ([15]) *Let f be analytic in $\mathbb{D}$ and suppose that*

$$\lim_{|z|\to 1} \left|f(z)\right| = 1.$$

Then f is a finite Blaschke product.

The complex *pseudo-hyperbolic distance* $[z, w]$ in $\mathbb{D}$ is defined by

$$[z, w] = \frac{z - w}{1 - \bar{w}z}.$$

For any $f : \mathbb{D} \to \mathbb{D}$, the hyperbolic difference quotient of f is given by

$$f^*(z, w) = \frac{[f(z), f(w)]}{[z, w]}.$$

Theorem 14 ([5, Theorem 2.4(d)]) *For a given $w \in \mathbb{D}$, $B(z)$ is a finite Blaschke product of degree k if and only if $B^*(z, w)$ is a finite Blaschke product of degree $k - 1$ with the variable z.*

Now we will state Carathéodory's theorem.

Theorem 15 ([21, p. 6]) *Let f be an analytic function on $\mathbb{D}$. If $|f| \le 1$ on $\mathbb{D}$, then there is a sequence $\{B_k\}$ of finite Blaschke products that converges to f pointwise on $\mathbb{C}$.*

2.2.2 Critical Points and Critical Values

Let B be a finite Blaschke product of degree n. The point $z \in \mathbb{C}$ is called a *critical point* of B if $B'(z) = 0$. And B is said to have a critical point at ∞ if either $B(1/z)$ or $\frac{1}{B(1/z)}$ has a critical point at 0. Moreover, w is called a *critical value* of B if $w = B(z)$ for some critical point z of B. Let $\{w_1, w_2, \ldots, w_k\}$ be the set of the critical values of B in $\mathbb{D}$. For each w_j, the inverse image of $B^{-1}(w_j)$ has less than n points, say $n - \delta_B^{\mathbb{D}}(w_j)$ points. We call $\delta_B^{\mathbb{D}}(w_j)$ the *deficiency* of B at w_j. It is well known that $\sum_{j=1}^{k} \delta_B^{\mathbb{D}}(w_j) = n - 1$ (for example, see [4, pp. 352–353]). Conversely, we have the following result.

Theorem 16 (For Instance, [4, p. 353]) *Let $w_1, w_2, \ldots, w_k$ be distinct points of $\mathbb{D}$ and let $\delta(w_1), \delta(w_2), \ldots, \delta(w_k) \in \mathbb{N}$ such that $\sum_{j=1}^{k} \delta(w_j) = n - 1$. Then there is a finite Blaschke product B of degree n with $\{w_1, w_2, \ldots, w_k\}$ as its set of the critical values such that $\delta_B^{\mathbb{D}}(w_j) = \delta(w_j)$ for all $j = 1, \ldots, k$.*

The following result shows how the set of critical points can characterize a finite Blaschke product.

Theorem 17 ([52, Corollary 2]) *Two finite Blaschke products f, g have the same critical points, counted with multiplicity, if and only if $f = \tau \circ g$ for some Möbius transformation τ.*

The following "mean value" inequality was proven by T. Sheil-Small in 2002.

Theorem 18 ([43, p. 366]) *Let B be a finite Blaschke product of degree $d \geq 2$ with critical points $\zeta_j \in \mathbb{D}$ of B such that $B(0) = 0$. If 0 is not a critical point of B, then*

$$\left| \frac{B(\zeta_j)}{\zeta_j} \right| \leq \frac{4}{(1 + |\zeta_j|)^2} \cdot \left| B'(0) \right| \quad \textit{for some } j. \tag{4}$$

It is natural to ask if one can replace 4 by something smaller and therefore consider the following problem.

Problem 1 What is the smallest constant so that (4) still holds?

J.L. Walsh [48] gave an analogous result of the Gauss-Lucas theorem for finite Blaschke products in terms of non-euclidean lines in $\mathbb{D}$, which are segments of circles contained in $\mathbb{D}$ orthogonal to $\partial\mathbb{D}$, or else diameters of $\mathbb{D}$.

Theorem 19 ([48], [43, p. 377]) *Let B be a finite Blaschke product of degree n with zeros $z_1, \ldots, z_n$ in $\mathbb{D}$. Then $B(z)$ has exactly $n - 1$ critical points in $\mathbb{D}$ and these all lie in the non-euclidean convex hull of the set $\{z_1, \ldots, z_n\}$. In particular the critical points in $\mathbb{D}$ lie in the (euclidean) convex hull of the set $\{0, z_1, \ldots, z_n\}$. The critical points of B outside $\mathbb{D}$ are the conjugates (relative to $\partial\mathbb{D}$ of those in $\mathbb{D}$.)*

2.2.3 Factorizations of Finite Blaschke Products

Let B be a finite Blaschke product with $\deg B > 1$. We say that B is *prime* if and only if there do not exist two finite Blaschke products B_1 and B_2 both with degree greater than one such that $B(z) = B_1(B_2(z))$. Otherwise, B is called *composite* or *factorized*.

Clearly, for a given finite Blaschke product B, one can always factorize it as a composition of prime finite Blaschke products only and this factorization will be called a *prime factorization*. The number of prime finite Blaschke products in a prime factorization is called the *length* of this factorization.

All the three results of Ritt on the factorizations of polynomials has been considered for finite Blaschke products and this has been done in [32] or [49].

Theorem R1′ *A finite Blaschke product B ($\deg B > 1$) is composite if and only if the monodromy group of B is imprimitive.*

Theorem R2′ *The length of a finite Blaschke product* B $(\deg B > 1)$ *is independent of its prime factorizations.*

Theorem R3′ *Given two prime factorizations of a finite Blaschke product* B $(\deg B > 1)$, *one can pass from one to the other by repeatedly use of the following operations*:

1. $\tilde{B} \circ \hat{B} = (\tilde{B} \circ M) \circ (M^{-1} \circ \hat{B})$, *with finite Blaschke products* $\tilde{B}$, $\hat{B}$ *and a Möbius transformation* M;
2. $f_{m,n\tau} \circ f_{n,\tau} = f_{n,m\tau} \circ f_{m,\tau}$, *where* $f_{n,\tau}$ *is the Chebyshev-Blaschke product of degree* n (definition of the Chebyshev-Blaschke products $f_{n,\tau}$ will be given in Sect. 3);
3. $z^r[B_0(z)]^k \circ z^k = z^k \circ [z^r B_0(z^k)]$, *with integers* r, k *and a finite Blaschke product* B_0.

We summarize the results in this section by the following table:

Polynomials	Finite Blaschke products
A finite map from $\mathbb{C}$ to $\mathbb{C}$ (Theorem 1)	A finite map from $\mathbb{D}$ to $\mathbb{D}$ (Theorems 10, 11)
Theorem 2, Remark 1	Theorem 12, Remark 2
Theorem 3	Theorem 13
Theorem 4	Theorem 14
Theorem 5	Theorem 15
Theorem 6	Theorem 16
Theorem 7	Theorem 17
Theorem 8, Conjecture 1	Theorem 18, Problem 1
Theorem 9	Theorem 19
Ritt's theory: Theorems R1, R2, R3	Ritt's theory: Theorems R1′, R2′, R3′

3 Chebyshev Polynomials and Chebyshev-Blaschke Products

3.1 Part One: Chebyshev Polynomials

In this section, we will look at a special kind of polynomials, that is, the Chebyshev polynomials and state some elementary properties of them. For more details, see for instance [27], [7, Chap. 2.1] and [42].

3.1.1 Definitions and Some Basic Properties

The *Chebyshev polynomial* T_n is a polynomial of degree n, defined by the relation

$$T_n(z) = \cos n\theta, \quad z = \cos\theta.$$

In other words,

$$T_n(z) = \cos(n \arccos z).$$

The zeros of T_n are precisely the points

$$z_k = \cos \frac{(2k-1)\pi}{2n}, \quad k = 1, 2, \ldots, n.$$

Proposition 1 *The Chebyshev polynomial T_n has the following properties*

1. $T_n^{-1}([-1, 1]) = [-1, 1]$.
2. *Nesting property*: $T_{mn} = T_m \circ T_n$.
3. *The critical points of T_n are contained in* $[-1, 1]$ *and they are precisely the points*

$$\zeta_k = \cos \frac{k\pi}{n}, \quad k = 1, \ldots, n-1.$$

4. *The critical values of T_n are* ± 1.

3.1.2 Monodromy

In this section, we shall study the monodromy of the Chebyshev polynomial T_n of degree n $(n > 2)$.

First let us define the monodromy for a surjective finite map $f : X \to Y$ of degree n, where X and Y are Riemann surfaces. Let Δ be the set of all the critical values of f. Select a point $y \in Y \setminus \Delta$, then $f^{-1}(y) = \{x_1, x_2, \ldots, x_n\}$. For any loop $\alpha \in \pi_1(Y \setminus \Delta, y)$ and any $i \in \{1, 2, \ldots, n\}$, α can be uniquely lifted by f^{-1} as a path with the initial point x_i and end point $x_{\alpha(i)}$, for some $\alpha(i) \in \{1, 2, \ldots, n\}$. Therefore, α can be considered as a permutation of n points, and we can naturally get a group homomorphism $\mu : \pi_1(Y \setminus \Delta) \to S_n$, which we will call the *monodromy* of f. The image of μ is said to be the *monodromy group* of f. So, by considering $X = Y = \mathbb{C}$, the monodromy of a polynomial can be defined.

Now let us find the monodromy of the Chebyshev polynomial T_n. The set of critical values of the Chebyshev polynomial is simply $\Delta = \{-1, 1\}$, and $\pi_1(\mathbb{C} \setminus \Delta) = \langle \sigma, \tau \rangle$, where σ is the loop around 1, and τ is the loop around -1, both with counterclockwise orientation. The monodromy representation μ of T_n is a map from $\pi_1(\mathbb{C} \setminus \Delta)$ to the symmetric group S_n, defined by

$$\begin{cases} \mu(\sigma) = (2, 2k)(3, 2k-1) \cdots (k, k+2), & \text{if } n = 2k, \\ \mu(\tau) = (2, 1)(3, 2k) \cdots (k+1, k+2), & \\ \mu(\sigma) = (2, 2k+1)(3, 2k) \cdots (k+1, k+2), & \text{if } n = 2k+1. \\ \mu(\tau) = (2, 1)(3, 2k+1) \cdots (k+1, k+3), & \end{cases} \tag{5}$$

The above monodromy of the Chebyshev polynomial T_n was first given by J.F. Ritt in [41] and its detailed explanation can also be found in [32].

3.1.3 Differential Equations

The Chebyshev polynomial T_n satisfies the following differential equations

$$n^2\left(w^2-1\right)=\left(w'\right)^2\left(z^2-1\right) \tag{6}$$

and

$$\left(1-z^2\right)w''-zw'+n^2w=0. \tag{7}$$

These results can be found in [27].

3.1.4 The Julia Set

Before looking at the Julia set of the Chebyshev polynomial, let us review some of the standard facts on the iteration theory of rational functions. For more details, we refer the reader to [3] or [46].

Let $f:\overline{\mathbb{C}}\to\overline{\mathbb{C}}$ be a rational function of degree greater than one. We will denote the n-th iterate of f by $f^{\circ n}$, that is, $f^{\circ 0}(z)=z$ and $f^{\circ n}(z)=f(f^{\circ(n-1)}(z))$ for $n\geq 1$. To study the Fatou and Julia sets of rational functions, we first introduce normal families of rational functions.

Definition 1 Let U be an open subset of $\overline{\mathbb{C}}$ and $\mathcal{F}=\{f_i : i\in I\}$ a family of rational functions defined on U with values in $\overline{\mathbb{C}}$ (I is any index set). The family $\mathcal{F}$ is a *normal family* if every sequence $\{f_n\}$ contains a subsequence $\{f_{n_j}\}$ which converges uniformly on compact subsets of U.

Let $f:\overline{\mathbb{C}}\to\overline{\mathbb{C}}$ be a rational function of degree greater than one. The *Fatou set*, $F(f)$ consists of point z in $\overline{\mathbb{C}}$ at which the sequence $\{f^{\circ n}\}_{n\in\mathbb{N}}$ is defined and normal in some neighborhood of z. The *Julia set*, $J(f)$ is defined as the complement of $F(f)$ in $\overline{\mathbb{C}}$.

Here are some properties of Fatou and Julia sets.

1. By definition, $F(f)$ is open and $J(f)$ is closed.
2. $F(f)$ and $J(f)$ are completely invariant (a set S is said to be *completely invariant* if $f(S)=S=f^{-1}(S)$).
3. (The minimality of Julia sets) $J(f)$ is the smallest closed, completely invariant set with at least three points.

Note that $[-1,1]$ is completely invariant under T_n. By the minimality of the Julia set, $J(T_n)$ is contained in $[-1,1]$. In fact, $J(T_n)$ is exactly $[-1,1]$. For the proof, we refer the reader to [3].

3.1.5 The Approximation Problems

Let $\mathcal{P}_n$ denote the set of all polynomials of degree n. Before stating the approximation problems related to Chebyshev polynomials, we first review some facts about the n-th polynomial of least deviation. Given any compact set E and any continuous function $\varphi : E \to \mathbb{C}$, there exists a polynomial $p^*_{n,\varphi} \in \mathcal{P}_n$ such that

$$\max_{z\in E}\left|\varphi(z) - p^*_{n,\varphi}(z)\right| = \min_{p_n\in\mathcal{P}_n}\max_{z\in E}\left|\varphi(z) - p_n(z)\right|.$$

Such a polynomial is called the *n-th polynomial of least deviation* from φ on E. The n-th polynomial of least deviation from φ is unique if E contains at least $n+1$ points. For more details, we refer the reader to [26].

Now we consider the problem when $E = [-1, 1]$ and $\varphi(z) = z^n$:

Problem I Find $p^*_{n-1} \in \mathcal{P}_{n-1}$ that attains the minimum

$$\sigma_n = \min_{p_{n-1}\in\mathcal{P}_{n-1}}\max_{z\in[-1,1]}\left|z^n - p_{n-1}(z)\right|.$$

The solution to Problem I is $p^*_{n-1}(z) = z^n - 2^{1-n}T_n(z)$ with $\sigma_n = 2^{1-n}$ (see for instance [7, Theorem 2.1.1]).

Remark 3 Such a polynomial p^*_{n-1} is actually the $(n-1)$-th polynomial of least deviation from z^n on $[-1, 1]$.

Let $\mathcal{P}^{mon}_n$ denote the set of all monic polynomials of degree n. We can rephrase the above problem as follows:

Problem II Find $q^*_n \in \mathcal{P}^{mon}_n$ that attains the minimum

$$\hat{\sigma}_n = \min_{q_n\in\mathcal{P}^{mon}_n}\max_{z\in[-1,1]}\left|q_n(z)\right|. \tag{8}$$

It follows from Problem I that the solution to Problem II is $q^*_n = 2^{1-n}T_n$ with $\hat{\sigma}_n = 2^{1-n}$.

3.2 *Part Two: Chebyshev-Blaschke Products*

3.2.1 Definition and Some Basic Properties

The construction of Chebyshev-Blaschke products was first considered in [49] by studying the monodromy of Chebyshev polynomials. Here we shall take a more direct approach by defining the Chebyshev-Blaschke products in terms of Jacobi elliptic functions. We therefore first review the properties of Jacobi elliptic functions (see [9] or [47]).

Jacobi Elliptic Functions For any $\tau \in \mathbb{H}$, write $q = e^{\pi i \tau}$ and define the four theta functions as follows:

$$\begin{aligned}
\vartheta_1(v,\tau) &= \sum_{n=-\infty}^{\infty} i^{2n-1} q^{(n+\frac{1}{2})^2} e^{(2n+1)vi} \\
&= 2q^{1/4}\sin\pi v - 2q^{9/4}\sin 3\pi v + 2q^{25/4}\sin 5\pi v - \cdots, \\
\vartheta_2(v,\tau) &= \sum_{n=-\infty}^{\infty} q^{(n+\frac{1}{2})^2} e^{(2n+1)vi} \\
&= 2q^{1/4}\cos\pi v + 2q^{9/4}\cos 3\pi v + 2q^{25/4}\cos 5\pi v + \cdots, \\
\vartheta_3(v,\tau) &= \sum_{n=-\infty}^{\infty} q^{n^2} e^{2nvi} \\
&= 1 + 2q\cos 2\pi v + 2q^4\cos 4\pi v + 2q^9\cos 6\pi v + \cdots, \\
\vartheta_0(v,\tau) &= \sum_{n=-\infty}^{\infty} (-1)^n q^{n^2} e^{2nvi} \\
&= 1 - 2q\cos 2\pi v + 2q^4\cos 4\pi v - 2q^9\cos 6\pi v + \cdots.
\end{aligned}$$

We further define:

$$\begin{aligned}
\omega_1(\tau) &= \pi\vartheta_3^2(0,\tau) = \pi\left(1 + 2q + 2q^4 + \cdots\right)^2, \\
\omega_2(\tau) &= \tau\omega_1(\tau), \\
k(\tau) &= \frac{\vartheta_2^2(0,\tau)}{\vartheta_3^2(0,\tau)}, \\
\sqrt{k(\tau)} &= \frac{\vartheta_2(0,\tau)}{\vartheta_3(0,\tau)}, \\
k'(\tau) &= \frac{\vartheta_0^2(0,\tau)}{\vartheta_3^2(0,\tau)}, \\
\sqrt{k'(\tau)} &= \frac{\vartheta_0(0,\tau)}{\vartheta_3(0,\tau)}.
\end{aligned}$$

Notice that if $\tau \in \mathbb{R}_+ i$, then $\sqrt{k(\tau)} \in \mathbb{R}_+$ and $\omega_1 \in \mathbb{R}_+$.

The Jacobi elliptic functions can be expressed as follows:

$$\begin{aligned}
\operatorname{sn} u &= \frac{\vartheta_3(0,\tau)}{\vartheta_2(0,\tau)} \cdot \frac{\vartheta_1(u/\omega_1,\tau)}{\vartheta_0(u/\omega_1,\tau)} = \frac{1}{\sqrt{k}} \cdot \frac{\vartheta_1(u/\omega_1,\tau)}{\vartheta_0(u/\omega_1,\tau)}, \\
\operatorname{cn} u &= \frac{\vartheta_0(0,\tau)}{\vartheta_2(0,\tau)} \cdot \frac{\vartheta_2(u/\omega_1,\tau)}{\vartheta_0(u/\omega_1,\tau)} = \frac{\sqrt{k'}}{\sqrt{k}} \cdot \frac{\vartheta_2(u/\omega_1,\tau)}{\vartheta_0(u/\omega_1,\tau)},
\end{aligned}$$

$$\operatorname{dn} u = \frac{\vartheta_0(0,\tau)}{\vartheta_3(0,\tau)} \cdot \frac{\vartheta_3(u/\omega_1,\tau)}{\vartheta_0(u/\omega_1,\tau)} = \sqrt{k'} \cdot \frac{\vartheta_3(u/\omega_1,\tau)}{\vartheta_0(u/\omega_1,\tau)}.$$

Notice that sn is an elliptic function of order 2 with primitive periods $2\omega_1$ and ω_2. Moreover,

$$\operatorname{sn}\left(\pm\frac{\omega_1}{2},\tau\right) = \pm 1, \qquad \operatorname{sn}(\omega_1 - u,\tau) = \operatorname{sn}(u,\tau).$$

Definition and Properties of Chebyshev-Blaschke Products Let $\tau \in \mathbb{R}_+ i$. Now we will consider the following Jacobi function

$$\operatorname{cd} = \frac{\operatorname{cn}}{\operatorname{dn}}.$$

The Jacobi cd function can be expressed in terms of sn:

$$\operatorname{cd}(u,\tau) = \operatorname{sn}\left(u + \frac{\omega_1}{2},\tau\right).$$

So cd is an elliptic function of order two with the primitive periods $2\omega_1$ and ω_2. Both cn and dn are even, so is cd. One of the properties of cd is

$$\sqrt{k(\tau)}\operatorname{cd}\left(u + \frac{\omega_2(\tau)}{2},\tau\right) = \frac{1}{\sqrt{k(\tau)}\operatorname{cd}(u,\tau)},$$

which tells us that cd has a similar property to the Schwarz reflection principle and therefore cd has a close relationship with finite Blaschke products.

Define $x_{n\tau}(u) = \sqrt{k(n\tau)}\operatorname{cd}(nu\omega_1(n\tau), n\tau)$, $n \in \mathbb{N}$. It is easy to show that $f_{n,\tau}(z) := x_{n\tau} \circ x_\tau^{-1}(z)$ is a rational function. In other words, we can define $f_{n,\tau}$ by the following parametric equations

$$f_{n,\tau}(z) = \sqrt{k(n\tau)}\operatorname{cd}\big(nu\omega_1(n\tau), n\tau\big), \quad z = \sqrt{k(\tau)}\operatorname{cd}\big(u\omega_1(\tau),\tau\big).$$

We can easily check that the zeros of $f_{n,\tau}$ are $z_p = \sqrt{k(\tau)}\operatorname{cd}(\frac{(2p-1)\omega_1(\tau)}{2n},\tau)$, $p = 1,\dots,n$. In particular, all the zeros are contained in $[-\sqrt{k(\tau)}, \sqrt{k(\tau)}]$. Moreover, it is clear that $f_{n,\tau}$ satisfies the symmetry property

$$\frac{1}{f_{n,\tau}(z)} = f_{n,\tau}\left(\frac{1}{z}\right).$$

So $f_{n,\tau}$ is a finite Blaschke product of degree n. We call $f_{n,\tau}$ a *Chebyshev-Blaschke product*. The proof of the following proposition is given in [30].

Proposition 2 *The Chebyshev-Blaschke product $f_{n,\tau}$ has the following properties*

1. $f_{n,\tau}^{-1}([-\sqrt{k(n\tau)}, \sqrt{k(n\tau)}]) = [-\sqrt{k(\tau)}, \sqrt{k(\tau)}]$.
2. *Nesting property*: $f_{mn,\tau} = f_{m,n\tau} \circ f_{n,\tau}$.

3. *The critical points of* $f_{n,\tau}$ *are contained in*

$$\left(-\infty, -\frac{1}{\sqrt{k(\tau)}}\right] \cup [-\sqrt{k(\tau)}, \sqrt{k(\tau)}] \cup \left[\frac{1}{\sqrt{k(\tau)}}, \infty\right) \cup \{\infty\}.$$

More precisely,

a. $f_{n,\tau}$ *has* $n-1$ *critical points in* $[-\sqrt{k(\tau)}, \sqrt{k(\tau)}]$ *and they are the points*

$$w_p = \sqrt{k(\tau)}\,\mathrm{cd}\left(\frac{p\omega_1(\tau)}{n}, \tau\right), \quad p = 1, \ldots, n-1$$

with

$$\begin{aligned} w_{n-1} &= -w_1 \\ w_{n-2} &= -w_2 \\ &\vdots \\ w_{n-i} &= -w_i, \quad \textit{where } 1 \le i \le \left\lfloor \frac{n}{2} \right\rfloor \\ &\vdots \end{aligned}$$

and for even n we have

$$w_{\frac{n}{2}} = 0;$$

b. $f_{n,\tau}$ *has* $n-1$ *critical points in* $(-\infty, -\frac{1}{\sqrt{k(\tau)}}] \cup [\frac{1}{\sqrt{k(\tau)}}, \infty) \cup \{\infty\}$ *and they are the points*

$$\frac{1}{w_p}, \quad p = 1, \ldots, n-1.$$

Moreover, all the critical points of $f_{n,\tau}$ *have multiplicity* 1.

4. *There are only* 4 *critical values* $\pm\sqrt{k(n\tau)}, \pm\frac{1}{\sqrt{k(n\tau)}}$ *and only* 4 *non-critical points* $\pm\sqrt{k(\tau)}, \pm\frac{1}{\sqrt{k(\tau)}}$, *whose images are critical values.*

Chebyshev-Blaschke Products of Small Degrees When $n = 2$, the Chebyshev-Blaschke product $f_{2,\tau}$ has the form

$$\frac{z^2 - a}{1 - az^2}.$$

To determine the number a, notice that $f_{2,\tau}$ has only one critical point 0 and

$$f(0) = -\sqrt{k(2\tau)}.$$

So

$$a = \sqrt{k(2\tau)} = \frac{\vartheta_2(0, 2\tau)}{\vartheta_3(0, 2\tau)}.$$

When $n = 3$, the Chebyshev-Blaschke product $f_{3,\tau}$ has the form

$$z\frac{z^2 - a}{1 - az^2}.$$

To determine the number a, notice that $f_{3,\tau}$ has three zeros 0 and $\pm\sqrt{k(\tau)} \times \mathrm{cd}(\frac{\omega_1(\tau)}{6}, \tau)$. So

$$a = k(\tau)\mathrm{cd}^2\left(\frac{\omega_1(\tau)}{6}, \tau\right) = \frac{\vartheta_2^2(1/6, \tau)}{\vartheta_3^2(1/6, \tau)}.$$

When $n = 2^i 3^j$, we can apply the nesting property to get

$$f_{2^i 3^j, \tau} = f_{2, 2^{i-1}3^j \tau} \circ \cdots \circ f_{2, 2\cdot 3^j \tau} \circ f_{2, 3^j \tau} \circ f_{3, 3^{j-1}\tau} \circ \cdots \circ f_{3, 3\tau} \circ f_{3,\tau}.$$

Chebyshev-Blaschke Products $f_{n,\tau}$ for $\tau \to \infty$ Let $\tau \in \mathbb{R}_+ i$ and $z = \mathrm{cd}(\theta, \tau)$. It is easily seen that

$$\lim_{\tau\to\infty} \mathrm{cd}(\theta, \tau) = \cos\theta \quad \text{and} \quad \lim_{\tau\to\infty} \omega_1(\tau) = \pi. \tag{9}$$

By the definition of $f_{n,\tau}$, we get

$$f_{n,\tau}\big(\sqrt{k(\tau)}z\big) = f_{n,\tau}\big(\sqrt{k(\tau)}\,\mathrm{cd}(\theta, \tau)\big) = \sqrt{k(n\tau)}\,\mathrm{cd}\left(\frac{n\theta\omega_1(n\tau)}{\omega_1(\tau)}, n\tau\right). \tag{10}$$

Let $\mathcal{T}_{n,\tau}(z) = \frac{f_{n,\tau}(\sqrt{k(\tau)}z)}{\sqrt{k(n\tau)}}$. Then it follows from (9) and (10) that for $z = \cos\theta$,

$$\lim_{\tau\to\infty} \mathcal{T}_{n,\tau}(\cos\theta) = \lim_{\tau\to\infty} \mathcal{T}_{n,\tau}\big(\mathrm{cd}(\theta, \tau)\big) = \lim_{\tau\to\infty} \mathrm{cd}\left(\frac{n\theta\omega_1(n\tau)}{\omega_1(\tau)}, n\tau\right) = \cos n\theta.$$

By the definition of the Chebyshev polynomial,

$$T_n(z) = \lim_{\tau\to\infty} \mathcal{T}_{n,\tau}(z) = \lim_{\tau\to\infty} \frac{f_{n,\tau}(\sqrt{k(\tau)}z)}{\sqrt{k(n\tau)}}. \tag{11}$$

3.2.2 Monodromy

The monodromy of the Chebyshev-Blaschke product of degree $n > 2$ is the same as that of the Chebyshev polynomial: note that the set of critical values of the Chebyshev-Blaschke product $f_{n,\tau}$ on the unit disk $\mathbb{D}$ is simply $\Delta = \{\pm\sqrt{k(n\tau)}\}$, and $\pi_1(\mathbb{D} - \Delta) = \langle\sigma, \tau\rangle$, where σ is the loop around $\sqrt{k(n\tau)}$, and τ is the loop around $-\sqrt{k(n\tau)}$, both with counterclockwise orientation. The monodromy representation μ of $f_{n,\tau}$ is $\pi_1(\mathbb{D}-\Delta)$ to the symmetric group S_n, defined by

$$\begin{cases} \mu(\sigma) = (2,2k)(3,2k-1)\cdots(k,k+2), \\ \mu(\tau) = (2,1)(3,2k)\cdots(k+1,k+2), \end{cases} \text{if } n=2k, \\ \begin{cases} \mu(\sigma) = (2,2k+1)(3,2k)\cdots(k+1,k+2), \\ \mu(\tau) = (2,1)(3,2k+1)\cdots(k+1,k+3), \end{cases} \text{if } n=2k+1.$$

In fact, the Chebyshev-Blaschke product can be recovered from this monodromy. Such a construction was shown in [32, 49] or [50].

3.2.3 Differential Equations

It follows from Proposition 2(iii), (iv) that the Chebyshev-Blaschke product $f_{n,\tau}$ satisfies the differential equation

$$c_{n,\tau}\cdot\left(w^2-k(n\tau)\right)\left(w^2-\frac{1}{k(n\tau)}\right)=\left(w'\right)^2\left(z^2-k(\tau)\right)\left(z^2-\frac{1}{k(\tau)}\right), \tag{12}$$

for some constant $c_{n,\tau}$. In fact, the constant $c_{n,\tau}$ can be found explicitly by solving (12):

$$c_{n,\tau}=\left[\frac{n\sqrt{k(n\tau)}\vartheta_3^2(0,n\tau)}{\sqrt{k(\tau)}\vartheta_3^2(0,\tau)}\right]^2=\left[\frac{n\vartheta_2(0,n\tau)\vartheta_3(0,n\tau)}{\vartheta_2(0,\tau)\vartheta_3(0,\tau)}\right]^2=\frac{n^2\vartheta_2^4(0,\frac{n\tau}{2})}{\vartheta_2^4(0,\frac{\tau}{2})},$$

where the last equality holds by using the theta constant identity (for example, see [14])

$$\vartheta_2^2(0,\tau)=2\vartheta_2(0,2\tau)\vartheta_3(0,2\tau).$$

Remark 4 The constant $c_{n,\tau}$ has the following property: $c_{mn,\tau}=c_{m,n\tau}c_{n,\tau}$.

3.2.4 The Julia Set

The Julia set of a finite Blaschke product of degree >1 is either the unit circle or a Cantor set on the circle [8]. The following theorem is a well known fact about the Julia set of a finite Blaschke product.

Theorem 20 ([28]) *If a finite Blaschke product B of degree >1 has a fixed point z_0 in $\mathbb{D}$, then the Julia set $J(B)$ of B is the unit circle.*

Since it is proven in [30] that the Chebyshev-Blaschke product $f_{n,\tau}$ has a fixed point in the interval $[-\sqrt{k(\tau)},\sqrt{k(\tau)}]\subset\mathbb{D}$, by Theorem 20, the Julia set $J(f_{n,\tau})$ is the unit circle $\partial\mathbb{D}$.

3.2.5 The Approximation Problems

In this section, we would like to point out that the Chebyshev-Blaschke product $f_{n,\tau}$ are solutions to some approximation problems which are related to Zolotarev's third problem (Problem B) and fourth problem (Problem C). For more details of these two problems, see for instance [2, 25].

Before looking at Zolotarev's problems, let us state the following problem in Gonchar's paper [22]:

Problem A Find $g^*_{n,\tau} \in R_{nn}$ that attains the minimum

$$\sigma_{A,n,\tau} = \min_{g_n \in R_{nn}} \frac{\max_{z\in E} |g_n(z)|}{\min_{z \in F} |g_n(z)|},$$

where $E = [-\sqrt{k(\tau)}, \sqrt{k(\tau)}]$, $F = (-\infty, -\frac{1}{\sqrt{k(\tau)}}] \cup [\frac{1}{\sqrt{k(\tau)}}, \infty)$ and R_{nn} is the set of irreducible rational functions whose numerator and denominator are real polynomials with degree at most n.

Remark 5 In fact, Problem A was originally stated in Gonchar's paper [22] for any disjoint compact sets E and F of $\mathbb{C}$.

It is obvious that Problem A can be rephrased as the following problem.

Problem B (Zolotarev's 3rd Problem) Find $h^*_{n,\tau} \in R_{nn}$ that attains the minimum

$$\sigma_{B,n,\tau} = \min_{h_n \in R_{nn}} \max_{z \in E} \left| h_n(z) \right|$$

subject to

$$\min_{z\in F} \left| h_n(z) \right| = 1,$$

where $E = [-\sqrt{k(\tau)}, \sqrt{k(\tau)}]$ and $F = (-\infty, -\frac{1}{\sqrt{k(\tau)}}] \cup [\frac{1}{\sqrt{k(\tau)}}, \infty)$.

In [2, Chap. 9], Akhiezer showed that Problem B is equivalent to the following problem.

Problem C (Zolotarev's 4th Problem) Find $r^*_{n,t} \in R_{nn}$ that attains the minimum

$$\sigma_{C,n,t} = \min_{r_n \in R_{nn}} \max_{z \in G} \left| r_n(x) - \operatorname{sgn} x \right|,$$

where $G = [-1/t, -1] \cup [1, 1/t]$ $(0 < t < 1)$ and

$$\operatorname{sgn} x = \begin{cases} -1, & \text{for } x < 0, \\ 1, & \text{for } x > 0. \end{cases}$$

Using the technique in [2], we obtained in [30] the solution (in terms of $f_{n,\tau}$) to Problem C and hence Problems A and B as well. More precisely, we have

1. $f_{n,\tau}$ is a solution to Problem A with $\sigma_{A,n,\tau} = k(n\tau)$;
2. $\sqrt{k(n\tau)}f_{n,\tau}$ is a solution to Problem B with $\sigma_{B,n,\tau} = k(n\tau)$;
3. $r_{n,t}(x) = \frac{1-k(n\tau)}{1+k(n\tau)} \cdot \frac{f_{n,\tau}(z)-1}{f_{n,\tau}(z)+1}$ is a solution to Problem C with $\sigma_{C,n,t} = \frac{2\sqrt{k(n\tau)}}{1+k(n\tau)}$, where

$$x = \frac{1+\sqrt{k(\tau)}}{1-\sqrt{k(\tau)}} \cdot \frac{z-1}{z+1} \quad \text{and} \quad t = \left(\frac{1-\sqrt{k(\tau)}}{1+\sqrt{k(\tau)}}\right)^2.$$

New Approximation Problems Let $\mathcal{B}_n$ denote the set of all finite Blaschke products of degree n and let $\mathcal{B}_{n,\tau}^{\mathbb{R}}$ denote the set of all finite Blaschke products of degrees n such that all the zeros are contained in $E = [-\sqrt{k(\tau)}, \sqrt{k(\tau)}]$.

Problem D Find $\tilde{B}_{n,\tau} \in \mathcal{B}_{n,\tau}^{\mathbb{R}}$ that attains the minimum

$$\sigma_{D,n,\tau} = \min_{B_n \in \mathcal{B}_{n,\tau}^{\mathbb{R}}} \max_{z \in E} \left|B_n(z)\right| = \min_{\alpha_1,\ldots,\alpha_n \in E} \max_{z \in E} \left|\frac{(z-\alpha_1)\cdots(z-\alpha_n)}{(1-\overline{\alpha_1}z)\cdots(1-\overline{\alpha_n}z)}\right|,$$

where $E = [-\sqrt{k(\tau)}, \sqrt{k(\tau)}]$.

By the symmetry property $B_n(\frac{1}{z}) = \frac{1}{B_n(z)}$, we have

$$\left(\max_{z \in E}\left|B_n(z)\right|\right)^2 = \frac{\max_{z\in E} |B_n(z)|}{\min_{z\in F} |B_n(z)|}.$$

On the other hand, $f_{n,\tau} \in \mathcal{B}_{n,\tau}^{\mathbb{R}}$ is a solution to Problem A and hence is also a solution to the following minimization problem

$$\min_{B_n \in \mathcal{B}_{n,\tau}^{\mathbb{R}}} \frac{\max_{z\in E} |B_n(z)|}{\min_{z\in F} |B_n(z)|},$$

with $E = [-\sqrt{k(\tau)}, \sqrt{k(\tau)}]$ and $F = (-\infty, -\frac{1}{\sqrt{k(\tau)}}] \cup [\frac{1}{\sqrt{k(\tau)}}, \infty)$.

Therefore, $f_{n,\tau}$ is also a solution to Problem D with $\sigma_{D,n,\tau} = \sqrt{\sigma_{A,n,\tau}} = \sqrt{k(n\tau)}$.

Finally, we consider the following problem which is an analogue of Problem II.

Problem II′ Find $B^*_{n,\tau} \in \mathcal{B}_n$ that attains the minimum

$$\sigma_{II',n,\tau} = \min_{B_n \in \mathcal{B}_n} \max_{z \in E} \left|B_n(z)\right| = \min_{\alpha_1,\ldots,\alpha_n \in \mathbb{D}} \max_{z \in E} \left|\frac{(z-\alpha_1)\cdots(z-\alpha_n)}{(1-\overline{\alpha_1}z)\cdots(1-\overline{\alpha_n}z)}\right|, \tag{13}$$

where $E = [-\sqrt{k(\tau)}, \sqrt{k(\tau)}]$.

It was shown in [30] that $f_{n,\tau}$ is a solution to Problem II′ with $\sigma_{II',n,\tau} = \sqrt{k(n\tau)}$.

We summarize the results in this section by the following table:

Chebyshev polynomials T_n	Chebyshev-Blaschke products $f_{n,\tau}$
$T_n(z) = \cos n\theta$, where $z = \cos\theta$	$f_{n,\tau}(z) = \sqrt{k(n\tau)}\,\mathrm{cd}(nu\omega_1(n\tau), n\tau)$, where $z = \sqrt{k(\tau)}\,\mathrm{cd}(u\omega_1(\tau), \tau)$
Zeros: $z_p = \cos\frac{(2p-1)\pi}{2n}$ $(p = 1, \dots, n)$	Zeros: $z_p = \sqrt{k(\tau)}\,\mathrm{cd}(\frac{(2p-1)\omega_1(\tau)}{2n}, \tau)$ $(p = 1, \dots, n)$
Critical points in $[-1, 1]$: $w_p = \cos\frac{p\pi}{n}$ $(p = 1, \dots, n-1)$	Critical points in $[-\sqrt{k(\tau)}, \sqrt{k(\tau)}]$: $w_p = \sqrt{k(\tau)}\,\mathrm{cd}(\frac{p\omega_1(\tau)}{n}, \tau)$ $(p = 1, \dots, n-1)$
Critical values in $\mathbb{C}$: ± 1	Critical values in $\mathbb{D}$: $\pm\sqrt{k(n\tau)}$
$T_n^{-1}([-1, 1]) = [-1, 1]$	$f_{n,\tau}^{-1}([-\sqrt{k(n\tau)}, \sqrt{k(n\tau)}]) = [-\sqrt{k(\tau)}, \sqrt{k(\tau)}]$
Nesting property: $T_{mn} = T_m \circ T_n$	Nesting property: $f_{mn,\tau} = f_{m,n\tau} \circ f_{n,\tau}$
Monodromy: (5)	Monodromy: same as that of T_n
DE: Equation (6)&(7)	DE: Equation (12)
Julia set: $J(T_n) = [-1, 1]$	Julia set: $J(f_{n,\tau}) = \partial\mathbb{D}$
Minimax problem: (8) Solution: $2^{1-n}T_n$	Minimax problem: (13) Solution: $f_{n,\tau}$
Equation (11): $T_n(z) = \lim_{\tau\to\infty}[f_{n,\tau}(\sqrt{k(\tau)}z)/\sqrt{k(n\tau)}]$	

4 Polynomials and Finite Blaschke Products That Share a Set

In this section, we discuss some problems about polynomials and finite Blaschke products sharing a set. In Sect. 4.1, some results for polynomials are reviewed. Some analogous results for finite Blaschke products are stated in Sect. 4.2.

4.1 Part One: Polynomials

4.1.1 Polynomials That Share Two Values in the Complex Plane

In 1971, W. Adams and E. Sraus [1] proved that the two nonconstant one variable polynomials p and q are identical if they share two distinct finite values a and b IM (ignoring multiplicities), that is, $p^{-1}(\{a\}) = q^{-1}(\{a\})$ and $p^{-1}(\{b\}) = q^{-1}(\{b\})$.

4.1.2 Polynomials That Share a Set

In 1978, C.C. Yang [51, p. 169] raised the following problem: what can be said if p and q are nonconstant polynomials of the same degree and share the set $\{0, 1\}$ IM, that is, $p^{-1}(\{0, 1\}) = q^{-1}(\{0, 1\})$?

F.B. Pakovich [34] solved this problem by using the uniqueness property of polynomials of least deviation, and proved that a polynomial of a given degree is uniquely determined up to the sign $\pm$ by the preimage of the set $\{-1, 1\}$. This problem was also solved and in fact the result was generalized to the preimage of any compact set containing at least 2 points in [35] and [33]. When two polynomials p, q of arbitrary degrees share a compact set K of positive (logarithmic) capacity, that is, $p^{-1}(K) = q^{-1}(K)$, T.C. Dinh gave a complete description of p and q in [12]. This result was extended to an arbitrary infinite compact set K in [13]. Later F.B. Pakovich [36] considered the more general case when $p^{-1}(K_1) = q^{-1}(K_2)$ for two arbitrary compact sets K_1 and K_2.

The sharing set problem is related to the functional equation

$$f \circ p = g \circ q, \tag{14}$$

where f, g, p, q are polynomials. In fact, T.C. Dinh [12] showed that if $p^{-1}(K) = q^{-1}(K)$ for a compact set K of positive (logarithmic) capacity, then there exist two polynomials f, g such that (14) holds. The idea of his proof is to make use of the uniqueness of equilibrium measures to obtain subharmonic functions ϕ and ψ such that $\phi \circ p = \psi \circ q$. Then by considering the germ of conformal map near ∞, there actually exist two polynomials f, g such that (14) holds. Finally the complete classification of p and q can be obtained by applying Ritt's theorem (Theorem R3) for polynomials. Later this result was also proved and in fact extended to the case $p^{-1}(K_1) = q^{-1}(K_2)$ (for any compact sets K_1, K_2) by F.B. Pakovich [36] who made use of the uniqueness of the least deviations from zero instead of the uniqueness of the equilibrium measures.

4.1.3 Dinh's Result

Theorem 21 ([12, Theorem 1]) *Let f_1 and f_2 be polynomials of degree $d_1 \geq 1$ and $d_2 \geq 1$, and let $K_0 \subset \mathbb{C}$ be a compact set of positive (logarithmic) capacity such that $K := f_1^{-1}(K_0) = f_2^{-1}(K_0)$ holds. Then there exist polynomials $\tilde{f}_1, \tilde{f}_2, Q$ with $\deg Q = d$, where $d = \gcd(d_1, d_2)$ such that*

$$f_1 = \tilde{f}_1 \circ Q, \qquad f_2 = \tilde{f}_2 \circ Q$$

and one of the following conditions is true:

1. *$\tilde{f}_1 = id$ or $\tilde{f}_2 = id$;*
2. *$d_1 > d$, $d_2 > d$ and $\tilde{f}_1 = \sigma \circ z^{d_1/d}$, $\tilde{f}_2 = \sigma \circ a z^{d_2/d}$, for some linear function σ and $a \in \mathbb{C} \setminus \{0\}$;*
3. *$d_1 > d$, $d_2 > d$ and $\tilde{f}_1 = \sigma \circ \pm T_{d_1/d}$, $\tilde{f}_2 = \sigma \circ \pm T_{d_2/d}$, for some linear function σ, where T_k is the Chebyshev polynomial of degree k.*

The above theorem was extended to an infinite compact set K_0 in [13] (see also [36, Theorem 2]).

4.1.4 Pakovich's Result

Theorem 22 ([36, Theorem 1]) *Let* f_1, f_2 *be polynomials*, $\deg f_1 = d_1$, $\deg f_2 = d_2$, $d_1 \le d_2$, *and* $K_1, K_2 \subset \mathbb{C}$ *be compact sets such that* $K := f_1^{-1}(K_1) = f_2^{-1}(K_2)$ *holds. Suppose that* $\mathrm{card}(K) \ge \mathrm{lcm}(d_1, d_2)$. *We have*

1. *if* d_1 *divides* d_2, *then there exists a polynomial* g_1 *such that* $f_2 = g_1 \circ f_1$ *and* $K_1 = g_1^{-1}(K_2)$;
2. *if* d_1 *does not divide* d_2, *then there exist polynomials* g_1, g_2, *with* $\deg g_1 = d_2/d$, $\deg g_2 = d_1/d$, *where* $d = \gcd(d_1, d_2)$ *such that*

$$g_1 \circ f_1 = g_2 \circ f_2$$

and a compact set $K_3 \subset \mathbb{C}$ *such that*

$$K_1 = g_1^{-1}(K_3) \quad \text{and} \quad K_2 = g_2^{-1}(K_3).$$

Furthermore, there exist polynomials $\tilde{f}_1, \tilde{f}_2, W$, *with* $\deg W = d$, *such that*

$$f_1 = \tilde{f}_1 \circ W, \qquad f_2 = \tilde{f}_2 \circ W$$

and there exist linear functions σ_1, σ_2 *such that either*

$$\begin{aligned} g_1 &= z^c\big[R(z)\big]^{d_1/d} \circ \sigma_1^{-1}, & \tilde{f}_1 &= \sigma_1 \circ z^{d_1/d}, \\ g_2 &= z^{d_1/d} \circ \sigma_2^{-1}, & \tilde{f}_2 &= \sigma_2 \circ z^c R\big(z^{d_1/d}\big) \end{aligned}$$

for some polynomial R *and for* c *being the remainder after division of* d_2/d *by* d_1/d, *or*

$$\begin{aligned} g_1 &= T_{d_2/d} \circ \sigma_1^{-1}, & \tilde{f}_1 &= \sigma_1 \circ T_{d_1/d}, \\ g_2 &= T_{d_1/d} \circ \sigma_2^{-1}, & \tilde{f}_2 &= \sigma_2 \circ T_{d_2/d} \end{aligned}$$

for the Chebyshev polynomials $T_{d_1/d}, T_{d_2/d}$.

4.2 Part Two: Finite Blaschke Products

4.2.1 Finite Blaschke Products That Share Two Values in the Unit Disk

In [1], W. Adams and E. Sraus also showed that two nonconstant rational functions that share four values a, b, c, d IM are identical. Now we consider two nonconstant finite Blaschke products B_1 and B_2, sharing two distinct finite values a and b IM on $\mathbb{D}$. By the symmetry property

$$B_i(1/\overline{z}) = 1/\overline{B_i(z)} \quad (i = 1, 2),$$

B_1 and B_2 also share $1/\bar{a}$ and $1/\bar{b}$. Therefore B_1 and B_2 are identical.

4.2.2 Finite Blaschke Products That Share a Set

For finite Blaschke products, we can also study the problem of sharing a compact set E. Similar to the case of polynomials, it makes sense to relate $B_1^{-1}(E) = B_2^{-1}(E)$ (or more generally $B_1^{-1}(E_1) = B_2^{-1}(E_2)$) to the functional equation $f \circ B_1 = g \circ B_2$. We study such a problem by using the hyperbolic equilibrium measure instead of the logarithmic equilibrium measure as well as Ritt's Theorem (Theorem R3′) for finite Blaschke products. In fact, we obtain the following results and refer the reader to [31] for the proofs.

Theorem 23 *Let B_1, B_2 be finite Blaschke products,* $\deg B_1 = d_1$, $\deg B_2 = d_2$, *$d_1 \le d_2$, and $E_1, E_2 \subset \mathbb{D}$ be compact sets such that*

$$E := B_1^{-1}(E_1) = B_2^{-1}(E_2)$$

holds. Suppose that E_1 and E_2 are connected sets of positive (hyperbolic) capacities. We have

1. *if d_1 divides d_2, then there exists a finite Blaschke product g_1 such that $B_2 = g_1 \circ B_1$ and $E_1 = g_1^{-1}(E_2)$;*
2. *if d_1 does not divide d_2, then there exist finite Blaschke products g_1, g_2, with $\deg g_1 = d_2/d, \deg g_2 = d_1/d$, where $d = \gcd(d_1, d_2)$ such that*

$$g_1 \circ B_1 = g_2 \circ B_2,$$

and a compact connected set $E_3 \subset \mathbb{D}$ such that

$$E_1 = g_1^{-1}(E_3) \quad \textit{and} \quad E_2 = g_2^{-1}(E_3).$$

Furthermore there exist finite Blaschke products $\tilde{B}_1, \tilde{B}_2, W$, with $\deg W = d$, such that

$$B_1 = \tilde{B}_1 \circ W, \qquad B_2 = \tilde{B}_2 \circ W$$

and there exist Möbius transformations τ_1, τ_2 such that either

$$\begin{aligned} g_1 &= z^c\big[R(z)\big]^{d_1/d} \circ \tau_1, & \tilde{B}_1 &= \tau_1^{-1} \circ z^{d_1/d}, \\ g_2 &= z^{d_1/d} \circ \tau_2, & \tilde{B}_2 &= \tau_2^{-1} \circ z^c R\big(z^{d_1/d}\big) \end{aligned}$$

for some finite Blaschke product R and for c being the remainder after division of d_2/d by d_1/d, or

$$\begin{aligned} g_1 &= f_{d_2/d, d_1\tau/d} \circ \tau_1, & \tilde{B}_1 &= \tau_1^{-1} \circ f_{d_1/d,\tau}, \\ g_2 &= f_{d_1/d, d_2\tau/d} \circ \tau_2, & \tilde{B}_2 &= \tau_2^{-1} \circ f_{d_2/d,\tau} \end{aligned}$$

for the Chebyshev-Blaschke products $f_{d_2/d,d_1\tau/d}, f_{d_1/d,d_2\tau/d}, f_{d_1/d,\tau}, f_{d_2/d,\tau}$.

For the case $E_1 = E_2$, we have the following

Theorem 24 *Let B_1, B_2 be finite Blaschke products,* $\deg B_1 = d_1$, $\deg B_2 = d_2$, $d_1 \le d_2$, *and* $E_0 \subset \mathbb{D}$ *be a compact set such that* $E := B_1^{-1}(E_0) = B_2^{-1}(E_0)$ *holds. Suppose that E_0 is a connected set of positive (hyperbolic) capacity. Then $d_1 = d_2$ and there exists a Möbius transformation g_1 such that $B_2 = g_1 \circ B_1$ and $E_0 = g_1^{-1}(E_0)$.*

Remark 6 In Theorem 24, if $E = E_0$, then B_1 and B_2 must be Möbius transformations. To see this, assume on the contrary that $\deg B_i \ge 2$ for $i = 1$ or 2. Since E_0 is completely invariant under B_i, the Julia set $J(B_i)$ is contained in E_0 by the minimality of Julia set (see Sect. 3.1.4). On the other hand, $J(B_i)$ is either the unit circle or a Cantor set on the circle (mentioned in Sect. 3.2.4), which is a contradiction.

Finally, in view of Theorems 22 and 23, we make the following conjecture.

Conjecture 2 *Theorem 23 still holds by replacing the condition that E_1 and E_2 are connected sets of positive (hyperbolic) capacities by* $\operatorname{card}(E) \ge \operatorname{lcm}(d_1, d_2)$.

We summarize the results in this section by the following table:

Polynomials	Finite Blaschke products
Theorem 22	Theorem 23, Conjecture 2
Theorem 21	Theorem 24

5 Final Remark

We hope the reader are convinced that there does exist a nice correspondence between polynomials and finite Blaschke products, and there is still a lot of work to be done in extending the dictionary given here. For example, after proving a very interesting result on products of polynomials in uniform norms in [37], I.E. Pritsker also considered and proved a corresponding version for finite Blaschke products in [38] very recently.

Acknowledgements The first author was partially supported by RGC grant HKU 704409P. The second author was partially supported by graduate studentship of HKU and RGC grant HKU 704409P.

References

1. Adams, W.W., Straus, E.G.: Non-Archimedean analytic functions taking the same values at the same points. Ill. J. Math. **15**, 418–424 (1971)
2. Akhiezer, N.I.: Elements of the Theory of Elliptic Functions. AMS, Providence (1990)
3. Beardon, A.F.: Iteration of Rational Functions. Springer, Berlin (1991)

4. Beardon, A.F., Carne, T.K., Ng, T.W.: The critical values of a polynomial. Constr. Approx. **18**, 343–354 (2002)
5. Beardon, A.F., Minda, D.: A multi-point Schwarz-Pick lemma. J. Anal. Math. **92**, 81–104 (2004)
6. Beardon, A.F., Minda, D., Ng, T.W.: Smale's mean value conjecture and the hyperbolic metric. Math. Ann. **322**, 623–632 (2002)
7. Borwein, P., Erdélyi, T.: Polynomials and Polynomial Inequalities. Graduate Texts in Mathematics, vol. 161. Springer, New York (1995)
8. Carleson, L., Gamelin, T.W.: Complex Dynamics. Springer, Berlin (1993)
9. Chandrasekharan, K.: Elliptic Functions. Springer, Berlin (1985)
10. Conte, A., Fujikawa, E., Lakic, N.: Smale's mean value conjecture and the coefficients of univalent functions. Proc. Am. Math. Soc. **135**, 3295–3300 (2007)
11. Crane, E.: A bound for Smale's mean value conjecture for complex polynomials. Bull. Lond. Math. Soc. **39**, 781–791 (2007)
12. Dinh, T.C.: Ensembles d'unicité pour les polynômes. Ergod. Theory Dyn. Syst. **22**(1), 171–186 (2002)
13. Dinh, T.C.: Distribution des préimages et des points périodiques d'une correspondance polynomiale. Bull. Soc. Math. Fr. **133**, 363–394 (2005)
14. Farkas, H.M., Kra, I.: Theta Constants, Riemann Surfaces and the Modular Group. Grad. Stud. Math., vol. 37. Am. Math. Soc., Providence (2001)
15. Fatou, P.: Sur les équations fonctionnelles. Bull. Soc. Math. Fr. **47**, 161–271 (1919)
16. Fatou, P.: Sur les équations fonctionnelles. Bull. Soc. Math. Fr. **48**, 33–94 (1920)
17. Fatou, P.: Sur les équations fonctionnelles. Bull. Soc. Math. Fr. **48**, 208–314 (1920)
18. Fatou, P.: Sur les fonctions holomorphes et bornées à l'intérieur d'uncercle. Bull. Soc. Math. Fr. **51**, 191–202 (1923)
19. Forster, O.: Lectures on Riemann Surfaces. Springer, New York (1991)
20. Fujikawa, E., Sugawa, T.: Geometric function theory and Smale's mean value conjecture. Proc. Jpn. Acad., Ser. A, Math. Sci. **82**, 97–100 (2006)
21. Garnett, J.B.: Bounded Analytic Functions. Grad. Texts in Math., vol. 236. Springer, New York (2007)
22. Gonchar, A.A.: Zolotarev problems connected with rational functions. Math. USSR Sb. **7**, 623–635 (1969)
23. Hinkkanen, A., Kayumov, I.: Smale's problem for critical points on certain two rays. J. Aust. Math. Soc. **88**, 183–191 (2010)
24. Horwitz, A.L., Rubel, L.A.: A uniqueness theorem for monic Blaschke products. Proc. Am. Math. Soc. **96**, 180–182 (1986)
25. Istace, M.P., Thiran, J.P.: On the third and fourth Zolotarev problems in the complex plane. SIAM J. Numer. Anal. **32**, 249–259 (1995)
26. Lorentz, G.G.: Approximation of Functions. Chelsea, New York (1986)
27. Mason, J.C., Handscomb, D.: Chebyshev Polynomials. Chapman and Hall, London (2003)
28. Milnor, J.: Dynamics in One Complex Variable. Princeton University Press, Princeton (2006)
29. Ng, T.W.: Smale's mean value conjecture for odd polynomials. J. Aust. Math. Soc. **75**, 409–411 (2003)
30. Ng, T.W., Tsang, C.Y.: Chebyshev-Blaschke products. Preprint (2012)
31. Ng, T.W., Tsang, C.Y.: On finite Blaschke products sharing preimages of sets. Preprint (2012)
32. Ng, T.W., Wang, M.X.: Ritt's theory on the unit disk. Forum Math. doi:10.1515/form.2011.136
33. Ostrovskii, I.V., Pakovich, F.B., Zaidenberg, M.G.: A remark on complex polynomials of least deviation. Int. Math. Res. Not. **14**, 699–703 (1996)
34. Pakovich, F.B.: Sur un problème d'unicité pour les polynômes, prépublication. Inst. Fourier Math. **324**, 1–4 (1995). Grenoble
35. Pakovich, F.B.: Sur un problème d'unicité pour les fouctions méromorphes. C.R. Acad. Sci. Paris Sr. I Math. **323**, 745–748 (1996)

36. Pakovich, F.B.: On polynomials sharing preimages of compact sets and related questions. Geom. Funct. Anal. **18**, 163–183 (2008)
37. Pritsker, I.E.: Products of polynomials in uniform norms. Trans. Am. Math. Soc. **353**, 3971–3993 (2001)
38. Pritsker, I.E.: Inequalities for sums of Green potentials and Blaschke products. Bull. Lond. Math. Soc. **43**, 561–575 (2011)
39. Radó, T.: Zur Theorie der mehrdeutigen konformen Abbildungen. Acta Litt. Sci. Univ. Hung. **1**, 55–64 (1922) (German)
40. Remmert, R.: Classical Topics in Complex Function Theory. Springer, New York (1998)
41. Ritt, J.F.: Prime and composite polynomials. Trans. Am. Math. Soc. **23**(1), 51–66 (1922)
42. Rivlin, T.J.: Chebyshev Polynomials: From Approximation Theory to Algebra and Number Theory, 2nd edn. Wiley, New York (1990)
43. Sheil-Small, T.: Complex Polynomials. Cambridge University Press, Cambridge (2002)
44. Singer, D.A.: The location of critical points of finite Blaschke products. Conform. Geom. Dyn. **10**, 117–124 (2006)
45. Smale, S.: The fundamental theorem of algebra and complexity theory. Bull. Am. Math. Soc. **4**(1), 1–36 (1981)
46. Steinmetz, N.: Rational Iteration: Complex Analytic Dynamical Systems. Studies in Mathematics (1993)
47. Walker, P.L.: Elliptic Functions: A Constructive Approach. Wiley, New York (1996)
48. Walsh, J.L.: Note on the location of zeros of extremal polynomials in the non-euclidean plane. Acad. Serbe Sci. Publ. Inst. Math. **4**, 157–160 (1952)
49. Wang, M.X.: Factorizations of finite mappings on Riemann surfaces. M.Phil. thesis, HKU (2008). http://hub.hku.hk/handle/123456789/51854
50. Wang, M.X.: Rational points and transcendental points. Ph.D. thesis, ETH (2011). http://e-collection.library.ethz.ch/view/eth:4704
51. Yang, C.C.: Open problems in complex analysis. In: Proceedings of the S.U.N.Y. Brockport Conference. Dekker, New York (1978)
52. Zakeri, S.: On critical points of proper holomorphic maps on the unit disk. Bull. Lond. Math. Soc. **30**, 62–66 (1996)

Recent Progress on Truncated Toeplitz Operators

Stephan Ramon Garcia and William T. Ross

Abstract This paper is a survey on the emerging theory of truncated Toeplitz operators. We begin with a brief introduction to the subject and then highlight the many recent developments in the field since Sarason's seminal paper (Oper. Matrices 1(4):491–526, 2007).

1 Introduction

Although the subject has deep classical roots, the systematic study of *truncated Toeplitz operators* for their own sake was only recently spurred by the seminal 2007 paper of Sarason [88]. In part motivated by several of the problems posed in the aforementioned article, the area has undergone vigorous development during the past several years [13, 14, 23, 25, 28, 35, 45, 49–52, 56, 59, 65, 88–90, 92, 93, 96]. While several of the initial questions raised by Sarason have now been resolved, the study of truncated Toeplitz operators has nevertheless proven to be fertile ground, spawning both new questions and unexpected results. In this survey, we aim to keep the interested reader up to date and to give the uninitiated reader a historical overview and a summary of the important results and major developments in this area.

Our survey commences in Sect. 2 with an extensive treatment of the basic definitions, theorems, and techniques of the area. Consequently, we shall be brief in this introduction and simply declare that a *truncated Toeplitz operator* is the compression $A^u_\varphi : \mathcal{K}_u \to \mathcal{K}_u$ of a classical Toeplitz operator T_φ to a shift coinvariant subspace $\mathcal{K}_u := H^2 \ominus uH^2$ of the classical Hardy space H^2. Here u denotes a nonconstant inner function and we write $A^u_\varphi f = P_u(\varphi f)$ where P_u denotes the orthog-

S.R. Garcia
Department of Mathematics, Pomona College, Claremont, CA 91711, USA
e-mail: Stephan.Garcia@pomona.edu

W.T. Ross (✉)
Department of Mathematics and Computer Science, University of Richmond, Richmond, VA 23173, USA
e-mail: wross@richmond.edu

J. Mashreghi, E. Fricain (eds.), *Blaschke Products and Their Applications*, Fields Institute Communications 65, DOI 10.1007/978-1-4614-5341-3_15,

onal projection from L^2 onto $\mathcal{K}_u$. Interestingly, the study of potentially unbounded truncated Toeplitz operators, having symbols φ in L^2 as opposed to L^∞, has proven to be spectacularly fruitful. Indeed, a number of important questions in the area revolve around this theme (e.g., the results of Sect. 5).

Before proceeding, let us first recall several instances where truncated Toeplitz operators have appeared in the literature. This will not only provide a historical perspective on the subject, but it will also illustrate the fact that truncated Toeplitz operators, in various guises, form the foundations of much of modern function-related operator theory.

Let us begin with the powerful Sz.-Nagy-Foiaş model theory for Hilbert space contractions, the simplest manifestation of which is the *compressed shift* A_z^u [16, 77–79, 82]. To be more specific, every Hilbert space contraction T having defect indices $(1,1)$ and such that $\lim_{n\to\infty} T^{*n} = 0$ (SOT) is unitarily equivalent to A_z^u for some inner function u. Natural generalizations of this result are available to treat contractions with arbitrary defect indices by employing the machinery of vector-valued model spaces and operator-valued inner functions.

In his approach to the *Gelfand problem*, that is the characterization of the invariant subspace lattice $\operatorname{Lat} V$ of the Volterra integration operator

$$[Vf](x) = \int_0^x f(y)\,dy \tag{1}$$

on $L^2[0,1]$, Sarason noted that the Volterra operator is unitarily equivalent to the Cayley transform of the compressed shift A_z^u corresponding to the atomic inner function $u(z) = \exp(\frac{z+1}{z-1})$ [84]. This equivalence was then used in conjunction with Beurling's Theorem to demonstrate the unicellularity of $\operatorname{Lat} V$ [78, 79, 84, 86]. Interestingly, it turns out that the Volterra operator, and truncated Toeplitz operators in general, are natural examples of *complex symmetric operators*, a large class of Hilbert space operators which has also undergone significant development in recent years [24, 35, 45–50, 54, 55, 61, 63, 71, 72, 74, 103–107]. This link between truncated Toeplitz operators and complex symmetric operators is explored in Sect. 9.

Sarason himself identified the commutant of A_z^u as the set $\{A_\varphi^u : \varphi \in H^\infty\}$ of all *analytic* truncated Toeplitz operators. He also obtained an H^∞ functional calculus for the compressed shift, establishing that $\varphi(A_z^u) = A_\varphi^u$ holds for all φ in H^∞ [85]. These seminal observations mark the beginning of the so-called *commutant lifting* theory, which has been developed to great effect over the ensuing decades [42, 80, 91]. Moreover, these techniques have given new perspectives on several classical problems in complex function theory. For instance, the Carathéodory and Pick problems lead one naturally to consider lower triangular Toeplitz matrices (i.e., analytic truncated Toeplitz operators on $\mathcal{K}_{z^n}$) and the backward shift on the span of a finite collection of Cauchy kernels (i.e., $A_{\overline{z}}^u$ on a finite dimensional model space $\mathcal{K}_u$). We refer the reader to the text [1] which treats these problems in greater detail and generality.

Toeplitz matrices, which can be viewed as truncated Toeplitz operators on $\mathcal{K}_{z^n}$, have long been the subject of intense research. We make no attempt to give even a superficial discussion of this immense topic. Instead, we merely refer the reader to several recent texts which analyze various aspects of this fascinating subject. The pseudospectral properties of Toeplitz matrices are explored in [100]. The asymptotic analysis of Toeplitz operators on H^2 via large truncated Toeplitz matrices is the focus of [19]. The role played by Toeplitz determinants in the study of orthogonal polynomials is discussed in [94, 95] and its relationship to random matrix theory is examined in [15, 70]. Finally, we should also remark that a special class of Toeplitz matrices, namely circulant matrices, are a crucial ingredient in many aspects of numerical computing [37].

We must also say a few words about the appearance of truncated Toeplitz operators in applications to control theory and electrical engineering. In such contexts, extremal problems posed over H^∞ often appear. It is well-known that the solution to many such problems can be obtained by computing the norm of an associated Hankel operator [43, 44]. However, it turns out that many questions about Hankel operators can be phrased in terms of analytic truncated Toeplitz operators and, moreover, this link has long been exploited [81, Eq. (2.9)]. Changing directions somewhat, we remark that the *skew Toeplitz operators* arising in H^∞ control theory are closely related to selfadjoint truncated Toeplitz operators [17, 18].

Among other things, Sarason's recent article [88] is notable for opening the general study of truncated Toeplitz operators, beyond the traditional confines of the analytic ($\varphi \in H^\infty$) and co-analytic ($\overline{\varphi} \in H^\infty$) cases and the limitations of the case $u = z^N$ (i.e., Toeplitz matrices), all of which are evidently well-studied in the literature. By permitting arbitrary symbols in L^∞, and indeed under some circumstances in L^2, an immense array of new theorems, novel observations, and difficult problems emerges. It is our aim in this article to provide an overview of the ensuing developments, with an eye toward promoting further research. In particular, we make an effort to highlight open problems and unresolved issues which we hope will spur further advances.

2 Preliminaries

In this section we gather together some of the standard results on model spaces and Aleksandrov-Clark measures which will be necessary for what follows. Since most of this material is familiar to those who have studied Sarason's article [88], the following presentation is somewhat terse. Indeed, it serves primarily as a review of the standard notations and conventions of the field.

2.1 Basic Notation

Let $\mathbb{D}$ be the open unit disk, $\partial\mathbb{D}$ the unit circle, $m = d\theta/2\pi$ *normalized* Lebesgue measure on $\partial\mathbb{D}$, and $L^p := L^p(\partial\mathbb{D}, m)$ be the standard Lebesgue spaces on $\partial\mathbb{D}$.

For $0 < p < \infty$ we use H^p to denote the classical Hardy spaces on $\mathbb{D}$ and H^∞ to denote the bounded analytic functions on $\mathbb{D}$. As is standard, we regard H^p as a closed subspace of L^p by identifying each $f \in H^p$ with its m-almost everywhere defined L^p boundary function

$$f(\zeta) := \lim_{r \to 1^-} f(r\zeta), \qquad m\text{-a.e. } \zeta \in \partial\mathbb{D}.$$

In the Hilbert space setting H^2 (or L^2) we denote the norm as $\|\cdot\|$ and the usual integral inner product by $\langle \cdot, \cdot \rangle$. On the rare occasions when we need to discuss L^p norms we will use $\|\cdot\|_p$. We let $\widehat{\mathbb{C}}$ denote the Riemann sphere $\mathbb{C} \cup \{\infty\}$ and, for a set $A \subseteq \mathbb{C}$, we let A^- denote the closure of A. For a subset $V \subset L^p$, we let $\overline{V} := \{\overline{f} : f \in V\}$. We interpret the Cauchy integral formula

$$f(\lambda) = \int_{\partial\mathbb{D}} \frac{f(\zeta)}{1 - \overline{\zeta}\lambda} dm(\zeta),$$

valid for all f in H^2, in the context of reproducing kernel Hilbert spaces by writing $f(\lambda) = \langle f, c_\lambda \rangle$ where

$$c_\lambda(z) := \frac{1}{1 - \overline{\lambda} z} \tag{2}$$

denotes the *Cauchy kernel* (also called the *Szegő kernel*). A short computation now reveals that the orthogonal projection P_+ from L^2 onto H^2 (i.e., the *Riesz projection*) satisfies

$$[P_+ f](\lambda) = \langle f, c_\lambda \rangle$$

for all $f \in L^2$ and $\lambda \in \mathbb{D}$.

2.2 *Model Spaces*

Let $S : H^2 \to H^2$ denote the unilateral shift

$$[Sf](z) = zf(z), \tag{3}$$

and recall that Beurling's Theorem asserts that the nonzero S-invariant subspaces of H^2 are those of the form uH^2 for some inner function u. Letting

$$\left[S^* f\right](z) = \frac{f(z) - f(0)}{z} \tag{4}$$

denote the backward shift operator, it follows that the proper S^*-invariant subspaces of H^2 are precisely those of the form

$$\mathcal{K}_u := H^2 \ominus uH^2. \tag{5}$$

The subspace (5) is called the *model space* corresponding to the inner function u, the terminology stemming from the important role that $\mathcal{K}_u$ plays in the model theory for Hilbert space contractions [79, Part C].

Although they will play only a small role in what follows, we should also mention that the backward shift invariant subspaces of the Hardy spaces H^p for $0 < p < \infty$ are also known. In particular, for $1 \leqslant p < \infty$ the proper backward shift invariant subspaces of H^p are all of the form

$$\mathcal{K}_u^p := H^p \cap u\overline{H_0^p}, \tag{6}$$

where H_0^p denotes the subspace of H^p consisting of those H^p functions which vanish at the origin and where the right-hand side of (6) is to be understood in terms of boundary functions on $\partial\mathbb{D}$. For further details and information on the more difficult case $0 < p < 1$, we refer the reader to the text [27] and the original article [4] of Aleksandrov. For $p = 2$, we often suppress the exponent and simply write $\mathcal{K}_u$ in place of $\mathcal{K}_u^2$.

2.3 Pseudocontinuations

Since the initial definition (6) of $\mathcal{K}_u^p$ is somewhat indirect, one might hope for a more concrete description of the functions belonging to $\mathcal{K}_u^p$. A convenient function-theoretic characterization of $\mathcal{K}_u^p$ is provided by the following important result.

Theorem 1 (Douglas-Shapiro-Shields [39]) *If $1 \leqslant p < \infty$, then f belongs to $\mathcal{K}_u^p$ if and only if there exists a $G \in H^p(\widehat{\mathbb{C}}\backslash\mathbb{D}^-)$ which vanishes at infinity*[1] *such that*

$$\lim_{r\to 1^+} G(r\zeta) = \lim_{r\to 1^-} \frac{f}{u}(r\zeta)$$

for almost every ζ on $\partial\mathbb{D}$.

The function G in the above theorem is called a *pseudocontinuation* of f/u to $\widehat{\mathbb{C}}\backslash\mathbb{D}^-$. We refer the reader to the references [5, 27, 39, 83] for further information about pseudocontinuations and their properties. An explicit function theoretic parametrization of the spaces $\mathcal{K}_u^p$ is discussed in detail in [45].

2.4 Kernel Functions, Conjugation, and Angular Derivatives

Letting P_u denote the orthogonal projection from L^2 onto $\mathcal{K}_u$, we see that

$$[P_u f](\lambda) = \langle f, k_\lambda \rangle \tag{7}$$

[1] Equivalently, $G(1/z) \in H_0^p$.

for each λ in $\mathbb{D}$. Here

$$k_\lambda(z) := \frac{1-\overline{u(\lambda)}u(z)}{1-\overline{\lambda}z} \tag{8}$$

denotes the *reproducing kernel* for $\mathcal{K}_u$. In particular, this family of functions has the property that $f(\lambda) = \langle f, k_\lambda \rangle$ for every $f \in \mathcal{K}_u$ and $\lambda \in \mathbb{D}$.

Each model space $\mathcal{K}_u$ carries a natural *conjugation* (an isometric, conjugate-linear involution) $C : \mathcal{K}_u \to \mathcal{K}_u$, defined in terms of boundary functions by

$$[Cf](\zeta) := \overline{f(\zeta)\zeta}u(\zeta). \tag{9}$$

For notational convenience, we sometimes denote the conjugate Cf of f by $\tilde{f}$. More information about conjugations in general, along with specific properties of the map (9) can be found in [45]. For the moment, we simply mention that the so-called *conjugate kernels*

$$[Ck_\lambda](z) = \frac{u(z)-u(\lambda)}{z-\lambda} \tag{10}$$

will be important in what follows. In particular, observe that each conjugate kernel is a difference quotient for the inner function u. We therefore expect that derivatives will soon enter the picture.

Definition 1 For an inner function u and a point ζ on $\partial\mathbb{D}$ we say that u has an *angular derivative in the sense of Carathéodory* (ADC) *at* ζ if the nontangential limits of u and u' exist at ζ and $|u(\zeta)| = 1$.

The following theorem provides several useful characterizations of ADCs.

Theorem 2 (Ahern-Clark [3]) *For an inner function $u = b_\Lambda s_\mu$, where b_Λ is a Blaschke product with zeros $\Lambda = \{\lambda_n\}_{n=1}^\infty$, repeated according to multiplicity, s_μ is a singular inner function with corresponding singular measure μ, and $\zeta \in \partial\mathbb{D}$, the following are equivalent*:

(i) *Every $f \in \mathcal{K}_u$ has a nontangential limit at ζ.*
(ii) *For every $f \in \mathcal{K}_u$, $f(\lambda)$ is bounded as $\lambda \to \zeta$ nontangentially.*
(iii) *u has an ADC at ζ.*
(iv) *The function*

$$k_\zeta(z) = \frac{1-\overline{u(\zeta)}u(z)}{1-\overline{\zeta}z}, \tag{11}$$

belongs to H^2.

(v) *The following condition holds*:

$$\sum_{n\geq 1} \frac{1-|\lambda_n|^2}{|\zeta-\lambda_n|^2} + \int_{\partial\mathbb{D}} \frac{d\mu(\xi)}{|\xi-\zeta|^2} < \infty. \tag{12}$$

In fact, the preceding is only a partial statement of the Ahern-Clark result, for there are several other additional conditions which can be appended to the preceding list. Moreover, they also characterized the existence of nontangential boundary limits of the derivatives (up to a given order) of functions in $\mathcal{K}_u$. An extension of Theorem 2 to the spaces $\mathcal{K}_u^p$ is due to Cohn [32].

Among other things, Theorem 2 tells us that whenever u has an ADC at a point ζ on $\partial\mathbb{D}$, then the functions (11), are reproducing kernels for $\mathcal{K}_u$ in the sense that the reproducing property $f(\zeta) = \langle f, k_\zeta \rangle$ holds for all f in $\mathcal{K}_u$. In a similar manner, the functions

$$[Ck_\zeta](z) = \frac{u(z) - u(\zeta)}{z - \zeta}$$

are also defined and belong to $\mathcal{K}_u$ whenever u has an ADC at ζ.

2.5 Two Results of Aleksandrov

Letting H^∞ denote the Banach algebra of all bounded analytic functions on $\mathbb{D}$, we observe that the set $\mathcal{K}_u^\infty := \mathcal{K}_u \cap H^\infty$ is dense in $\mathcal{K}_u$ since $\operatorname{span}\{S^{*n}u : n = 1, 2, \ldots\}$ is dense in $\mathcal{K}_u$. Another way to see that $\mathcal{K}_u^\infty$ is dense in $\mathcal{K}_u$ is to observe that each reproducing kernel (8) belongs to $\mathcal{K}_u^\infty$ whence $\operatorname{span}\{k_\lambda : \lambda \in \Lambda\}$ is dense in $\mathcal{K}_u$ whenever Λ is a uniqueness set for $\mathcal{K}_u$.

For many approximation arguments, the density of $\mathcal{K}_u^\infty$ in $\mathcal{K}_u$ is sufficient. In other circumstances, however, one requires continuity up to the boundary. Unfortunately, for many inner functions (e.g., singular inner functions) it is difficult to exhibit a single nonconstant function in $\mathcal{K}_u$ which is continuous on $\mathbb{D}^-$ (i.e., which belongs to the intersection of $\mathcal{K}_u$ with the *disk algebra* $\mathcal{A}$). The following surprising result asserts that $\mathcal{K}_u \cap \mathcal{A}$, far from being empty, is actually dense in $\mathcal{K}_u$.

Theorem 3 (Aleksandrov [6]) *For $p \in (1, \infty)$, $\mathcal{K}_u^p \cap \mathcal{A}$ is dense in $\mathcal{K}_u^p$.*

A detailed exposition of Aleksandrov's density theorem can be found in [26, p. 186]. For related results concerning whether or not $\mathcal{K}_u$ contains functions of varying degrees of smoothness, the reader is invited to consult [41]. One consequence of Theorem 3 is that it allows us to discuss whether or not $\mathcal{K}_u^p$ can be embedded in $L^p(\mu)$ where μ is a measure on $\partial\mathbb{D}$.

Theorem 4 (Aleksandrov [6]) *Let u be an inner function, μ be a positive Borel measure on $\partial\mathbb{D}$, and $p \in (1, \infty)$. If there exists a $C > 0$ such that*

$$\|f\|_{L^p(\mu)} \leqslant C\|f\|_p, \quad \forall f \in \mathcal{A} \cap \mathcal{K}_u^p, \tag{13}$$

then every function in $\mathcal{K}_u^p$ has a finite nontangential limit μ-almost everywhere and (13) *holds for all $f \in \mathcal{K}_u^p$.*

It is clear that every measure on $\partial\mathbb{D}$ which is also Carleson measure (see [62]) satisfies (13). However, there are generally many other measures which also satisfy (13). For example, if u has an ADC at ζ, then the point mass δ_ζ satisfies (13) with $p=2$.

2.6 The Compressed Shift

Before introducing truncated Toeplitz operators in general in Sect. 2.9, we should first introduce and familiarize ourselves with the most important and well-studied example. The so-called *compressed shift* operator is simply the compression of the unilateral shift (3) to a model space $\mathcal{K}_u$:

$$A_z^u := P_u S|_{\mathcal{K}_u}. \tag{14}$$

The adjoint of A_z^u is the restriction of the backward shift (4) to $\mathcal{K}_u$. Being the compression of a contraction, it is clear that A_z^u is itself a contraction and in fact, such operators and their vector-valued analogues can be used to model certain types of contractive operators [16, 77–79]. The following basic properties of A_z^u are well-known and can be found, for instance, in [78, 88].

Theorem 5 *If u is a nonconstant inner function, then*

(i) *The invariant subspaces of A_z^u are $vH^2 \cap (uH^2)^\perp$, where v is an inner function which divides u (i.e., $u/v \in H^\infty$).*
(ii) *A_z^u is cyclic with cyclic vector k_0. That is to say, the closed linear span of $\{(A_z^u)^n k_0 : n = 0, 1, 2, \ldots\}$ is equal to $\mathcal{K}_u$. Moreover, $f \in \mathcal{K}_u$ is cyclic for A_z^u if and only if u and the inner factor of f are relatively prime.*
(iii) *A_z^u is irreducible (i.e., has no proper, nontrivial reducing subspaces).*

To discuss the spectral properties of A_z^u we require the following definition.

Definition 2 If u is an inner function, then the *spectrum* $\sigma(u)$ of u is the set

$$\sigma(u) := \left\{\lambda \in \mathbb{D}^- : \liminf_{z\to\lambda} |u(z)| = 0\right\}.$$

If $u = b_\Lambda s_\mu$, where b is a Blaschke product with zero sequence $\Lambda = \{\lambda_n\}$ and s_μ is a singular inner function with corresponding singular measure μ, then

$$\sigma(u) = \Lambda^- \cup \operatorname{supp}\mu.$$

The following related lemma is well-known result of Moeller and we refer the reader to [78, p. 65] or [27, p. 84] for its proof.

Lemma 1 *Each function in $\mathcal{K}_u$ can be analytically continued across $\partial\mathbb{D}\backslash\sigma(u)$.*

An explicit description of the spectrum of the compressed shift A_z^u can be found in Sarason's article [88, Lemma 2.5], although portions of it date back to the work of Livšic and Moeller [78, Lec. III.1].

Theorem 6 *If u is an inner function, then*

(i) *The spectrum $\sigma(A_z^u)$ of A_z^u is equal to $\sigma(u)$.*
(ii) *The point spectrum of $\sigma_p(A_z^u)$ of A_z^u is equal to $\sigma(u) \cap \mathbb{D}$.*
(iii) *The essential spectrum $\sigma_e(A_z^u)$ of A_z^u is equal to $\sigma(u) \cap \partial\mathbb{D}$.*

2.7 Clark Unitary Operators and Their Spectral Measures

Maintaining the notation and conventions of the preceding subsection, let us define, for each $\alpha \in \partial\mathbb{D}$, the following operator on $\mathcal{K}_u$:

$$U_\alpha := A_z^u + \frac{\alpha}{1 - \overline{u(0)}\alpha} k_0 \otimes Ck_0. \tag{15}$$

In the above, the operator $f \otimes g$, for $f, g \in H^2$, is given by the formula

$$(f \otimes g)(h) = \langle h, g \rangle f.$$

A seminal result of Clark [29] asserts that each U_α is a cyclic unitary operator and, moreover, that every unitary, rank-one perturbation of A_z^u is of the form (15). Furthermore, Clark was even able to concretely identify the corresponding spectral measures σ_α for these so-called *Clark operators*. We discuss these results below (much of this material is presented in greater detail in the recent text [26]).

Theorem 7 (Clark) *For each $\alpha \in \partial\mathbb{D}$, U_α is a cyclic unitary operator on $\mathcal{K}_u$. Moreover, any unitary rank-one perturbation of A_z^u is equal to U_α for some $\alpha \in \partial\mathbb{D}$.*

The spectral theory for the *Clark operators* U_α is well-developed and explicit. For instance, if $u(0) = 0$, then a point ζ on $\partial\mathbb{D}$ is an eigenvalue of U_α if and only if u has an ADC at ζ and $u(\zeta) = \alpha$. The corresponding eigenvector is

$$k_\zeta(z) = \frac{1 - \overline{\alpha}u(z)}{1 - \overline{\zeta}z},$$

which is simply a boundary kernel (11).

Since each U_α is cyclic, there exists a measure σ_α, supported on $\partial\mathbb{D}$, so that U_α is unitarily equivalent to the operator $M_z : L^2(\sigma_\alpha) \to L^2(\sigma_\alpha)$ of multiplication by the independent variable on the Lebesgue space $L^2(\sigma_\alpha)$, i.e., $M_z f = zf$. To concretely identify the spectral measure σ_α, we require an old theorem of Herglotz, which states that any positive harmonic function on $\mathbb{D}$ can be written as the Poisson integral

$$(\mathfrak{P}\sigma)(z) = \int_{\partial\mathbb{D}} \frac{1-|z|^2}{|\zeta - z|^2}\, d\sigma(\zeta)$$

of some unique finite, positive, Borel measure σ on $\partial\mathbb{D}$ [40, Theorem 1.2].

Theorem 8 (Clark) *If σ_α is the unique measure on $\partial\mathbb{D}$ satisfying*

$$\frac{1-|u(z)|^2}{|\alpha - u(z)|^2} = \int_{\partial\mathbb{D}} \frac{1-|z|^2}{|\zeta - z|^2}\, d\sigma_\alpha(\zeta), \tag{16}$$

then U_α is unitarily equivalent to the operator $M_z : L^2(\sigma_\alpha) \to L^2(\sigma_\alpha)$ defined by $M_z f = zf$.

The *Clark measures* $\{\sigma_\alpha : \alpha \in \partial\mathbb{D}\}$ corresponding to an inner function u have many interesting properties. We summarize some of these results in the following theorem. The reader may consult [26] for further details.

Theorem 9

(i) *σ_α is singular with respect to Lebesgue measure for each $\alpha \in \partial\mathbb{D}$.*
(ii) *$\sigma_\alpha \perp \sigma_\beta$ when $\alpha \neq \beta$.*
(iii) *(Nevanlinna) $\sigma_\alpha(\{\zeta\}) > 0$ if and only if $u(\zeta) = \alpha$ and u has an ADC at ζ. Moreover,*

$$\sigma_\alpha\big(\{\zeta\}\big) = \frac{1}{|u'(\zeta)|}.$$

(iv) *(Aleksandrov) For any $f \in C(\partial\mathbb{D})$ we have*

$$\int_{\partial\mathbb{D}} \left(\int_{\partial\mathbb{D}} g(\zeta)\, d\sigma_\alpha(\zeta) \right) dm(\alpha) = \int_{\partial\mathbb{D}} g(\zeta)\, dm(\zeta). \tag{17}$$

Condition (iv) of the preceding theorem is a special case of the *Aleksandrov disintegration theorem*: If g belongs to L^1, then the map

$$\alpha \to \int_{\partial\mathbb{D}} g(\zeta)\, d\sigma_\alpha(\zeta)$$

is defined for m-almost every α in $\partial\mathbb{D}$ and, as a function of α, it belongs to L^1 and satisfies the natural analogue of (17). In fact, the Clark measures σ_α are often called *Aleksandrov-Clark* measures in light of Aleksandrov's deep work on the subject, which actually generalizes to measures μ_α on $\partial\mathbb{D}$ satisfying

$$\frac{1-|u(z)|^2}{|\alpha - u(z)|^2} = \int_{\partial\mathbb{D}} \frac{1-|z|^2}{|\zeta - z|^2}\, d\mu_\alpha(\zeta)$$

for arbitrary functions u belonging to the unit ball of H^∞. The details and ramifications of this remarkable result are discussed in detail in [26].

2.8 Finite Dimensional Model Spaces

It is not hard to show that the model space $\mathcal{K}_u$ is finite dimensional if and only if u is a finite Blaschke product. In fact, if u is a finite Blaschke product with zeros $\lambda_1, \lambda_2, \ldots, \lambda_N$, repeated according to multiplicity, then $\dim \mathcal{K}_u = N$ and

$$\mathcal{K}_u = \left\{ \frac{\sum_{j=0}^{N-1} a_j z^j}{\prod_{j=1}^{N}(1 - \overline{\lambda_j} z)} : a_j \in \mathbb{C} \right\}. \tag{18}$$

With respect to the representation (18), the conjugation (9) on $\mathcal{K}_u$ assumes the simple form

$$C\left(\frac{\sum_{j=0}^{N-1} a_j z^j}{\prod_{j=1}^{N}(1 - \overline{\lambda_j} z)} \right) = \frac{\sum_{j=0}^{N-1} \overline{a_{N-1-j}} z^j}{\prod_{j=1}^{N}(1 - \overline{\lambda_j} z)}.$$

If the zeros of u are distinct, then the Cauchy kernels c_{λ_i} from (2) corresponding to the λ_i form a basis for $\mathcal{K}_u$ whence

$$\mathcal{K}_u = \operatorname{span}\{c_{\lambda_1}, c_{\lambda_2}, \ldots, c_{\lambda_N}\}.$$

If some of the λ_i are repeated, then one must include the appropriate derivatives of the c_{λ_i} to obtain a basis for $\mathcal{K}_u$.

Although the natural bases for $\mathcal{K}_u$ described above are not orthogonal, a particularly convenient orthonormal basis exists. For $\lambda \in \mathbb{D}$, let

$$b_\lambda(z) = \frac{z - \lambda}{1 - \overline{\lambda} z}$$

be a disk automorphism with a zero at λ and for each $1 \leq n \leq N$ let

$$\gamma_n(z) = \frac{\sqrt{1 - |\lambda_n|^2}}{1 - \overline{\lambda_n} z} \prod_{k=1}^{n-1} b_{\lambda_k}(z). \tag{19}$$

The following important fact was first observed by Takenaka [98] in 1925, although it has been rediscovered many times since.

Theorem 10 (Takenaka) $\{\gamma_1, \gamma_2, \ldots, \gamma_N\}$ *is an orthonormal basis for* $\mathcal{K}_u$.

If u is an infinite Blaschke product, then an extension, due to Walsh [78], of the preceding result tells us that $\{\gamma_1, \gamma_2, \ldots\}$ is an orthonormal basis for $\mathcal{K}_u$.

Let us return now to the finite dimensional setting. Suppose that u is a finite Blaschke product with N zeros, repeated according to multiplicity. To avoid some needless technical details, we assume that $u(0) = 0$. If $\{\zeta_1, \zeta_2, \ldots, \zeta_N\}$ are the eigenvalues of the Clark unitary operator

$$U_\alpha := A_z^u + \alpha k_0 \otimes C k_0$$

(i.e., the N distinct solutions on $\partial\mathbb{D}$ to $u(\zeta) = \alpha$), then the corresponding eigenvectors $\{k_{\zeta_1}, k_{\zeta_2}, \ldots, k_{\zeta_N}\}$ are orthogonal. A routine computation shows that $\|k_{\zeta_j}\| = \sqrt{|u'(\zeta)|}$ so that

$$\left\{ \frac{k_{\zeta_1}}{\sqrt{|u'(\zeta_1)|}}, \frac{k_{\zeta_2}}{\sqrt{|u'(\zeta_2)|}}, \ldots, \frac{k_{\zeta_N}}{\sqrt{|u'(\zeta_N)|}} \right\} \tag{20}$$

is an orthonormal basis for $\mathcal{K}_u$. This is called a *Clark basis* for $\mathcal{K}_u$. Letting $w_j = \exp(-\frac{1}{2}(\arg\zeta_j - \arg\alpha))$, it turns out that

$$\left\{ \frac{w_1 k_{\zeta_1}}{\sqrt{|u'(\zeta_1)|}}, \frac{w_2 k_{\zeta_2}}{\sqrt{|u'(\zeta_2)|}}, \ldots, \frac{w_N k_{\zeta_N}}{\sqrt{|u'(\zeta_N)|}} \right\} \tag{21}$$

is an orthonormal basis for $\mathcal{K}_u$, each vector of which is fixed by the conjugation (9) on $\mathcal{K}_u$ (i.e., in the terminology of [49], (21) is a C-*real* basis for $\mathcal{K}_u$). We refer to a basis of the form (21) as a *modified Clark basis* for $\mathcal{K}_u$ (see [45] and [51] for further details).

2.9 Truncated Toeplitz Operators

The *truncated Toeplitz operator* A_φ^u on $\mathcal{K}_u$ having *symbol* φ in L^2 is the closed, densely defined operator

$$A_\varphi^u f := P_u(\varphi f)$$

having domain[2]

$$\mathscr{D}(A_\varphi^u) = \{ f \in \mathcal{K}_u : P_u(\varphi f) \in \mathcal{K}_u \}.$$

When there is no danger of confusion, we sometimes write A_φ in place of A_φ^u. A detailed discussion of unbounded truncated Toeplitz operators and their properties can be found in Sect. 10. For the moment, we focus on those truncated Toeplitz operators which can be extended to bounded operators on $\mathcal{K}_u$.

Definition 3 Let $\mathscr{T}_u$ denote the set of all *bounded* truncated Toeplitz operators on $\mathcal{K}_u$.

[2] Written as an integral transform, P_u can be regarded as an operator from L^1 into $\mathrm{Hol}(\mathbb{D})$.

For Toeplitz operators, recall that $\|T_\varphi\| = \|\varphi\|_\infty$ holds for each φ in L^∞. In contrast, we can say little more than

$$0 \leqslant \|A_\varphi^u\| \leqslant \|\varphi\|_\infty \tag{22}$$

for general truncated Toeplitz operators. In fact, computing, or at least estimating, the norm of a truncated Toeplitz operator is a difficult problem. This topic is discussed in greater detail in Sect. 4. However, a complete characterization of those symbols which yield the zero operator has been obtained by Sarason [88, Theorem 3.1].

Theorem 11 (Sarason) *A truncated Toeplitz operator A_φ^u is identically zero if and only if $\varphi \in uH^2 + \overline{uH^2}$.*

In particular, the preceding result tells us that there are always infinitely many symbols (many of them unbounded) which represent the same truncated Toeplitz operator. On the other hand, since $A_\varphi^u = A_\psi^u$ if and only if $\psi - \varphi$ belongs to $uH^2 + \overline{uH^2}$, we actually enjoy some freedom in specifying the symbol of a truncated Toeplitz operator. The following corollary makes this point concrete.

Corollary 1 *If A belongs to $\mathscr{T}_u$, then there exist φ_1 and φ_2 in $\mathcal{K}_u$ such that $A = A_{\varphi_1 + \overline{\varphi_2}}$. Furthermore, φ_1 and φ_2 are uniquely determined if we fix the value of one of them at the origin.*

To some extent, the preceding corollary can be reversed. As noted in [14], if we assume that $A \in \mathscr{T}_u$ has a symbol $\varphi_1 + \overline{\varphi_2}$, where φ_1 and φ_2 belong to $\mathcal{K}_u$ and $\varphi_2(0) = 0$, then we can recover φ_1 and φ_2 by knowing the action of A on the reproducing kernels k_λ and the conjugate reproducing kernels Ck_λ. Indeed, one just needs to solve the following linear system in the variables $\varphi_1(\lambda)$ and $\overline{\varphi_2(\lambda)}$:

$$\begin{aligned} \varphi_1(\lambda) - \overline{u(0)}u(\lambda)\overline{\varphi_2(\lambda)} &= \langle Ak_0, k_\lambda \rangle, \\ \overline{\varphi_2(\lambda)} - u(0)\overline{u(\lambda)}\varphi_1(\lambda) &= \langle ACk_0, Ck_0 \rangle - \langle Ak_0, k_0 \rangle. \end{aligned}$$

With more work, one can even obtain an estimate of $\max\{\|\varphi_1\|, \|\varphi_2\|\}$ [14].

Letting C denote the conjugation (9) on $\mathcal{K}_u$, a direct computation confirms the following result from [49]:

Theorem 12 (Garcia-Putinar) *For any $A \in \mathscr{T}_u$, we have $A = CA^*C$.*

In particular, Theorem 12 says each truncated Toeplitz operator is a *complex symmetric operator*, a class of Hilbert space operators which has undergone much recent study [24, 35, 45–50, 54, 55, 61, 63, 71, 72, 74, 103–107]. In fact, it is suspected that truncated Toeplitz operators might serve as some sort of model operator for various classes of complex symmetric operators (see Sect. 9). For the moment, let us simply note that the matrix representation of a truncated Toeplitz operator A_φ^u with respect to a modified Clark basis (21) is *complex symmetric* (i.e., self-transpose). This was first observed in [49] and developed further in [45].

An old theorem of Brown and Halmos [20] says that a bounded operator T on H^2 is a Toeplitz operator if and only if $T = STS^*$. Sarason recently obtained a version of this theorem for truncated Toeplitz operators [88, Theorem 4.1].

Theorem 13 (Sarason) *A bounded operator A on $\mathcal{K}_u$ belongs to $\mathscr{T}_u$ if and only if there are functions $\varphi, \psi \in \mathcal{K}_u$ such that*

$$A = A_z^u A \big(A_z^u\big)^* + \varphi \otimes k_0 + k_0 \otimes \psi.$$

When $\mathcal{K}_u$ is finite dimensional, one can get more specific results using matrix representations. For example, if $u = z^N$, then $\{1, z, \ldots, z^{N-1}\}$ is an orthonormal basis for $\mathcal{K}_{z^N}$. Any operator in $\mathscr{T}_{z^N}$ represented with respect to this basis yields a Toeplitz matrix and, conversely, any $N \times N$ Toeplitz matrix gives rise to a truncated Toeplitz operator on $\mathcal{K}_{z^N}$. Indeed the matrix representation of $A_\varphi^{z^N}$ with respect to $\{1, z, \ldots, z^{N-1}\}$ is the Toeplitz matrix $(\widehat{\varphi}(j-k))_{j,k=0}^{N-1}$. For more general finite Blaschke products we have the following result from [28].

Theorem 14 (Cima-Ross-Wogen) *Let u be a finite Blaschke product of degree n with distinct zeros $\lambda_1, \lambda_2, \ldots, \lambda_n$ and let A be any linear transformation on the n-dimensional space $\mathcal{K}_u$. If $M_A = (r_{i,j})_{i,j=1}^n$ is the matrix representation of A with respect to the basis $\{k_{\lambda_1}, k_{\lambda_2}, \ldots, k_{\lambda_n}\}$, then $A \in \mathscr{T}_u$ if and only if*

$$r_{i,j} = \overline{\left(\frac{u'(\lambda_1)}{u'(\lambda_i)}\right)} \left(\frac{r_{1,i}\overline{(\lambda_1 - \lambda_i)} + r_{1,j}\overline{(\lambda_j - \lambda_1)}}{\overline{\lambda_j - \lambda_i}}\right),$$

for $1 \leqslant i, j \leqslant n$ and $i \neq j$.

Although the study of general truncated Toeplitz operators appears to be difficult, there is a distinguished subset of these operators which are remarkably tractable. We say that A_φ^u is an *analytic* truncated Toeplitz operator if the symbol φ belongs to H^∞, or more generally, to H^2. It turns out that the natural polynomial functional calculus $p(A_z^u) = A_p^u$ can be extended to H^∞ in such a way that the symbol map $\varphi \mapsto \varphi(A_z^u) := A_\varphi^u$ is linear, contractive, and multiplicative. As a broad generalization of Theorem 6, we have the following spectral mapping theorem [78, p. 66], the proof of which depends crucially on the famous Corona Theorem of L. Carleson [21].

Theorem 15 *If $\varphi \in H^\infty$, then*

(i) $\sigma(A_\varphi^u) = \{\lambda : \inf_{z \in \mathbb{D}}(|u(z)| + |\varphi(z) - \lambda|) = 0\}$.
(ii) *If $\varphi \in H^\infty \cap C(\partial\mathbb{D})$, then $\sigma(A_\varphi^u) = \varphi(\sigma(u))$.*

We conclude this section by remarking that vector-valued analogues are available for most of the preceding theorems. However, these do not concern us here and we refer the reader to [78] for further details.

3 $\mathscr{T}_u$ as a Linear Space

Recent work of Baranov, Bessonov, and Kapustin [13] has shed significant light on the structure of $\mathscr{T}_u$ as a linear space. Before describing these results, let us first recount a few important observations due to Sarason. The next theorem is [88, Theorem 4.2].

Theorem 16 (Sarason) *$\mathscr{T}_u$ is closed in the weak operator topology.*

It is important to note that $\mathscr{T}_u$ is not an operator algebra, for the product of truncated Toeplitz operators is rarely itself a truncated Toeplitz operator (the precise conditions under which this occurs were found by Sedlock [92, 93]). On the other hand, $\mathscr{T}_u$ contains a number of interesting subsets which are algebras. The details are discussed in Sect. 7, followed in Sect. 8 by a brief discussion about C^*-algebras generated by truncated Toeplitz operators.

In order to better frame the following results, first recall that there are no nonzero compact Toeplitz operators on H^2 [20]. In contrast, there are many examples of finite rank (hence compact) truncated Toeplitz operators. In fact, the rank-one truncated Toeplitz operators were first identified by Sarason [88, Theorem 5.1].

Theorem 17 (Sarason) *For an inner function u, the operators*

(i) $k_\lambda \otimes Ck_\lambda = A^u_{\frac{\overline{u}}{\overline{z}-\overline{\lambda}}}$ *for* $\lambda \in \mathbb{D}$,
(ii) $Ck_\lambda \otimes k_\lambda = A^u_{\frac{u}{z-\lambda}}$ *for* $\lambda \in \mathbb{D}$,
(iii) $k_\zeta \otimes k_\zeta = A^u_{k_\zeta + \overline{k_\zeta} - 1}$ *where u has an ADC at* $\zeta \in \partial\mathbb{D}$,

are truncated Toeplitz operators having rank one. Moreover, any truncated Toeplitz operator of rank one is a scalar multiple of one of the above.

We should also mention the somewhat more involved results of Sarason [88, Theorems 6.1 & 6.2] which identify a variety of natural finite rank truncated Toeplitz operators. Furthermore, the following linear algebraic description of $\mathscr{T}_u$ has been obtained [88, Theorem 7.1] in the finite dimensional setting.

Theorem 18 (Sarason) *If* $\dim \mathcal{K}_u = n$, *then*

(i) $\dim \mathscr{T}_u = 2n - 1$,
(ii) *If* $\lambda_1, \lambda_2, \ldots, \lambda_{2n-1}$ *are distinct points of* $\mathbb{D}$, *then the operators* $k^u_{\lambda_j} \otimes \widetilde{k}^u_{\lambda_j}$ *for* $j = 1, 2, \ldots, 2n-1$ *form a basis for* $\mathscr{T}_u$.[3]

[3]Recall that we are using the notation $\widetilde{f} := Cf$ for $f \in \mathcal{K}_u$.

When confronted with a novel linear space, the first questions to arise concern duality. Baranov, Bessonov, and Kapustin recently identified the predual of $\mathscr{T}_u$ and discussed the weak-$*$ topology on $\mathscr{T}_u$ [13]. Let us briefly summarize some of their major results. First consider the space

$$\mathcal{X}_u := \left\{ F = \sum_{n=1}^{\infty} f_n \overline{g_n} \ : \ f_n, g_n \in \mathcal{K}_u, \ \sum_{n=1}^{\infty} \|f_n\| \|g_n\| < \infty \right\}$$

with norm

$$\|F\|_{\mathcal{X}_u} := \inf \left\{ \sum_{n=1}^{\infty} \|f_n\| \|g_n\| : F = \sum_{n=1}^{\infty} f_n \overline{g_n} \right\}.$$

It turns out that

$$\mathcal{X}_u \subseteq \overline{u} z H^1 \cap \overline{u z H^1},$$

and that each element of $\mathcal{X}_u$ can be written as a linear combination of four elements of the form $f\overline{g}$, where f and g belong to $\mathcal{K}_u$. The importance of the space $\mathcal{X}_u$ lies in the following important theorem and its corollaries.

Theorem 19 (Baranov-Bessonov-Kapustin [13]) *For any inner function u, $\mathcal{X}_u^*$, the dual space of $\mathcal{X}_u$, is isometrically isomorphic to $\mathscr{T}_u$ via the dual pairing*

$$(F, A) := \sum_{n=1}^{\infty} \langle A f_n, g_n \rangle, \quad F = \sum_{n=1}^{\infty} f_n \overline{g_n}, \ A \in \mathscr{T}_u.$$

Furthermore, if $\mathscr{T}_u^c$ denotes the compact truncated Toeplitz operators, then $(\mathscr{T}_u^c)^*$, the dual of $\mathscr{T}_u^c$, is isometrically isomorphic to $\mathcal{X}_u$.

Corollary 2 (Baranov-Bessonov-Kapustin)

(i) *The weak topology and the weak-$*$ topology on $\mathscr{T}_u$ are the same.*
(ii) *The norm closed linear span of the rank-one truncated Toeplitz operators is $\mathscr{T}_u^c$.*
(iii) *$\mathscr{T}_u^c$ is weakly dense in $\mathscr{T}_u$.*

For a general inner function u, we will see below that not every bounded truncated Toeplitz operator on $\mathcal{K}_u$ has a bounded symbol (see Sect. 5). On the other hand, the following corollary holds in general.

Corollary 3 (Baranov-Bessonov-Kapustin) *The truncated Toeplitz operators with bounded symbols are weakly dense in $\mathscr{T}_u$.*

This leaves open the following question.

Question 1 Are the truncated Toeplitz operators with bounded symbols *norm* dense in $\mathscr{T}_u$?

4 Norms of Truncated Toeplitz Operators

Recall that for φ in L^∞ we have the trivial estimates (22) on the norm of a truncated Toeplitz operator, but little other general information concerning this quantity. For φ in L^2, we may also consider the potentially unbounded truncated Toeplitz operator A_φ^u. Of interest is the quantity

$$\|A_\varphi^u\| := \sup\{\|A_\varphi^u f\| : f \in \mathcal{K}_u \cap H^\infty, \|f\| = 1\}, \tag{23}$$

which we regard as being infinite if A_φ^u is unbounded. For A_φ^u bounded, (23) is simply the operator norm of A_φ^u in light of Theorem 3. Evaluation or estimation of (23) is further complicated by the fact that the representing symbol φ for A_φ^u is never unique (Theorem 11).

If u is a finite Blaschke product (so that the corresponding model space $\mathcal{K}_u$ is finite dimensional) and φ belongs to H^∞, then straightforward residue computations allow us to represent A_φ^u with respect to any of the orthonormal bases mentioned earlier (i.e., the Takenaka (19), Clark (20), or modified Clark (21) bases). For $\mathcal{K}_{z^n}$, the Takenaka basis is simply the monomial basis $\{1, z, z^2, \ldots, z^{n-1}\}$ and the matrix representation of A_φ^u is just a lower triangular Toeplitz matrix. In any case, one can readily compute the norm of A_φ^u by computing the norm of one of its matrix representations. This approach was undertaken by the authors in [51]. One can also approach this problem using the theory of Hankel operators (see [81, Eq. (2.9)] and the method developed in [23]).

Let us illustrate the general approach with a simple example. If φ belongs to H^∞ and u is the finite Blaschke product with distinct zeros $\lambda_1, \lambda_2, \ldots, \lambda_n$, then the matrix representation for A_φ^u with respect to the modified Clark basis (21) is

$$\left(\frac{w_k}{\sqrt{|u'(\zeta_k)|}} \frac{w_j}{\sqrt{|u'(\zeta_j)|}} \sum_{i=1}^{n} \frac{\varphi(\lambda_i)}{u'(\lambda_i)(1-\overline{\zeta_k}\lambda_i)(1-\overline{\zeta_j}\lambda_i)} \right)_{j,k=1}^{n}. \tag{24}$$

In particular, observe that this matrix is complex symmetric, as predicted by Theorem 12. As a specific example, consider the Blaschke product

$$u(z) = z\frac{z-\frac{1}{2}}{1-\frac{1}{2}z}$$

and the H^∞ function

$$\varphi(z) = \frac{2z-\frac{1}{2}}{1-\frac{1}{2}z}.$$

The parameters in (24) are

$$\alpha = 1, \qquad \lambda_1 = 0, \qquad \lambda_2 = \frac{1}{2}, \qquad \zeta_1 = 1,$$
$$\zeta_2 = -1, \qquad w_1 = 1, \qquad w_2 = -i,$$

which yields

$$\|A^u_\varphi\| = \left\| \begin{pmatrix} \frac{5}{4} & -\frac{7i}{4\sqrt{3}} \\ -\frac{7i}{4\sqrt{3}} & -\frac{13}{12} \end{pmatrix} \right\| = \frac{1}{6}(7+\sqrt{37}) \approx 2.1805.$$

On a somewhat different note, it is possible to obtain lower estimates of $\|A^u_\varphi\|$ for general φ in L^2. This can be helpful, for instance, in determining whether a given truncated Toeplitz operator is unbounded. Although a variety of lower bounds on $\|A^u_\varphi\|$ are provided in [52], we focus here on perhaps the most useful of these. We first require the *Poisson integral*

$$(\mathfrak{P}\varphi)(z) := \int_{\partial\mathbb{D}} \frac{1-|z|^2}{|\zeta - z|^2}\varphi(\zeta)\,dm(\zeta)$$

of a function φ in L^1. In particular, recall that $\lim_{r\to 1^-}(\mathfrak{P}\varphi)(r\zeta) = \varphi(\zeta)$ whenever φ is continuous at a point $\zeta \in \partial\mathbb{D}$ [69, p. 32] or more generally, when ζ is a Lebesgue point of φ.

Theorem 20 (Garcia-Ross) *If $\varphi \in L^2$ and u is inner, then*

$$\left\|A^u_\varphi\right\| \geq \sup\left\{\left|(\mathfrak{P}\varphi)(\lambda)\right| : \lambda \in \mathbb{D} : u(\lambda) = 0\right\},$$

where the supremum is regarded as 0 *if u never vanishes on $\mathbb{D}$.*

Corollary 4 *If φ belongs to $C(\partial\mathbb{D})$ and u is an inner function whose zeros accumulate almost everywhere on $\partial\mathbb{D}$, then $\|A^u_\varphi\| = \|\varphi\|_\infty$.*

A related result on *norm attaining symbols* can be found in [51].

Theorem 21 (Garcia-Ross) *If u is inner, $\varphi \in H^\infty$, and A^u_φ is compact, then $\|A^u_\varphi\| = \|\varphi\|_\infty$ if and only if φ is a scalar multiple of the inner factor of a function from $\mathcal{K}_u$.*

It turns out that the norm of a truncated Toeplitz operator can be related to certain classical extremal problems from function theory. For the following discussion we require a few general facts about complex symmetric operators [45, 49, 50]. Recall that a *conjugation* on a complex Hilbert space $\mathcal{H}$ is a map $C : \mathcal{H} \to \mathcal{H}$ which is conjugate-linear, involutive (i.e., $C^2 = I$), and isometric (i.e., $\langle Cx, Cy\rangle = \langle y, x\rangle$). A bounded linear operator $T : \mathcal{H} \to \mathcal{H}$ is called *C-symmetric* if $T = CT^*C$ and *complex symmetric* if there exists a conjugation C with respect to which T is C-symmetric. Theorem 12 asserts that each operator in $\mathscr{T}_u$ is C-symmetric with respect to the conjugation C on $\mathcal{K}_u$ defined by (9). The following general result from [51] relates the norm of a given C-symmetric operator to a certain extremal problem (as is customary, $|T|$ denotes the positive operator $\sqrt{T^*T}$).

Theorem 22 (Garcia-Ross) *If $T : \mathcal{H} \to \mathcal{H}$ is a bounded C-symmetric operator, then*

(i) $\|T\| = \sup_{\|x\|=1} |\langle Tx, Cx\rangle|$.
(ii) *If* $\|x\| = 1$, *then* $\|T\| = |\langle Tx, Cx\rangle|$ *if and only if* $Tx = \omega\|T\|Cx$ *for some unimodular constant* ω.
(iii) *If* T *is compact, then the equation* $Tx = \|T\|Cx$ *has a unit vector solution. Furthermore, this unit vector solution is unique, up to a sign, if and only if the kernel of the operator* $|T| - \|T\|I$ *is one-dimensional.*

Applying the Theorem 22 to A_φ^u we obtain the following result.

Corollary 5 *For inner* u *and* $\varphi \in L^\infty$

$$\|A_\varphi^u\| = \sup\left\{\left|\frac{1}{2\pi i}\oint_{\partial\mathbb{D}} \frac{\varphi f^2}{u} dz\right| : f \in \mathcal{K}_u, \|f\| = 1\right\}.$$

For φ in H^∞, the preceding supremum can be taken over H^2.

Corollary 6 *For inner* u *and* $\varphi \in H^\infty$

$$\|A_\varphi^u\| = \sup\left\{\left|\frac{1}{2\pi i}\oint_{\partial\mathbb{D}} \frac{\varphi f^2}{u} dz\right| : f \in H^2, \|f\| = 1\right\}.$$

The preceding corollary relates the norm of a truncated Toeplitz operator to a certain *quadratic* extremal problem on H^2. We can relate this to a classical *linear* extremal problem in the following way. For a rational function ψ with no poles on $\partial\mathbb{D}$ we have the well-studied classical H^1 extremal problem [40, 62]:

$$\Lambda(\psi) := \sup\left\{\left|\frac{1}{2\pi i}\oint_{\partial\mathbb{D}} \psi F\, dz\right| : F \in H^1, \|F\|_1 = 1\right\}. \tag{25}$$

On the other hand, basic functional analysis tells us that

$$\Lambda(\psi) = \operatorname{dist}\big(\psi, H^\infty\big).$$

Following [51], we recall that the extremal problem $\Lambda(\psi)$ has an *extremal function* F_e (not necessarily unique). It is also known that F_e can be taken to be outer and hence $F_e = f^2$ for some f in H^2. Therefore the linear extremal problem $\Lambda(\psi)$ and the quadratic extremal problem

$$\Gamma(\psi) := \sup\left\{\left|\frac{1}{2\pi i}\oint_{\partial\mathbb{D}} \psi f^2 dz\right| : f \in H^2, \|f\| = 1\right\} \tag{26}$$

have the same value. The following result from [51], combined with the numerical recipes discussed at the beginning of this section, permit one to explicitly evaluate many specific extremal problems. Before doing so, we remark that many of these problems can be attacked using the theory of Hankel operators, although in that case one must compute the norm of a finite-rank Hankel operator acting on an infinite-dimensional space. In contrast, the truncated Toeplitz approach employs only $n \times n$ matrices.

Corollary 7 *Suppose that* ψ *is a rational function having no poles on* $\partial\mathbb{D}$ *and poles* $\lambda_1, \lambda_2, \ldots, \lambda_n$ *lying in* $\mathbb{D}$*, counted according to multiplicity. Let* u *denote the associated Blaschke product whose zeros are precisely* $\lambda_1, \lambda_2, \ldots, \lambda_n$ *and note that* $\varphi = u\psi$ *belongs to* H^∞*. We then have the following*:

(i) $\|A^u_\varphi\| = \Gamma(\psi) = \Lambda(\psi)$.
(ii) *There is a unit vector* $f \in \mathcal{K}_u$ *satisfying* $A^u_\varphi f = \|A^u_\varphi\| Cf$ *and any such* f *is an extremal function for* $\Gamma(\psi)$*. In other words,*

$$\left|\frac{1}{2\pi i}\oint_{\partial\mathbb{D}} \psi f^2\, dz\right| = \|A^u_\varphi\|.$$

(iii) *Every extremal function* f *for* $\Gamma(\psi)$ *belongs to* $\mathcal{K}_u$ *and satisfies*

$$A^u_\varphi f = \|A_\varphi\| Cf.$$

(iv) *An extremal function for* $\Gamma(\psi)$ *is unique, up to a sign, if and only if the kernel of the operator* $|A^u_\varphi| - \|A^u_\varphi\| I$ *is one-dimensional.*

We refer the reader to [51] for several worked examples of classical extremal problems $\Lambda(\psi)$ along with a computation of several extremal functions F_e. For rational functions ψ with no poles on $\partial\mathbb{D}$, we have seen that the linear (25) and the quadratic (26) extremal problems have the same value. Recent work of Chalendar, Fricain, and Timotin shows that this holds in much greater generality.

Theorem 23 (Chalendar-Fricain-Timotin [23]) *For each* ψ *in* L^∞, $\Gamma(\psi) = \Lambda(\psi)$.

It is important to note that for general ψ in L^∞ an extremal function for $\Lambda(\psi)$ need not exist (see [51] for a relevant discussion). Nevertheless, for ψ in L^∞, Chalendar, Fricain, and Timotin prove that $\Lambda(\psi) = \Gamma(\psi)$ by using the fact that $\Lambda(\psi) = \|H_\psi\|$, where $H_\psi : H^2 \to L^2 \ominus H^2$ is the corresponding *Hankel operator* $H_\psi f = P_-(\psi f)$. Here P_- denotes the orthogonal projection from L^2 onto $L^2 \ominus H^2$. We certainly have the inequality

$$\Gamma(\psi) = \Lambda(\psi) = \|H_\psi\| = \operatorname{dist}\big(\psi, H^\infty\big) \leqslant \|\psi\|_\infty.$$

When equality holds in the preceding, we say that the symbol ψ is *norm attaining*. The authors of [23] prove the following.

Theorem 24 (Chalendar-Fricain-Timotin) *If* $\psi \in L^\infty$ *is norm attaining then* ψ *has constant modulus and there exists an extremal outer function for* $\Lambda(\psi)$.

Before proceeding, we should also mention the fact that computing the norm of certain truncated Toeplitz operators and solving their related extremal problems have been examined for quite some time in the study of H^∞ control theory and skew-Toeplitz operators [17, 18, 43, 44]. In the scalar setting, a *skew-Toeplitz operator* is a truncated Toeplitz operator A_φ^u, where the symbol takes the form

$$\varphi(\zeta) = \sum_{j,k=0}^{n} a_{j,k}\zeta^{j-k}, \quad a_{j,k} \in \mathbb{R},$$

making A_φ^u self-adjoint. In H^∞ control theory, the extremal problem

$$\operatorname{dist}\big(\psi, uH^\infty\big),$$

where ψ is a rational function belonging to H^∞, plays an important role. From the preceding results, we observe that $\|A_\psi^u\| = \operatorname{dist}(\psi, uH^\infty)$.

5 The Bounded Symbol and Related Problems

Recall that $\mathscr{T}_u$ denotes the set of all truncated Toeplitz operators A_φ^u, densely defined on $\mathcal{K}_u$ and having symbols φ in L^2, that can be extended to bounded operators on all of $\mathcal{K}_u$. As a trivial example, if φ belongs to L^∞, then clearly A_φ^u belongs to $\mathscr{T}_u$. A major open question involved the converse. In other words, if A_φ^u is a bounded truncated Toeplitz operator, does there exist a symbol φ_0 in L^∞ such that $A_\varphi^u = A_{\varphi_0}^u$? This question was recently resolved in the negative by Baranov, Chalendar, Fricain, Mashreghi, and Timotin [14]. We describe this groundbreaking work, along with important related contributions by Baranov, Bessonov, and Kapustin [13], below.

For symbols φ in H^2, a complete and elegant answer to the bounded symbol problem is available. In the following theorem, the implication (i) $\Leftrightarrow$ (ii) below is due to Sarason [85]. Condition (iii) is often referred to as the *reproducing kernel thesis* for A_φ^u.

Theorem 25 (Baranov, Chalendar, Fricain, Mashreghi, Timotin [14]) *For $\varphi \in H^2$, the following are equivalent.*

(i) $A_\varphi^u \in \mathscr{T}_u$.
(ii) $A_\varphi^u = A_{\varphi_0}^u$ *for some* $\varphi_0 \in H^\infty$.
(iii) $\sup_{\lambda\in\mathbb{D}} \|A_\varphi^u \frac{k_\lambda}{\|k_\lambda\|}\| < \infty$.

Furthermore, there exists a universal constant $C > 0$ so that any $A_\varphi^u \in \mathscr{T}_u$ with $\varphi \in H^2$, has a bounded symbol φ_0 such that

$$\|\varphi_0\|_\infty \leqslant C \sup_{\lambda\in\mathbb{D}} \left\| A_\varphi^u \frac{k_\lambda}{\|k_\lambda\|} \right\|.$$

The following result demonstrates the existence of bounded truncated Toeplitz operators with no bounded symbol (a small amount of function theory, discussed below, is required to exhibit concrete examples). Recall from Theorem 17 that for

ζ in $\partial\mathbb{D}$, the rank one operator $k_\zeta \otimes k_\zeta$ belongs to $\mathscr{T}_u$ if and only if u has an ADC at ζ. Using two technical lemmas from [14] (Lemmas 5.1 and 5.2), they prove the following theorem.

Theorem 26 (Baranov, Chalendar, Fricain, Mashreghi, Timotin) *If u has an ADC at $\zeta \in \partial\mathbb{D}$ and $p \in (2,\infty)$, then the following are equivalent*:

(i) *$k_\zeta \otimes k_\zeta$ has a symbol in L^p.*
(ii) *$k_\zeta \in L^p$.*

Consequently, if $k_\zeta \notin L^p$ for some $p \in (2,\infty)$, then $k_\zeta \otimes k_\zeta$ belongs to $\mathscr{T}_u$ and has no bounded symbol.

From Theorem 2 we know that if $u = b_\Lambda s_\mu$, where b is a Blaschke product with zeros $\Lambda = \{\lambda_n\}_{n=1}^\infty$ (repeated according to multiplicity) and s_μ is a singular inner function with corresponding singular measure μ, then

$$k_\zeta \in H^2 \quad \Longleftrightarrow \quad \sum_{n=1}^{\infty} \frac{1-|\lambda|^2}{|\zeta - \lambda_n|^2} + \int \frac{d\mu(\xi)}{|\xi - \zeta|^2} < \infty. \tag{27}$$

This was extended [3, 32] to $p \in (1,\infty)$ as follows:

$$k_\zeta \in H^p \quad \Longleftrightarrow \quad \sum_{n=1}^{\infty} \frac{1-|\lambda|^2}{|\zeta - \lambda_n|^p} + \int \frac{d\mu(\xi)}{|\xi - \zeta|^p} < \infty. \tag{28}$$

Based upon this, one can arrange it so that k_ζ belongs to H^2 but not to H^p for any $p > 2$. This yields the desired example of a bounded truncated Toeplitz operator which cannot be represented using a bounded symbol. We refer the reader to the article [14] where the details and further examples are considered.

Having seen that there exist bounded truncated Toeplitz operators which lack bounded symbols, it is natural to ask if there exist inner functions u so that *every* operator in $\mathscr{T}_u$ has a bounded symbol? Obviously this holds when u is a finite Blaschke product. Indeed, in this case the symbol can be taken to be a polynomial in z and $\overline{z}$. A more difficult result is the following (note that the initial symbol φ belongs to L^2, as opposed to H^2, as was the case in Theorem 25).

Theorem 27 (Baranov, Chalendar, Fricain, Mashreghi, Timotin) *If $a > 0$, $\zeta \in \partial\mathbb{D}$, and*

$$u(z) = \exp\left(a\frac{z+\zeta}{z-\zeta}\right),$$

then the following are equivalent for $\varphi \in L^2$:

(i) $A^u_\varphi \in \mathscr{T}_u$.
(ii) $A^u_\varphi = A^u_{\varphi_0}$ *for some* $\varphi_0 \in L^\infty$.
(iii) $\sup_{\lambda\in\mathbb{D}} \|A^u_\varphi \frac{k_\lambda}{\|k_\lambda\|}\| < \infty$.

Furthermore, there exists a universal constant $C > 0$ so that any $A_\varphi^u \in \mathscr{T}_u$ with $\varphi \in L^2$, has a bounded symbol φ_0 such that

$$\|\varphi_0\|_\infty \leqslant C \sup_{\lambda \in \mathbb{D}} \left\| A_\varphi^u \frac{k_\lambda}{\|k_\lambda\|} \right\|.$$

In light of Theorems 25 and 27, one might wonder whether condition (iii) (the reproducing kernel thesis) is always equivalent to asserting that A_φ^u belongs to $\mathscr{T}_u$. Unfortunately, the answer is again negative [14, Sect. 5].

On a positive note, Baranov, Bessonov, and Kapustin recently discovered a condition on the inner function u which ensures that *every* operator in $\mathscr{T}_u$ has a bounded symbol [13]. After a few preliminary details, we discuss their work below.

Definition 4 For $p > 0$, let $\mathcal{C}_p(u)$ denote the finite complex Borel measures μ on $\partial\mathbb{D}$ such that $\mathcal{K}_u^p$ embeds continuously into $L^p(|\mu|)$.

Since S^*u belongs to $\mathcal{K}_u$, it follows from Aleksandrov's embedding theorem (Theorem 4) that for each μ in $\mathcal{C}_2(u)$, the boundary values of u are defined $|\mu|$-almost everywhere. Moreover, it turns out that $|u| = 1$ holds $|\mu|$-almost everywhere [6, 13]. For $\mu \in \mathcal{C}_2(u)$ the quadratic form

$$(f, g) \mapsto \int_{\partial\mathbb{D}} f\overline{g}\, d\mu, \qquad f, g \in \mathcal{K}_u$$

is continuous and so, by elementary functional analysis, there is a bounded operator $\mathcal{A}_\mu : \mathcal{K}_u \to \mathcal{K}_u$ such that

$$\langle \mathcal{A}_\mu f, g\rangle = \int_{\partial\mathbb{D}} f\overline{g}\, d\mu.$$

The following important result of Sarason [88, Theorem 9.1] asserts that each such $\mathcal{A}_\mu$ is a truncated Toeplitz operator.

Theorem 28 (Sarason) *$\mathcal{A}_\mu \in \mathscr{T}_u$ whenever $\mu \in \mathcal{C}_2(u)$.*

A natural question, posed by Sarason [88, p. 513], is whether the converse holds. In other words, does every bounded truncated Toeplitz operator arise from a so-called *u-compatible measure* [88, Sect. 9]? This question was recently settled in the affirmative by Baranov, Bessonov, and Kapustin [13].

Theorem 29 (Baranov-Bessonov-Kapustin [13]) *$A \in \mathscr{T}_u$ if and only if $A = \mathcal{A}_\mu$ for some $\mu \in \mathcal{C}_2(u)$.*

The measure μ above is called the *quasi-symbol* for the truncated Toeplitz operator. For $\varphi \in L^\infty$ we adopt the convention that $\mathcal{A}_{\varphi\, dm}^u := A_\varphi^u$ so that every bounded symbol is automatically a quasi-symbol.

It turns out that $\mathcal{C}_1(u^2) \subseteq \mathcal{C}_2(u) = \mathcal{C}_2(u^2)$ always holds. Baranov, Bessonov, and Kapustin showed that equality is the precise condition which ensures that every $A \in \mathscr{T}_u$ can be represented using a bounded symbol.

Theorem 30 (Baranov-Bessonov-Kapustin) *An operator $A \in \mathscr{T}_u$ has a bounded symbol if and only if $A = A_\mu$ for some $\mu \in \mathcal{C}_1(u^2)$. Consequently, every operator in $\mathscr{T}_u$ has a bounded symbol if and only if $\mathcal{C}_1(u^2) = \mathcal{C}_2(u)$.*

Recall that each function F in H^1 can be written as the product $F = fg$ of two functions in H^2. Conversely, the product of any pair of functions in H^2 lies in H^1. For each $f, g \in \mathcal{K}_u$ we note that

$$H^1 \ni fg = \tilde{\tilde{f}}\tilde{\tilde{g}} = \overline{\tilde{f}\tilde{g}z^2u^2} \in \overline{z}u^2\overline{H_0^1},$$

whence $fg \in H^1 \cap \overline{z}u^2\overline{H_0^1}$. Moreover, one can show that finite linear combinations of pairs of products of functions from $\mathcal{K}_u$ form a dense subset of $H^1 \cap \overline{z}u^2\overline{H_0^1}$. As a consequence, this relationship between $\mathcal{K}_u$ and $H^1 \cap \overline{z}u^2\overline{H_0^1}$ is sometimes denoted

$$\mathcal{K}_u \odot \mathcal{K}_u = H^1 \cap \overline{z}u^2\overline{H_0^1}.$$

For certain inner functions, one can say much more.

Theorem 31 (Baranov-Bessonov-Kapustin) *For an inner function u the following statements are equivalent.*

(i) $\mathcal{C}_1(u^2) = \mathcal{C}_2(u)$.
(ii) *For each $f \in H^1 \cap \overline{z}u^2\overline{H_0^1}$ there exists sequences g_j, h_j in $\mathcal{K}_u$ such that $\sum_j \|g_j\|\|h_j\| < \infty$ and*

$$f = \sum_j g_j h_j.$$

Moreover, there exists a universal $C > 0$, independent of f, such that the g_j, h_j can be chosen to satisfy $\sum_j \|g_j\|\|h_j\| \leqslant C\|f\|_1$.

As we have seen, the condition $\mathcal{C}_1(u^2) = \mathcal{C}_2(u)$ is of primary importance. Unfortunately, it appears difficult to test whether or not a given inner function has this property. On the other hand, the following related geometric condition appears somewhat more tractable.

Definition 5 An inner function u is called a *one-component inner function* if the set $\{z \in \mathbb{D} : |u(z)| < \epsilon\}$ is connected for some $\epsilon > 0$.

One can show, for instance, that for the atomic inner functions s_{δ_ζ} (where δ_ζ denotes the point mass at a point ζ on $\partial\mathbb{D}$), the set $\{|u| < \epsilon\}$ is a disk internally tangent to $\partial\mathbb{D}$ at ζ. In other words, these inner functions are one-component inner functions. The relevance of one-component inner functions lies in the following result of Aleksandrov.

Theorem 32 (Aleksandrov [7]) *If u is inner, then*

(i) *u is a one-component inner function if and only if*

$$\sup_{\lambda \in \mathbb{D}} \frac{\|k_\lambda\|_\infty}{\|k_\lambda\|} < \infty.$$

(ii) *If u is a one-component inner function, then $\mathcal{C}_{p_1}(u) = \mathcal{C}_{p_2}(u)$ for all $p_1, p_2 > 0$.*

There is also a related result of Treil and Volberg [102]. We take a moment to mention that W. Cohn [32] has described the set $\mathcal{C}_2(u)$ in the case where u is a one component inner function.

Theorem 33 (Cohn) *If u is a one component inner function and μ is a positive measure on $\partial\mathbb{D}$ for which $\mu(\sigma(u) \cap \partial\mathbb{D}) = 0$, then $\mu \in \mathcal{C}_2(u)$ if and only if there is a constant $C > 0$ such that*

$$\int_{\partial\mathbb{D}} \frac{1-|z|^2}{|\zeta - z|^2}\, d\mu(\zeta) \leqslant \frac{C}{1-|u(z)|^2}$$

for all z in $\mathbb{D}$.

Note that if u is a one-component inner function, then so is u^2. Combining the preceding results we obtain the following.

Corollary 8 *If u is a one-component inner function, then every operator in $\mathscr{T}_u$ has a bounded symbol.*

Of course the natural question now (and conjectured in [14]) is the following:

Question 2 Is the converse of the preceding corollary true?

It turns out that there is an interesting and fruitful interplay between the material discussed above and the family of Clark measures $\{\sigma_\alpha : \alpha \in \partial\mathbb{D}\}$, defined by (16), associated with an inner function u. To be more specific, Aleksandrov showed in [6] that if $\mathcal{C}_1(u^2) = \mathcal{C}_2(u^2)$ (the equivalent condition for every operator in $\mathscr{T}_u$ to have a bounded symbol), then every Clark measure is discrete. This leads to the following corollary.

Corollary 9 *Let u be an inner function. If for some $\alpha \in \partial\mathbb{D}$ the Clark measure σ_α is not discrete, then there is an operator in $\mathscr{T}_u$ without a bounded symbol.*

Since *any* singular measure μ (discrete or not) is equal to σ_1 for some inner function u [26], it follows that if we let μ be a continuous singular measure, then the corollary above yields an example of a truncated Toeplitz operator space $\mathscr{T}_u$ which contains operators without a bounded symbol.

6 The Spatial Isomorphism Problem

For two inner functions u_1 and u_2, when is $\mathscr{T}_{u_1}$ *spatially isomorphic* to $\mathscr{T}_{u_2}$? In other words, when does there exist a unitary operator $U : \mathcal{K}_{u_1} \to \mathcal{K}_{u_2}$ such that $U\mathscr{T}_{u_1}U^* = \mathscr{T}_{u_2}$? This is evidently a stronger condition than isometric isomorphism since one insists that the isometric isomorphism is implemented in a particularly restrictive manner.

A concrete solution to the spatial isomorphism problem posed above was given in [25]. Before discussing the solution, let us briefly introduce three basic families of spatial isomorphisms between truncated Toeplitz operator spaces. If $\psi : \mathbb{D} \to \mathbb{D}$ is a disk automorphism, then one can check that the weighted composition operator

$$U_\psi : \quad \mathcal{K}_u \to \mathcal{K}_{u\circ\psi}, \qquad U_\psi f = \sqrt{\psi'}(f \circ \psi),$$

is unitary. In particular, this implies that the map

$$\Lambda_\psi : \quad \mathscr{T}_u \to \mathscr{T}_{u\circ\psi}, \qquad \Lambda_\psi(A) = U_\psi A U_\psi^*,$$

which satisfies the useful relationship $\Lambda_\psi(A^u_\varphi) = A^{u\circ\psi}_{\varphi\circ\psi}$, implements a spatial isomorphism between $\mathscr{T}_u$ and $\mathscr{T}_{u\circ\psi}$.

Another family of spatial isomorphisms arises from the so-called *Crofoot transforms* [34] (see also [88, Sect. 13]). For $a \in \mathbb{D}$ and $\psi_a = \frac{z-a}{1-\overline{a}z}$, one can verify that the operator

$$U_a : \quad \mathcal{K}_u \to \mathcal{K}_{\psi_a \circ u}, \qquad U_a f = \frac{\sqrt{1-|a|^2}}{1-\overline{a}u} f,$$

is unitary. In particular, the corresponding map

$$\Lambda_a : \quad \mathscr{T}_u \to \mathscr{T}_{\psi_a \circ u}, \qquad \Lambda_a(A) = U_a A U_a^*,$$

implements a spatial isomorphism between $\mathscr{T}_u$ and $\mathscr{T}_{\psi_a \circ u}$.

Finally, let us define

$$[U_\# f](\zeta) = \overline{\zeta} f(\overline{\zeta}) u^\#(\zeta), \quad u^\#(z) := \overline{u(\overline{z})}.$$

The operator $U_\# : \mathcal{K}_u \to \mathcal{K}_{u^\#}$ is unitary and if

$$\Lambda_\# : \quad \mathscr{T}_u \to \mathscr{T}_{u^\#}, \qquad \Lambda_\#(A) = U_\# A U_\#^*,$$

then $\Lambda_{\#}(A_{\varphi}^{u}) = A_{\varphi^{\#}}^{u^{\#}}$ whence $\mathscr{T}_u$ is spatially isomorphic to $\mathscr{T}_{u^{\#}}$. Needless to say, the three classes of unitary operators U_{ψ}, U_a, and $U_{\#}$ introduced above should not be confused with the Clark operators U_{α} (15), which play no role here.

It turns out that any spatial isomorphism between truncated Toeplitz operator spaces can be written in terms of the three basic types described above [25].

Theorem 34 (Cima-Garcia-Ross-Wogen) *For two inner functions u_1 and u_2 the spaces $\mathscr{T}_{u_1}$ and $\mathscr{T}_{u_2}$ are spatially isomorphic if and only if either $u_1 = \psi \circ u_2 \circ \varphi$ or $u_1 = \psi \circ u_2^{\#} \circ \varphi$ for some disk automorphisms φ, ψ. Moreover, any spatial isomorphism $\Lambda : \mathscr{T}_{u_1} \to \mathscr{T}_{u_2}$ can be written as $\Lambda = \Lambda_a \Lambda_{\psi}$ or $\Lambda_a \Lambda_{\#} \Lambda_{\psi}$, where we allow $a = 0$ or $\psi(z) = z$.*

The preceding theorem leads immediately to the following question.

Question 3 Determine practical conditions on inner functions u_1 and u_2 which ensure that $u_1 = \psi \circ u_2 \circ \varphi$ or $u_1 = \psi \circ u_2^{\#} \circ \varphi$ for some disk automorphisms φ, ψ. For instance, do this when u_1 and u_2 are finite Blaschke products having the same number of zeros, counted according to multiplicity.

In the case where one of the inner functions is z^n, there is a complete answer [25].

Corollary 10 *For a finite Blaschke product u of order n, $\mathscr{T}_u$ is spatially isomorphic to $\mathscr{T}_{z^n}$ if and only if either u has one zero of order n or u has n distinct zeros all lying on a circle Γ in $\mathbb{D}$ with the property that if these zeros are ordered according to increasing argument on Γ, then adjacent zeros are equidistant in the hyperbolic metric.*

7 Algebras of Truncated Toeplitz Operators

Recall that $\mathscr{T}_u$ is a weakly closed subspace of $\mathcal{B}(\mathcal{H})$ (see Sect. 3). Although $\mathscr{T}_u$ is not an algebra, there are many interesting algebras contained within $\mathscr{T}_u$. In fact, the recent thesis [92] and subsequent paper of Sedlock [93] described them all. We discuss the properties of these so-called *Sedlock algebras* below, along with several further results from [59].

To begin with, we require the following generalization (see [88, Sect. 10]) of the Clark unitary operators (15):

$$S_u^a = A_z^u + \frac{a}{1 - \overline{u(0)}a} k_0 \otimes C k_0, \tag{29}$$

where the parameter a is permitted to vary over the closed unit disk $\mathbb{D}^-$ (we prefer to reserve the symbol α to denote complex numbers of unit modulus). The operators

S_u^a turn out to be fundamental to the study of Sedlock algebras. Before proceeding, let us recall a few basic definitions.

For $A \in \mathscr{T}_u$, the commutant $\{A\}'$ of A is defined to be the set of all bounded operators on $\mathcal{K}_u$ which commute with A. The weakly closed linear span of $\{A^n : n \geq 0\}$ will be denoted by $\mathcal{W}(A)$. Elementary operator theory says that $\mathcal{W}(A) \subseteq \{A\}'$ holds and that $\{A\}'$ is a weakly closed subset of $\mathcal{B}(\mathcal{K}_u)$. The relevance of these concepts lies in the following two results from [88, p. 515] and [59], respectively.

Theorem 35 (Sarason) *For each $a \in \mathbb{D}^-$, $\{S_u^a\}' \subseteq \mathscr{T}_u$.*

Theorem 36 (Garcia-Ross-Wogen) *For each $a \in \mathbb{D}^-$, $\{S_u^a\}' = \mathcal{W}(S_u^a)$.*

The preceding two theorems tell us that $\mathcal{W}(S_u^a)$ and $\mathcal{W}((S_u^b)^*)$, where a, b belong to $\mathbb{D}^-$, are algebras contained in $\mathscr{T}_u$. We adopt the following notation introduced by Sedlock [93]:

$$\mathscr{B}_u^a := \begin{cases} \mathcal{W}(S_u^a) & \text{if } a \in \mathbb{D}^-, \\ \mathcal{W}((S_u^{1/\overline{a}})^*) & \text{if } a \in \widehat{\mathbb{C}} \setminus \mathbb{D}^-. \end{cases}$$

Note that $\mathscr{B}_u^0$ is the algebra of analytic truncated Toeplitz operators (i.e., $\mathscr{B}_u^0 = \mathcal{W}(A_z^u)$) and that $\mathscr{B}_u^\infty$ is the algebra of co-analytic truncated Toeplitz operators (i.e., $\mathscr{B}_u^\infty = \mathcal{W}(A_{\overline{z}}^u)$). The following theorem of Sedlock asserts that the algebras $\mathscr{B}_u^a$ for $a \in \widehat{\mathbb{C}}$ are the only maximal abelian algebras in $\mathscr{T}_u$.

Theorem 37 (Sedlock [93]) *If $A, B \in \mathscr{T}_u \setminus \{\mathbb{C}I, 0\}$, then $AB \in \mathscr{T}_u$ if and only if $A, B \in \mathscr{B}_u^a$ for some $a \in \widehat{\mathbb{C}}$. Consequently, every weakly closed algebra in $\mathscr{T}_u$ is abelian and is contained in some $\mathscr{B}_u^a$.*

Let us gather together a few facts about the Sedlock algebras $\mathscr{B}_u^a$, all of which can be found in Sedlock's paper [93]. First we note that

$$\left(\mathscr{B}_u^a\right)^* = \mathscr{B}_u^{1/\overline{a}},$$

and

$$\mathscr{B}_u^a \cap \mathscr{B}_u^b = \mathbb{C}I, \quad a \neq b.$$

Most importantly, we have the following concrete description of $\mathscr{B}_u^a$.

Theorem 38 (Sedlock [93]) *If $a \in \mathbb{D}$, then*

$$\mathscr{B}_u^a = \left\{A^u_{\frac{\varphi}{1-\overline{a}u}} : \varphi \in H^\infty\right\}.$$

Furthermore, if $\varphi, \psi \in H^\infty$, then we have the following product formula

$$A^u_{\frac{\varphi}{1-\overline{a}u}} A^u_{\frac{\psi}{1-\overline{a}u}} = A^u_{\frac{\varphi\psi}{1-\overline{a}u}}.$$

In particular, if $a \in \widehat{\mathbb{C}} \setminus \partial\mathbb{D}$, then every operator in $\mathscr{B}_u^a$ is a truncated Toeplitz operator which can be represented using a bounded symbol. On the other hand, if a belongs to $\partial\mathbb{D}$, then there may be operators in $\mathscr{B}_u^a$ which do not have bounded symbols. In fact, if u has an ADC at some ζ in $\partial\mathbb{D}$, then $k_\zeta \otimes k_\zeta$ belongs to $\mathscr{B}_u^{u(\zeta)}$. From here, one can use the example from the remarks after Theorem 26 to produce an operator in $\mathscr{B}_u^{u(\zeta)}$ which has no bounded symbol.

Let us now make a few remarks about normal truncated Toeplitz operators. For a in $\partial\mathbb{D}$ the Sedlock algebra $\mathscr{B}_u^a$ is generated by a unitary operator (i.e., a Clark operator) and is therefore an abelian algebra of normal operators. When $a \in \widehat{\mathbb{C}} \setminus \partial\mathbb{D}$, the situation drastically changes [59].

Theorem 39 (Garcia-Ross-Wogen) *If $a \in \widehat{\mathbb{C}} \setminus \partial\mathbb{D}$, then $A \in \mathscr{B}_u^a$ is normal if and only if $A \in \mathbb{C}I$.*

In Sect. 6, we characterized all possible spatial isomorphisms between truncated Toeplitz operator spaces. In particular, recall that the basic spatial isomorphisms Λ_ψ, $\Lambda_\#$, Λ_a played a key role. Let us examine their effect on Sedlock algebras. The following result is from [59].

Theorem 40 (Garcia-Ross-Wogen) *For u inner, $a \in \widehat{\mathbb{C}}$, and $c \in \mathbb{D}$, we have*

$$\Lambda_\psi\left(\mathscr{B}_u^a\right) = \mathscr{B}_{u\circ\psi}^a, \qquad \Lambda_\#\left(\mathscr{B}_u^a\right) = \mathscr{B}_{u^\#}^{1/a}, \qquad \Lambda_c\left(\mathscr{B}_u^a\right) = \mathscr{B}_{u_c}^{\ell_c(a)},$$

where

$$u_c = \frac{u-c}{1-\overline{c}u}, \qquad \ell_c(a) = \begin{cases} \frac{a-c}{1-\overline{c}a} & \text{if } a \neq \frac{1}{\overline{c}}, \\ \infty & \text{if } a = \frac{1}{\overline{c}}. \end{cases}$$

For a in $\widehat{\mathbb{C}} \setminus \partial\mathbb{D}$, the preceding theorem follows from direct computations based upon Theorem 38. When a belongs to $\partial\mathbb{D}$, however, a different proof is required. One can go even further and investigate when two Sedlock algebras are spatially isomorphic. The following results are from [59].

Theorem 41 (Garcia-Ross-Wogen) *If $u(z) = z^n$ and $a, a' \in \mathbb{D}$, then $\mathscr{B}_u^a$ is spatially isomorphic to $\mathscr{B}_u^{a'}$ if and only if $|a| = |a'|$.*

Theorem 42 (Garcia-Ross-Wogen) *If*

$$u(z) = \exp\left(\frac{z+1}{z-1}\right)$$

and $a, a' \in \mathbb{D}$, then $\mathscr{B}_u^a$ is spatially isomorphic to $\mathscr{B}_u^{a'}$ if and only if $a = a'$.

Before moving on, let us take a moment to highlight an interesting operator integral formula from [88, Sect. 12] which is of some relevance here. Recall that for

α in $\partial\mathbb{D}$, the operator S_u^α given by (29) is unitary, whence $\varphi(S_u^\alpha)$ is defined by the functional calculus for φ in L^∞. By Theorem 35 we see that $\varphi(S_u^\alpha)$ belongs to $\mathscr{T}_u$. Using the Aleksandrov disintegration theorem (Theorem 9), one can then prove that

$$\langle A_\varphi f, g\rangle = \int_{\partial\mathbb{D}} \langle \varphi(S_u^\alpha) f, g\rangle \, dm(\alpha), \quad f, g \in \mathcal{K}_u,$$

which can be written in the more compact and pleasing form

$$A_\varphi = \int_{\partial\mathbb{D}} \varphi(S_u^\alpha) \, dm(\alpha).$$

A similar formula exists for symbols φ in L^2, but the preceding formulae must be interpreted carefully since the operators $\varphi(S_u^\alpha)$ may be unbounded.

Recall that for each φ in L^∞, the Cesàro means of φ are trigonometric polynomials φ_n which approximate φ in the weak-$*$ topology of L^∞. From the discussion above and Corollary 3 we know that

$$\left\{q\left(S_u^\alpha\right) : q \text{ is a trigonometric polynomial}, \alpha \in \partial\mathbb{D}\right\}$$

is weakly dense in $\mathscr{T}_u$. When u is a finite Blaschke product, it turns out that we can do much better. The following result can be found in [28], although it can be gleaned from [88, Sect. 12].

Theorem 43 *Let u be a Blaschke product of degree N and let $\alpha_1, \alpha_2 \in \partial\mathbb{D}$ with $\alpha_1 \neq \alpha_2$. Then for any $\varphi \in L^2$, there are polynomials p, q of degree at most N so that $A_\varphi = p(S_u^{\alpha_1}) + q(S_u^{\alpha_2})$.*

8 Truncated Toeplitz C^*-Algebras

In the following, we let $\mathcal{H}$ denote a separable complex Hilbert space. For each $\mathscr{X} \subseteq \mathcal{B}(\mathcal{H})$, let $C^*(\mathscr{X})$ denote the unital C^*-algebra generated by $\mathscr{X}$. In other words, $C^*(\mathscr{X})$ is the closure, in the norm of $\mathcal{B}(\mathcal{H})$, of the unital algebra generated by the operators in $\mathscr{X}$ and their adjoints. Since we are frequently interested in the case where $\mathscr{X} = \{A\}$ is a singleton, we often write $C^*(A)$ in place of $C^*(\{A\})$ in order to simplify our notation.

Recall that the *commutator ideal* $\mathscr{C}(C^*(\mathscr{X}))$ of $C^*(\mathscr{X})$, is the smallest closed two-sided ideal which contains the *commutators*

$$[A, B] := AB - BA$$

where A and B range over all elements of $C^*(\mathscr{X})$. Since the quotient algebra $C^*(\mathscr{X})/\mathscr{C}(C^*(\mathscr{X}))$ is an abelian C^*-algebra, it is isometrically $*$-isomorphic to

$C(Y)$, the set of all continuous functions on some compact Hausdorff space Y [33, Theorem 1.2.1]. We denote this relationship

$$\frac{C^*(\mathscr{X})}{\mathscr{C}(C^*(\mathscr{X}))} \cong C(Y). \tag{30}$$

Putting this all together, we have the short exact sequence

$$0 \longrightarrow \mathscr{C}\big(C^*(\mathscr{X})\big) \stackrel{\iota}{\longrightarrow} C^*(\mathscr{X}) \stackrel{\pi}{\longrightarrow} C(Y) \longrightarrow 0, \tag{31}$$

where $\iota : \mathscr{C}(C^*(\mathscr{X})) \to C^*(\mathscr{X})$ is the inclusion map and $\pi : C^*(\mathscr{X}) \to C(Y)$ is the composition of the quotient map with the isometric $*$-isomorphism which implements (30).

The *Toeplitz algebra* $C^*(T_z)$, where T_z denotes the unilateral shift on the classical Hardy space H^2, has been extensively studied since the seminal work of Coburn in the late 1960s [30, 31]. Indeed, the Toeplitz algebra is now one of the standard examples discussed in many well-known texts (e.g., [10, Sect. 4.3], [36, Chap. V.1], [38, Chap. 7]). In this setting, we have $\mathscr{C}(C^*(T_z)) = \mathscr{K}$, the ideal of compact operators on H^2, and $Y = \partial\mathbb{D}$, so that the short exact sequence (31) takes the form

$$0 \longrightarrow \mathscr{K} \stackrel{\iota}{\longrightarrow} C^*(T_z) \stackrel{\pi}{\longrightarrow} C(\partial\mathbb{D}) \longrightarrow 0. \tag{32}$$

In other words, $C^*(T_z)$ is an *extension* of $\mathscr{K}$ by $C(\partial\mathbb{D})$. It also follows that

$$C^*(T_z) = \big\{T_\varphi + K : \varphi \in C(\partial\mathbb{D}),\, K \in \mathscr{K}\big\},$$

and, moreover, that each element of $C^*(T_z)$ enjoys a unique decomposition of the form $T_\varphi + K$ [10, Theorem 4.3.2]. As a consequence, we see that the map $\pi : C^*(T_z) \to C(\partial\mathbb{D})$ is given by $\pi(T_\varphi + K) = \varphi$.

Needless to say, the preceding results have spawned numerous generalizations and variants over the years. For instance, one can consider C^*-algebras generated by matrix-valued Toeplitz operators or Toeplitz operators acting on other Hilbert function spaces (e.g., the Bergman space [11]). As another example, if $\mathscr{X}$ denotes the truncated Toeplitz operators whose symbols are both piecewise and left continuous on $\partial\mathbb{D}$, then Gohberg and Krupnik proved that $\mathscr{C}(C^*(\mathscr{X})) = \mathscr{K}$ and obtained the short exact sequence

$$0 \longrightarrow \mathscr{K} \stackrel{\iota}{\longrightarrow} C^*(\mathscr{X}) \stackrel{\pi}{\longrightarrow} C(Y) \longrightarrow 0,$$

where Y is the cylinder $\partial\mathbb{D} \times [0, 1]$, endowed with a nonstandard topology [64].

In the direction of truncated Toeplitz operators, we have the following analogue of Coburn's work.

Theorem 44 *If u is an inner function, then*

(i) $\mathscr{C}(C^*(A_z^u)) = \mathscr{K}^u$*, the algebra of compact operators on* $\mathcal{K}_u$*,*
(ii) $C^*(A_z^u)/\mathscr{K}^u$ *is isometrically $*$-isomorphic to* $C(\sigma(u) \cap \partial\mathbb{D})$*,*

(iii) *If* $\varphi \in C(\partial\mathbb{D})$, *then* A^u_φ *is compact if and only if* $\varphi(\sigma(u) \cap \partial\mathbb{D}) = \{0\}$,
(iv) $C^*(A^u_z) = \{A^u_\varphi + K : \varphi \in C(\partial\mathbb{D}), K \in \mathcal{K}^u\}$,
(v) *If* $\varphi \in C(\partial\mathbb{D})$, *then* $\sigma_e(A^u_\varphi) = \varphi(\sigma_e(A^u_z))$,
(vi) *For* $\varphi \in C(\partial\mathbb{D})$, $\|A^u_\varphi\|_e = \sup\{|\varphi(\zeta)| : \zeta \in \sigma(u) \cap \partial\mathbb{D}\}$.

In recent work [60], the authors and W. Wogen were able to provide operator algebraic proofs of the preceding results, utilizing an approach similar in spirit to the original work of Coburn. However, it should also be noted that many of the statements in Theorem 44 can be obtained using the explicit triangularization theory developed by Ahern and Clark in [2] (see the exposition in [78, Lec. V]).

9 Unitary Equivalence to a Truncated Toeplitz Operator

A significant amount of evidence is mounting that truncated Toeplitz operators may play a significant role in some sort of model theory for complex symmetric operators [25, 56, 96]. At this point, however, it is still too early to tell what exact form such a model theory should take. On the other hand, a surprising array of complex symmetric operators can be concretely realized in terms of truncated Toeplitz operators (or direct sums of such operators), without yet even venturing to discuss vector-valued truncated Toeplitz operators.

Before discussing unitary equivalence, however, we should perhaps say a few words about similarity. A number of years ago, D.S. Mackey, N. Mackey, and Petrovic asked whether or not the inverse Jordan structure problem can be solved in the class of Toeplitz matrices [75]. In other words, given any Jordan canonical form, can one find a Toeplitz matrix which is similar to this form? A negative answer to this question was subsequently provided by Heinig in [67]. On the other hand, it turns out that the inverse Jordan structure problem is always solvable in the class of truncated Toeplitz operators, for we have the following theorem [25, Theorem 6.2].

Theorem 45 (Cima-Garcia-Ross-Wogen) *Every operator on a finite dimensional space is similar to a co-analytic truncated Toeplitz operator.*

In light of the preceding theorem, it is clear that simple, purely algebraic, tools will be insufficient to settle the question of whether every complex symmetric operator can be represented in some fashion using truncated Toeplitz operators. We turn our attention now toward unitary equivalence.

Let us begin by recalling an early result of Sarason, who observed that the Volterra integration operator (1), a standard example of a complex symmetric operator [45, 49, 50], is unitarily equivalent to a truncated Toeplitz operator acting on the $\mathcal{K}_u$ space corresponding to the atomic inner function $u(z) = \exp(\frac{z+1}{z-1})$ [84] (although the term "truncated Toeplitz operator" was not yet coined). Detailed computations using the theory of model operators and characteristic functions can be found in [79, p. 41].

What was at first only an isolated result has recently begun to be viewed as a seminal observation. More recently, a number of standard classes of complex symmetric operators have been identified as being unitarily equivalent to truncated Toeplitz operators. Among the first observed examples are

(i) rank-one operators [25, Theorem 5.1],
(ii) 2×2 matrices [25, Theorem 5.2],
(iii) normal operators [25, Theorem 5.6],
(iv) for $k \in \mathbb{N} \cup \{\infty\}$, the k-fold inflation of a finite Toeplitz matrix [25, Theorem 5.7].

This last item was greatly generalized by Strouse, Timotin, and Zarrabi [96], who proved that a remarkable array of inflations of truncated Toeplitz operators are themselves truncated Toeplitz operators. In addition, a variety of related results concerning tensor products, inflations, and direct sums are given in [96]. The key to many of these results lies in the fact that if B is an inner function, then

$$h \otimes f \mapsto h(f \circ B)$$

extends linearly to a unitary operator $\Omega_B : \mathcal{K}_B \otimes L^2 \to L^2$ and, moreover, this operator maps $\mathcal{K}_B \otimes H^2$ onto H^2. Letting $\omega_B : \mathcal{K}_B \otimes \mathcal{K}_u \to \mathcal{K}_{u \otimes B}$ denote the restriction of Ω_B to $\mathcal{K}_B \otimes \mathcal{K}_u$, one can obtain the following general theorem.

Theorem 46 (Strouse-Timotin-Zarrabi [96]) *Let B and u be inner functions, and suppose that ψ, φ belong to L^2 and satisfy the following conditions:*

(i) *The operators $A^B_{\overline{B}^j \psi}$ are bounded, and nonzero only for a finite number of $j \in \mathbb{Z}$.*
(ii) *A^u_φ is bounded.*
(iii) *$\psi(\varphi \circ B) \in L^2$.*

Then $A^{u \circ B}_{\psi(\varphi \circ B)}$ is bounded and

$$A^{u \circ B}_{\psi(\varphi \circ B)} \omega_B = \omega_B \left(\sum_j \left(A^B_{\overline{B}^j \psi} \otimes A^u_{z^j \varphi} \right) \right).$$

We state explicitly only a few more results from [96], hoping to give the reader the general flavor of this surprising work. In the following, we say that the inner function B is of order n if B is a finite Blaschke product of degree n, and of order infinity otherwise.

Theorem 47 (Strouse-Timotin-Zarrabi) *Suppose that u is an inner function, that $\varphi \in L^2$, and that B is an inner function of order k for some $k \in \mathbb{N} \cup \{\infty\}$. Assume also that A^u_φ is bounded. Then $A^{u \circ B}_{\varphi \circ B}$ is bounded and unitarily equivalent to $I_k \otimes A^u_\varphi$.*

Theorem 48 (Strouse-Timotin-Zarrabi) *If ψ is an analytic function, A_{ψ}^{B} is bounded, and R is a non-selfadjoint operator of rank one, then $A_{\psi}^{B} \otimes R$ is unitarily equivalent to a truncated Toeplitz operator.*

Theorem 49 (Strouse-Timotin-Zarrabi) *Suppose u is inner, $\varphi \in H^{\infty}$, and $(A_{\varphi}^{u})^2 = 0$. If $k = \dim \mathcal{K}_u \ominus \ker A_{\varphi}^{u}$, then $A_{\varphi}^{u} \oplus 0_k$ is unitarily equivalent to a truncated Toeplitz operator.*

Although a few results concerning matrix representations of truncated Toeplitz operators have been obtained [25, 28, 96], the general question of determining whether a given matrix represents a truncated Toeplitz operator, with respect to some orthonormal basis of some $\mathcal{K}_u$ space, appears difficult. On the other hand, it is known that every truncated Toeplitz operator is unitarily equivalent to a complex symmetric matrix [45, 49], a somewhat more general issue which has been studied for its own independent interest [12, 57, 58, 99, 101].

The main result of [56] is the following simple criterion for determining whether or not a given matrix is unitarily equivalent to a truncated Toeplitz operator having an analytic symbol.

Theorem 50 (Garcia-Poore-Ross) *Suppose $M \in \mathbf{M}_n(\mathbb{C})$ has distinct eigenvalues $\lambda_1, \lambda_2, \ldots, \lambda_n$ with corresponding unit eigenvectors $\mathbf{x}_1, \mathbf{x}_2, \ldots, \mathbf{x}_n$. Then M is unitarily equivalent to an analytic truncated Toeplitz operator, on some model space $\mathcal{K}_u$ if and only if there are distinct points $z_1, z_2, \ldots, z_{n-1}$ in $\mathbb{D}$ such that*

$$\langle \mathbf{x}_n, \mathbf{x}_i \rangle \langle \mathbf{x}_i, \mathbf{x}_j \rangle \langle \mathbf{x}_j, \mathbf{x}_n \rangle = \frac{(1-|z_i|^2)(1-|z_j|^2)}{1-\overline{z_j} z_i} \tag{33}$$

holds for $1 \leqslant i \leqslant j < n$.

The method of Theorem 50 is constructive, in the sense that if (33) is satisfied, then one can construct an inner function u and a polynomial φ such that M is unitarily equivalent to A_{φ}^{u}. In fact, u is the Blaschke product having zeros at $z_1, z_2, \ldots, z_{n-1}$ and $z_n = 0$. Using Theorem 50 and other tools, one can prove the following result from [56].

Theorem 51 (Garcia-Poore-Ross) *Every complex symmetric operator on a 3-dimensional Hilbert space is unitarily equivalent to a direct sum of truncated Toeplitz operators.*

Taken together, these results from [25, 56, 96] yield a host of open questions, many of which are still open, even in the finite dimensional setting.

Question 4 Is every complex symmetric matrix $M \in \mathbf{M}_n(\mathbb{C})$ unitarily equivalent to a direct sum of truncated Toeplitz operators?

Question 5 Let $n \geq 4$. Is every irreducible complex symmetric matrix $M \in \mathbf{M}_n(\mathbb{C})$ unitarily equivalent to a truncated Toeplitz operator?

Recently, the first author and J. Tener [53] showed that every finite complex symmetric matrix is unitarily equivalent to a direct sum of (i) irreducible complex symmetric matrices or (ii) matrices of the form $A \oplus A^T$ where A is irreducible and not unitarily equivalent to a complex symmetric matrix (such matrices are necessarily 6×6 or larger). This immediately suggests the following question.

Question 6 For $A \in \mathbf{M}_n(\mathbb{C})$, is the matrix $A \oplus A^T \in \mathbf{M}_{2n}(\mathbb{C})$ unitarily equivalent to a direct sum of truncated Toeplitz operators?

One method for producing complex symmetric matrix representations of a given truncated Toeplitz operator is to use *modified Clark bases* (21) for $\mathcal{K}_u$.

Question 7 Suppose that $M \in \mathbf{M}_n(\mathbb{C})$ is complex symmetric. If M is unitarily equivalent to a truncated Toeplitz operator, does there exist an inner function u, a symbol $\varphi \in L^\infty$, and a modified Clark basis for $\mathcal{K}_u$ such that M is the matrix representation of A_φ^Θ with respect to this basis? In other words, do all such unitary equivalences between complex symmetric matrices and truncated Toeplitz operators arise from Clark representations?

10 Unbounded Truncated Toeplitz Operators

As we mentioned earlier (Sect. 2.9), for a symbol φ in L^2 and an inner function u, the truncated Toeplitz operator A_φ^u is closed and densely defined on the domain

$$\mathscr{D}(A_\varphi^u) = \{f \in \mathcal{K}_u : P_u(\varphi f) \in \mathcal{K}_u\}$$

in $\mathcal{K}_u$. In particular, the analytic function $P_u(\varphi f)$ can be defined on $\mathbb{D}$ by writing the formula (7) as an integral. In general, we actually have $CA_\varphi^u C = A_{\overline{\varphi}}^u$, and, when A_φ^u is bounded (i.e., $A_\varphi^u \in \mathscr{T}_u$), we have $A_{\overline{\varphi}}^u = (A_\varphi^u)^*$. Let us also recall an old and important result of Sarason [85] which inspired the commutant lifting theorem [97].

Theorem 52 (Sarason) *The bounded operators on $\mathcal{K}_u$ which commute with A_z^u are $\{A_\varphi^u : \varphi \in H^\infty\}$.*

In the recent papers [89, 90], Sarason studied unbounded Toeplitz operators (recall that a Toeplitz operator on H^2 is bounded if and only if the symbol is bounded) as well as unbounded truncated Toeplitz operators. We give a brief survey of these results.

In the following, N denotes the *Nevanlinna class*, the set of all quotients f/g where f and g belong to H^∞ and g is non-vanishing on $\mathbb{D}$. The *Smirnov class* N^+ denotes the subset of N for which the denominator g is not only non-vanishing on

$\mathbb{D}$ but outer. By [90] each φ in N^+ can be written uniquely as $\varphi = \frac{b}{a}$ where a and b belong to H^∞, a is outer, $a(0) > 0$, and $|a|^2 + |b|^2 = 1$ almost everywhere on $\partial\mathbb{D}$. Sarason calls this the *canonical representation* of φ.

For φ in N^+ define the Toeplitz operator T_φ as multiplication by φ on its domain $\mathscr{D}(T_\varphi) = \{f \in H^2 : \varphi f \in H^2\}$. In particular, observe that there is no projection involved in the preceding definition.

Theorem 53 (Sarason [90]) *For $\varphi = b/a \in N^+$, written in canonical form, T_φ is a closed operator on H^2 with dense domain $\mathscr{D}(T_\varphi) = aH^2$.*

There is no obvious way to define the co-analytic Toeplitz operator $T_{\overline{\varphi}}$ on H^2 for $\varphi \in N^+$. Of course we can always *define* $T_{\overline{\varphi}}$ to be T_φ^* and this makes sense when φ belongs to H^∞. In order to justify the definition $T_{\overline{\varphi}} := T_\varphi^*$ for $\varphi \in N^+$, however, we need to take care of some technical details.

As the preceding theorem shows, if $\varphi \in N^+$, then T_φ is a closed operator with dense domain aH^2. Basic functional analysis tells us that its adjoint T_φ^* is also closed and densely defined. In fact, one can show that $\mathscr{D}(T_\varphi^*)$ is the associated deBranges-Rovnyak space $\mathcal{H}(b)$ [90]. In order to understand $T_{\overline{\varphi}}$ we proceed, at least formally, as we do when examining $T_{\overline{\varphi}}$ when φ is bounded. Let φ and f have Fourier expansions

$$\varphi \sim \sum_{n=0}^{\infty} \varphi_n \zeta^n, \qquad f \sim \sum_{n=0}^{\infty} f_n \zeta^n.$$

Formal series manipulations show that

$$\begin{aligned}
T_{\overline{\varphi}} f &= P_+(\overline{\varphi} f) \\
&= P_+\left(\left(\sum_{n=0}^{\infty} \overline{\varphi_n}\,\overline{\zeta}^n\right)\left(\sum_{m=0}^{\infty} f_m \zeta^m\right)\right) \\
&= P_+\left(\sum_{n,m=0}^{\infty} \zeta^{m-n}\overline{\varphi_n} f_m\right) \\
&= P_+\left(\sum_{k=-\infty}^{\infty} \zeta^k \sum_{m=0}^{\infty} \overline{\varphi_m} f_{k+m}\right) \\
&= \sum_{k=0}^{\infty} \zeta^k \sum_{m=0}^{\infty} \overline{\varphi_m} f_{k+m}.
\end{aligned}$$

This suggests that if $\varphi = \sum_{n=0}^{\infty} \varphi_n z^n$ is the power series representation for φ in N^+, then we should define, for each function $f(z) = \sum_{n=0}^{\infty} f_n z^n$ analytic in a neighborhood of $\mathbb{D}^-$,

$$(t_{\overline{\varphi}}f)(z) := \sum_{k=0}^{\infty} z^k \left(\sum_{m=0}^{\infty} \overline{\varphi_m} f_{k+m} \right).$$

It turns out that $t_{\overline{\varphi}}f$, so defined, is an analytic function on $\mathbb{D}$. The following result indicates that this is indeed the correct approach to defining T_{φ}^*.

Theorem 54 (Sarason [90]) *If $\varphi \in N^+$, then $t_{\overline{\varphi}}$ is closable and T_{φ}^* is its closure.*

In light of the preceding theorem, for φ in N^+ we may define $T_{\overline{\varphi}}$ to be T_{φ}^*. More generally, we can define, for each φ in N^+ and u inner, the truncated Toeplitz operator $A_{\overline{\varphi}}^u$ by

$$A_{\overline{\varphi}}^u := T_{\overline{\varphi}}|_{\mathscr{D}(T_{\overline{\varphi}}) \cap \mathcal{K}_u}.$$

We gather up some results about $A_{\overline{\varphi}}^u$ from [90].

Theorem 55 (Sarason) *If $\varphi = b/a \in N^+$ is in canonical form and u is inner, then*

(i) *$A_{\overline{\varphi}}^u$ is closed and densely defined.*
(ii) *$A_{\overline{\varphi}}^u$ is bounded if and only if* $\operatorname{dist}(b, uH^\infty) < 1$.

From Theorem 12 we know that $A = CA^*C$ whenever $A \in \mathscr{T}_u$. Therefore it makes sense for us to define

$$A_{\varphi}^u := CA_{\overline{\varphi}}C$$

for $\varphi \in N^+$. It turns out that $\mathscr{D}(A_{\varphi}^u) = C\mathscr{D}(A_{\varphi}^u)$ and that the operator A_{φ}^u is closed and densely defined. Fortunately this definition makes sense in terms of adjoints.

Theorem 56 (Sarason) *For inner u and $\varphi \in N^+$, the operators A_{φ} and $A_{\overline{\varphi}}$ are adjoints of each other.*

What is the analog of Theorem 52 for of unbounded truncated Toeplitz operators? In [89] Sarason showed that

$$A_{\overline{\varphi}}^u A_{\overline{z}}^u f = A_{\overline{z}}^u A_{\overline{\varphi}}^u f,$$

holds for f in $\mathscr{D}(A_{\overline{\varphi}}^u)$ and thus one might be tempted to think that the closed densely defined operators which commute with $A_{\overline{z}}^u$ are simply $\{A_{\overline{\varphi}} : \varphi \in N^+\}$. Unfortunately the situation is more complicated and one needs to define $A_{\overline{\varphi}}^u$ for a slightly larger class of symbols than N^+. Sarason works out the details in [89] and identifies the closed densely defined operators on $\mathcal{K}_u$ which commute with $A_{\overline{z}}^u$ as the operators $A_{\overline{\varphi}}^u$ where the symbols φ come from a so-called *local Smirnov class* N_u^+. The details are somewhat technical and so we therefore leave it to the reader to explore this topic further in Sarason paper [89].

11 Smoothing Properties of Truncated Toeplitz Operators

Let us return to Theorem 2, an important result of Ahern and Clark which characterizes those functions in the model space $\mathcal{K}_u$ which have a finite angular derivative in the sense of Carathéodory (ADC) at some point ζ on $\partial\mathbb{D}$. In particular, recall that every function in $\mathcal{K}_u$ has a finite nontangential limit at ζ precisely when u has an ADC at ζ. The proof of this ultimately relies on the fact that this statement is equivalent to the condition that $(I - \overline{\lambda} A_z^u)^{-1} P_u 1$ is bounded as λ approaches ζ nontangentially. One can see this by observing that

$$\left\langle f, \left(I - \overline{\lambda} A_z^u\right)^{-1} P_u 1 \right\rangle = f(\lambda)$$

holds for all f in $\mathcal{K}_u$. If one replaces $P_u 1$ in the formula above with $P_u h$ for some h in H^∞, then a routine calculation shows that

$$\left\langle f, \left(I - \overline{\lambda} A_z^u\right)^{-1} P_u h \right\rangle = \left(A_{\overline{h}}^u f\right)(\lambda)$$

for all f in $\mathcal{K}_u$. An argument similar to that employed by Ahern and Clark shows that $(A_{\overline{h}}^u f)(\lambda)$ has a finite nontangential limit at ζ for each f in $\mathcal{K}_u$ if and only if $(I - \overline{\lambda} A_z^u)^{-1} P_u h$ is bounded as λ approaches ζ nontangentially.

Let us examine the situation when u is an infinite Blaschke product with zeros $\{\lambda_n\}_{n\geq 1}$, repeated according to multiplicity. Recall that the Takenaka basis $\{\gamma_n\}_{n\geq 1}$, defined by (19), is an orthonormal basis for $\mathcal{K}_u$. For each ζ in $\partial\mathbb{D}$, a calculation from [65] shows that $A_{\overline{h}}^u \gamma_n$ is a rational function (and so can be defined at any $\zeta \in \partial\mathbb{D}$). From this one can obtain the following analogue of the Ahern-Clark result [65].

Theorem 57 (Hartmann-Ross) *If u is a Blaschke product with zeros $\{\lambda_n\}_{n\geq 1}$ and $h \in H^\infty$, then every function in* ran $A_{\overline{h}}^u$ *has a finite nontangential limit at $\zeta \in \partial\mathbb{D}$ if and only if*

$$\sum_{n=1}^{\infty} \left|\left(A_{\overline{h}}^u \gamma_n\right)(\zeta)\right|^2 < \infty.$$

From here one can see the smoothing properties of the co-analytic truncated Toeplitz operator $A_{\overline{h}}^u$. If u happens to be an interpolating Blaschke product, then the condition in the above theorem reduces to

$$\sum_{n=1}^{\infty} (1 - |\lambda_n|^2) \left|\frac{h(\lambda_n)}{\zeta - \lambda_n}\right|^2 < \infty.$$

The following open problem now suggests itself.

Question 8 Obtain extensions of Theorem 57 to general inner functions u and symbols $h \in L^2$.

12 Nearly Invariant Subspaces

We conclude this survey with a few remarks about truncated Toeplitz operators which act on a family of spaces that are closely related to the model spaces $\mathcal{K}_u$. To be more precise, we say that a (norm closed) subspace $\mathcal{M}$ of H^2 is *nearly invariant* if the following divisibility condition holds

$$f \in \mathcal{M}, \qquad f(0) = 0 \quad \Longrightarrow \quad \frac{f}{z} \in \mathcal{M}. \tag{34}$$

These spaces were first considered and characterized in [68, 87] and they continue to be the focus of intense study [8, 9, 22, 66, 73, 76].

The link between nearly invariant subspaces of H^2 and model spaces is supplied by a crucial result of Hitt [68], which asserts that there is a unique solution g to the extremal problem

$$\sup\{\operatorname{Re} g(0) : g \in \mathcal{M},\ \|g\| = 1\}, \tag{35}$$

and moreover, that there is an inner function u so that

$$\mathcal{M} = g\mathcal{K}_u$$

and such that the map $W_g : \mathcal{K}_u \to \mathcal{M}$ defined by

$$W_g f = gf \tag{36}$$

is unitary. The function g is called the *extremal function* for the nearly invariant subspace $\mathcal{M}$. It is important to observe that since g belongs to $\mathcal{M} = g\mathcal{K}_u$, the inner function u must satisfy $u(0) = 0$. We remark that the content of these observations is nontrivial, for the set $g\mathcal{K}_u$, for arbitrary g in H^2 and u inner, is not necessarily even a subspace of H^2 since it may fail to be closed.

In the other direction, Sarason showed that if u is an inner function which satisfies $u(0) = 0$, then every isometric multiplier from $\mathcal{K}_u$ into H^2 takes the form

$$g = \frac{a}{1 - ub}, \tag{37}$$

where a and b are in the unit ball of H^∞ satisfy $|a|^2 + |b|^2 = 1$ a.e. on $\partial\mathbb{D}$ [87]. Consequently, one sees that $\mathcal{M} = g\mathcal{K}_u$ is a (closed) nearly invariant subspace of H^2 with extremal function g as in (35).

The next natural step towards defining truncated Toeplitz operators on the nearly invariant subspace $\mathcal{M} = g\mathcal{K}_u$ is to understand $P_{\mathcal{M}}$, the orthogonal projection of L^2 onto $\mathcal{M}$. The following lemma from [65] provides an explicit formula relating $P_{\mathcal{M}}$ and P_u.

Lemma 2 *If $\mathcal{M} = g\mathcal{K}_u$ is a nearly invariant subspace with extremal function g and associated inner function u satisfying $u(0) = 0$, then*

$$P_{\mathcal{M}} f = g P_u(\overline{g} f)$$

for all f in $\mathcal{M}$. Consequently, the reproducing kernel for $\mathcal{M}$ is given by

$$k_\lambda^{\mathcal{M}}(z) = \overline{g(\lambda)}g(z)\frac{1-\overline{u(\lambda)}u(z)}{1-\overline{\lambda}z}.$$

Now armed with the preceding lemma, we are in a position to introduce truncated Toeplitz operators on nearly invariant subspaces. Certainly whenever φ is a bounded function we can use Lemma 2 to see that the operator $A_\varphi^{\mathcal{M}} : \mathcal{M} \to \mathcal{M}$

$$A_\varphi^{\mathcal{M}} f := P_{\mathcal{M}}(\varphi f) = g P_u(\overline{g}\varphi f)$$

is well-defined and bounded. More generally, we may consider symbols φ such that $|g|^2\varphi$ belongs to L^2. In this case, for each h in $\mathcal{K}_u^\infty := \mathcal{K}_u \cap H^\infty$, the function $|g|^2\varphi h$ is in L^2 whence $P_u(|g|^2\varphi h)$ belongs to $\mathcal{K}_u$. By the isometric multiplier property of g on $\mathcal{K}_u$, we see that

$$P_{\mathcal{M}}(\varphi h) = g P_u\big(|g|^2\varphi h\big) \in g\mathcal{K}_u = \mathcal{M}.$$

Note that by the isometric property of g, the set $g\mathcal{K}_u^\infty$ is dense in $g\mathcal{K}_u$ by Theorem 3. Thus in this setting the operator $A_\varphi^{\mathcal{M}}$ is densely defined. We refer to any such operator as a *truncated Toeplitz operator* on $\mathcal{M}$. We denote by $\mathscr{T}_{\mathcal{M}}$ the set of all such densely defined truncated Toeplitz operators which have bounded extensions to $\mathcal{M}$. The following theorem from [65], which relies heavily upon the unitarity of the map (36), furnishes the explicit link between $\mathscr{T}_{\mathcal{M}}$ and $\mathscr{T}_u$.

Theorem 58 (Hartmann-Ross) *If $\mathcal{M} = g\mathcal{K}_u$ is a nearly invariant subspace with extremal function g and associated inner function u satisfying $u(0) = 0$, then for any Lebesgue measurable φ on $\partial\mathbb{D}$ with $|g|^2\varphi \in L^2$ we have*

$$W_g^* A_\varphi^{\mathcal{M}} W_g = A_{|g|^2\varphi}^u.$$

In light of the preceding theorem, we see that the map

$$A_\varphi^{\mathcal{M}} \mapsto A_{|g|^2\varphi}^u,$$

is a spatial isomorphism between $\mathscr{T}^M$ and $\mathscr{T}_u$. In particular, we have

$$\mathscr{T}^M = W_g \mathscr{T}_u W_g^*.$$

One can use the preceding results to prove the following facts about $\mathscr{T}^M$, all of which are direct analogues of the corresponding results on $\mathscr{T}_u$.

(i) $\mathscr{T}_{\mathcal{M}}$ is a weakly closed linear subspace of $\mathcal{B}(\mathcal{M})$.
(ii) $A_\varphi^{\mathcal{M}} \equiv 0$ if and only if $|g|^2\varphi \in uH^2 + \overline{uH^2}$.
(iii) $C_g := W_g C W_g^*$ defines a conjugation on $\mathcal{M}$ and $A = C_g A^* C_g$ for every $A \in \mathscr{T}^{\mathcal{M}}$.

(iv) If $S_g := W_g A_z W_g^*$, then a bounded operator A on $\mathcal{M}$ belongs to $\mathscr{T}^{\mathcal{M}}$ if and only if there are functions $\varphi_1, \varphi_2 \in \mathcal{M}$ so that

$$A - S_g A S_g^* = (\varphi_1 \otimes k_0^{\mathcal{M}}) + (k_0^{\mathcal{M}} \otimes \varphi_2).$$

(v) The rank-one operators in $\mathscr{T}^{\mathcal{M}}$ are constant multiples of

$$gk_\lambda \otimes gCk_\lambda, \qquad gCk_\lambda \otimes gk_\lambda, \qquad gk_\zeta \otimes gk_\zeta.$$

(vi) $\mathscr{T}^{\mathcal{M}_1}$ is spatially isomorphic to $\mathscr{T}^{\mathcal{M}_2}$ if and only if either $u_1 = \psi \circ u_2 \circ \varphi$ or $u_1 = \psi \circ \overline{u_2(\overline{z})} \circ \varphi$ for some disk automorphisms φ, ψ. In particular, this is completely independent of the corresponding extremal functions g_1 and g_2 for $\mathcal{M}_1$ and $\mathcal{M}_2$.

Acknowledgements First author partially supported by National Science Foundation Grant DMS-1001614. Second author partially supported by National Science Foundation Grant DMS-1001614.

References

1. Agler, J., McCarthy, J.E.: Pick Interpolation and Hilbert Function Spaces. Graduate Studies in Mathematics, vol. 44. Am. Math. Soc., Providence (2002)
2. Ahern, P.R., Clark, D.N.: On functions orthogonal to invariant subspaces. Acta Math. **124**, 191–204 (1970)
3. Ahern, P.R., Clark, D.N.: Radial limits and invariant subspaces. Amer. J. Math. **92**, 332–342 (1970)
4. Aleksandrov, A.B.: Invariant subspaces of the backward shift operator in the space H^p $(p \in (0, 1))$. Zap. Nauchn. Sem. Leningr. Otdel. Mat. Inst. Steklov. (LOMI) **92**, 7–29 (1979), also see p. 318. Investigations on linear operators and the theory of functions, IX
5. Aleksandrov, A.B.: Invariant subspaces of shift operators. An axiomatic approach. Zap. Nauchn. Sem. Leningr. Otdel. Mat. Inst. Steklov. (LOMI), **113**, 7–26 (1981), also see p. 264. Investigations on linear operators and the theory of functions, XI
6. Aleksandrov, A.B.: On the existence of angular boundary values of pseudocontinuable functions. Zap. Nauchn. Sem. S.-Peterburg. Otdel. Mat. Inst. Steklov. (POMI) **222**, 5–17 (1995), also see p. 307 (Issled. po, Linein. Oper. i Teor. Funktsii. 23)
7. Aleksandrov, A.B.: Embedding theorems for coinvariant subspaces of the shift operator. II. Zap. Nauchn. Sem. S.-Peterburg. Otdel. Mat. Inst. Steklov. (POMI) **262**, 5–48 (1999), also see p. 231 (Issled. po, Linein. Oper. i Teor. Funkts. 27)
8. Aleman, A., Korenblum, B.: Derivation-invariant subspaces of C^∞. Comput. Methods Funct. Theory **8**(1–2), 493–512 (2008)
9. Aleman, A., Richter, S.: Simply invariant subspaces of H^2 of some multiply connected regions. Integral Equ. Oper. Theory **24**(2), 127–155 (1996)
10. Arveson, W.: A Short Course on Spectral Theory. Graduate Texts in Mathematics, vol. 209. Springer, New York (2002)
11. Axler, S., Conway, J.B., McDonald, G.: Toeplitz operators on Bergman spaces. Can. J. Math. **34**(2), 466–483 (1982)
12. Balayan, L., Garcia, S.R.: Unitary equivalence to a complex symmetric matrix: geometric criteria. Oper. Matrices **4**(1), 53–76 (2010)
13. Baranov, A., Bessonov, R., Kapustin, V.: Symbols of truncated Toeplitz operators. J. Funct. Anal. **261**, 3437–3456 (2011)

14. Baranov, A., Chalendar, I., Fricain, E., Mashreghi, J.E., Timotin, D.: Bounded symbols and reproducing kernel thesis for truncated Toeplitz operators. J. Funct. Anal. **259**(10), 2673–2701 (2010)
15. Basor, E.L.: Toeplitz determinants, Fisher-Hartwig symbols, and random matrices. In: Recent Perspectives in Random Matrix Theory and Number Theory. London Math. Soc. Lecture Note Ser., vol. 322, pp. 309–336. Cambridge University Press, Cambridge (2005)
16. Bercovici, H.: Operator Theory and Arithmetic in H^∞. Mathematical Surveys and Monographs, vol. 26. Am. Math. Soc., Providence (1988)
17. Bercovici, H., Foias, C., Tannenbaum, A.: On skew Toeplitz operators. I. In: Topics in Operator Theory and Interpolation. Oper. Theory Adv. Appl., vol. 29, pp. 21–43. Birkhäuser, Basel (1988)
18. Bercovici, H., Foias, C., Tannenbaum, A.: On skew Toeplitz operators. II. In: Nonselfadjoint Operator Algebras, Operator Theory, and Related Topics, Oper. Theory Adv. Appl., vol. 104, pp. 23–35. Birkhäuser, Basel (1998)
19. Böttcher, A., Silbermann, B.: Introduction to Large Truncated Toeplitz Matrices. Universitext. Springer, New York (1999)
20. Brown, A., Halmos, P.R.: Algebraic properties of Toeplitz operators. J. Reine Angew. Math. **213**, 89–102 (1963/1964)
21. Carleson, L.: Interpolations by bounded analytic functions and the corona problem. Ann. Math. (2) **76**, 547–559 (1962)
22. Chalendar, I., Chevrot, N., Partington, J.R.: Nearly invariant subspaces for backwards shifts on vector-valued Hardy spaces. J. Oper. Theory **63**(2), 403–415 (2010)
23. Chalendar, I., Fricain, E., Timotin, D.: On an extremal problem of Garcia and Ross. Oper. Matrices **3**(4), 541–546 (2009)
24. Chevrot, N., Fricain, E., Timotin, D.: The characteristic function of a complex symmetric contraction. Proc. Am. Math. Soc. **135**(9), 2877–2886 (2007) (electronic)
25. Cima, J.A., Garcia, S.R., Ross, W.T., Wogen, W.R.: Truncated Toeplitz operators: spatial isomorphism, unitary equivalence, and similarity. Indiana Univ. Math. J. **59**(2), 595–620 (2010)
26. Cima, J.A., Matheson, A.L., Ross, W.T.: The Cauchy Transform. Mathematical Surveys and Monographs, vol. 125. Am. Math. Soc., Providence (2006)
27. Cima, J.A., Ross, W.T.: The Backward Shift on the Hardy Space. Mathematical Surveys and Monographs, vol. 79. Am. Math. Soc., Providence (2000)
28. Cima, J.A., Ross, W.T., Wogen, W.R.: Truncated Toeplitz operators on finite dimensional spaces. Oper. Matrices **2**(3), 357–369 (2008)
29. Clark, D.N.: One dimensional perturbations of restricted shifts. J. Anal. Math. **25**, 169–191 (1972)
30. Coburn, L.A.: The C^*-algebra generated by an isometry. Bull. Am. Math. Soc. **73**, 722–726 (1967)
31. Coburn, L.A.: The C^*-algebra generated by an isometry. II. Trans. Am. Math. Soc. **137**, 211–217 (1969)
32. Cohn, W.: Radial limits and star invariant subspaces of bounded mean oscillation. Amer. J. Math. **108**(3), 719–749 (1986)
33. Conway, J.B.: A Course in Operator Theory. Graduate Studies in Mathematics, vol. 21. Am. Math. Soc., Providence (2000)
34. Crofoot, R.B.: Multipliers between invariant subspaces of the backward shift. Pac. J. Math. **166**(2), 225–246 (1994)
35. Danciger, J., Garcia, S.R., Putinar, M.: Variational principles for symmetric bilinear forms. Math. Nachr. **281**(6), 786–802 (2008)
36. Davidson, K.R.: C^*-Algebras by Example. Fields Institute Monographs, vol. 6. Am. Math. Soc., Providence (1996)
37. Davis, P.J.: Circulant Matrices. Wiley, New York (1979). A Wiley-Interscience Publication, Pure and Applied Mathematics

38. Douglas, R.G.: Banach Algebra Techniques in Operator Theory, 2nd edn. Graduate Texts in Mathematics, vol. 179. Springer, New York (1998)
39. Douglas, R.G., Shapiro, H.S., Shields, A.L.: Cyclic vectors and invariant subspaces for the backward shift operator. Ann. Inst. Fourier (Grenoble) **20**, 37–76 (1970)
40. Duren, P.L.: Theory of H^p Spaces. Academic Press, New York (1970)
41. Dyakonov, K., Khavinson, D.: Smooth functions in star-invariant subspaces. In: Recent Advances in Operator-Related Function Theory. Contemp. Math., vol. 393, pp. 59–66. Am. Math. Soc., Providence (2006)
42. Foias, C., Frazho, A.E.: The Commutant Lifting Approach to Interpolation Problems. Operator Theory: Advances and Applications, vol. 44. Birkhäuser, Basel (1990)
43. Foias, C., Tannenbaum, A.: On the Nehari problem for a certain class of L^∞-functions appearing in control theory. J. Funct. Anal. **74**(1), 146–159 (1987)
44. Foias, C., Tannenbaum, A.: On the Nehari problem for a certain class of L^∞ functions appearing in control theory. II. J. Funct. Anal. **81**(2), 207–218 (1988)
45. Garcia, S.R.: Conjugation and Clark operators. In: Recent Advances in Operator-Related Function Theory. Contemp. Math., vol. 393, pp. 67–111. Am. Math. Soc., Providence (2006)
46. Garcia, S.R.: Aluthge transforms of complex symmetric operators. Integral Equ. Oper. Theory **60**(3), 357–367 (2008)
47. Garcia, S.R., Poore, D.E.: On the closure of the complex symmetric operators: compact operators and weighted shifts. Preprint. arXiv:1106.4855
48. Garcia, S.R., Poore, D.E.: On the norm closure problem for complex symmetric operators. Proc. Am. Math. Soc., to appear. arXiv:1103.5137
49. Garcia, S.R., Putinar, M.: Complex symmetric operators and applications. Trans. Am. Math. Soc. **358**(3), 1285–1315 (2006) (electronic)
50. Garcia, S.R., Putinar, M.: Complex symmetric operators and applications. II. Trans. Am. Math. Soc. **359**(8), 3913–3931 (2007) (electronic)
51. Garcia, S.R., Ross, W.T.: A nonlinear extremal problem on the Hardy space. Comput. Methods Funct. Theory **9**(2), 485–524 (2009)
52. Garcia, S.R., Ross, W.T.: The norm of a truncated Toeplitz operator. CRM Proc. Lect. Notes **51**, 59–64 (2010)
53. Garcia, S.R., Tener, J.E.: Unitary equivalence of a matrix to its transpose. J. Oper. Theory **68**(1), 179–203 (2012)
54. Garcia, S.R., Wogen, W.R.: Complex symmetric partial isometries. J. Funct. Anal. **257**(4), 1251–1260 (2009)
55. Garcia, S.R., Wogen, W.R.: Some new classes of complex symmetric operators. Trans. Am. Math. Soc. **362**(11), 6065–6077 (2010)
56. Garcia, S.R., Poore, D.E., Ross, W.: Unitary equivalence to a truncated Toeplitz operator: analytic symbols. Proc. Am. Math. Soc. **140**, 1281–1295 (2012)
57. Garcia, S.R., Poore, D.E., Tener, J.E.: Unitary equivalence to a complex symmetric matrix: low dimensions. Lin. Alg. Appl. **437**, 271–284 (2012)
58. Garcia, S.R., Poore, D.E., Wyse, M.K.: Unitary equivalence to a complex symmetric matrix: a modulus criterion. Oper. Matrices **4**(1), 53–76 (2010)
59. Garcia, S.R., Ross, W., Wogen, W.: Spatial isomorphisms of algebras of truncated Toeplitz operators. Indiana Univ. Math. J. **59**, 1971–2000 (2010)
60. Garcia, S.R., Ross, W., Wogen, W.: C^*-algebras generated by truncated Toeplitz operators. Oper. Theory. Adv. Appl., to appear
61. Garcia, S.R.: The eigenstructure of complex symmetric operators. In: Recent Advances in Matrix and Operator Theory. Oper. Theory Adv. Appl., vol. 179, pp. 169–183. Birkhäuser, Basel (2008)
62. Garnett, J.: Bounded Analytic Functions, 1st edn. Graduate Texts in Mathematics, vol. 236. Springer, New York (2007)
63. Gilbreath, T.M., Wogen, W.R.: Remarks on the structure of complex symmetric operators. Integral Equ. Oper. Theory **59**(4), 585–590 (2007)

64. Gohberg, I.C., Krupnik, N.Ja.: The algebra generated by the Toeplitz matrices. Funkc. Anal. Ego Prilož. **3**(2), 46–56 (1969)
65. Hartmann, A., Ross, W.T.: Boundary values in range spaces of co-analytic truncated Toeplitz operators. Publ. Mat. **56**, 191–223 (2012)
66. Hartmann, A., Sarason, D., Seip, K.: Surjective Toeplitz operators. Acta Sci. Math. (Szeged) **70**(3–4), 609–621 (2004)
67. Heinig, G.: Not every matrix is similar to a Toeplitz matrix. In: Proceedings of the Eighth Conference of the International Linear Algebra Society, Barcelona, 1999, vol. 332/334, pp. 519–531 (2001)
68. Hitt, D.: Invariant subspaces of H^2 of an annulus. Pac. J. Math. **134**(1), 101–120 (1988)
69. Hoffman, K.: Banach Spaces of Analytic Functions. Prentice-Hall Series in Modern Analysis. Prentice-Hall, Englewood Cliffs (1962)
70. Johansson, K.: Toeplitz determinants, random growth and determinantal processes. In: Proceedings of the International Congress of Mathematicians, Beijing, 2002, vol. III, pp. 53–62. Higher Ed. Press, Beijing (2002)
71. Jung, S., Ko, E., Lee, J.: On scalar extensions and spectral decompositions of complex symmetric operators. J. Math. Anal. Appl. **379**, 325–333 (2011)
72. Jung, S., Ko, E., Lee, M., Lee, J.: On local spectral properties of complex symmetric operators. J. Math. Anal. Appl. **379**, 325–333 (2011)
73. Kiselev, A.V., Naboko, S.N.: Nonself-adjoint operators with almost Hermitian spectrum: matrix model. I. J. Comput. Appl. Math. **194**(1), 115–130 (2006)
74. Li, C.G., Zhu, S., Zhou, T.: Foguel operators with complex symmetry. Preprint
75. Mackey, D.S., Mackey, N., Petrovic, S.: Is every matrix similar to a Toeplitz matrix? Linear Algebra Appl. **297**(1–3), 87–105 (1999)
76. Makarov, N., Poltoratski, A.: Meromorphic inner functions, Toeplitz kernels and the uncertainty principle. In: Perspectives in Analysis. Math. Phys. Stud., vol. 27, pp. 185–252. Springer, Berlin (2005)
77. Nikolski, N.: Operators, Functions, and Systems: An Easy Reading. Vol. 1. Mathematical Surveys and Monographs, vol. 92
78. Nikolski, N.: Treatise on the Shift Operator. Springer, Berlin (1986)
79. Nikolski, N.: Operators, Functions, and Systems: An Easy Reading. Vol. 2 Mathematical Surveys and Monographs, vol. 93. Am. Math. Soc., Providence (2002). Model operators and systems, Translated from the French by Andreas Hartmann and revised by the author
80. Partington, J.R.: Linear Operators and Linear Systems: An Analytical Approach to Control Theory. London Mathematical Society Student Texts, vol. 60. Cambridge University Press, Cambridge (2004)
81. Peller, V.V.: Hankel Operators and Their Applications. Springer Monographs in Mathematics. Springer, New York (2003)
82. Rosenblum, M., Rovnyak, J.: Hardy Classes and Operator Theory. Oxford Mathematical Monographs. The Clarendon Press/Oxford University Press, New York (1985). Oxford Science Publications
83. Ross, W.T., Shapiro, H.S.: Generalized Analytic Continuation. University Lecture Series, vol. 25. Am. Math. Soc., Providence (2002)
84. Sarason, D.: A remark on the Volterra operator. J. Math. Anal. Appl. **12**, 244–246 (1965)
85. Sarason, D.: Generalized interpolation in H^∞. Trans. Am. Math. Soc. **127**, 179–203 (1967)
86. Sarason, D.: Invariant Subspaces. Topics in Operator Theory, pp. 1–47. Am. Math. Soc., Providence (1974). Math. Surveys, No. 13
87. Sarason, D.: Nearly invariant subspaces of the backward shift. In: Contributions to Operator Theory and Its Applications, Mesa, AZ, 1987. Oper. Theory Adv. Appl., vol. 35, pp. 481–493. Birkhäuser, Basel (1988)
88. Sarason, D.: Algebraic properties of truncated Toeplitz operators. Oper. Matrices **1**(4), 491–526 (2007)
89. Sarason, D.: Unbounded operators commuting with restricted backward shifts. Oper. Matrices **2**(4), 583–601 (2008)

90. Sarason, D.: Unbounded Toeplitz operators. Integral Equ. Oper. Theory **61**(2), 281–298 (2008)
91. Sarason, D.: Commutant lifting. In: A Glimpse at Hilbert Space Operators. Oper. Theory Adv. Appl., vol. 207, pp. 351–357. Birkhäuser, Basel (2010)
92. Sedlock, N.: Properties of truncated Toeplitz operators. Ph.D. Thesis, Washington University in St. Louis, ProQuest LLC, Ann Arbor, MI (2010)
93. Sedlock, N.: Algebras of truncated Toeplitz operators. Oper. Matrices **5**(2), 309–326 (2011)
94. Simon, B.: Orthogonal Polynomials on the Unit Circle. Part 1 American Mathematical Society Colloquium Publications, vol. 54. Am. Math. Soc., Providence (2005). Classical theory. MR 2105088 (2006a:42002a)
95. Simon, B.: Orthogonal Polynomials on the Unit Circle. Part 2 American Mathematical Society Colloquium Publications, vol. 54. Am. Math. Soc., Providence (2005). Spectral theory. MR 2105089 (2006a:42002b)
96. Strouse, E., Timotin, D., Zarrabi, M.: Unitary equivalence to truncated Toeplitz operators. Indiana U. Math. J., to appear. http://arxiv.org/abs/1011.6055
97. Sz.-Nagy, B., Foias, C., Bercovici, H., Kérchy, L.: Harmonic Analysis of Operators on Hilbert Space, 2nd edn. Universitext. Springer, New York (2010)
98. Takenaka, S.: On the orthonormal functions and a new formula of interpolation. Jpn. J. Math. **2**, 129–145 (1925)
99. Tener, J.E.: Unitary equivalence to a complex symmetric matrix: an algorithm. J. Math. Anal. Appl. **341**(1), 640–648 (2008)
100. Trefethen, L.N., Embree, M.: Spectra and Pseudospectra: The Behavior of Nonnormal Matrices and Operators. Princeton University Press, Princeton (2005)
101. Vermeer, J.: Orthogonal similarity of a real matrix and its transpose. Linear Algebra Appl. **428**(1), 382–392 (2008)
102. Volberg, A.L., Treil, S.R.: Embedding theorems for invariant subspaces of the inverse shift operator. Zap. Nauchn. Sem. Leningr. Otdel. Mat. Inst. Steklov. (LOMI) **149**, 38–51 (1986), also see pp. 186–187 (Issled. Linein. Teor. Funktsii. XV)
103. Wang, X., Gao, Z.: A note on Aluthge transforms of complex symmetric operators and applications. Integral Equ. Oper. Theory **65**(4), 573–580 (2009)
104. Wang, X., Gao, Z.: Some equivalence properties of complex symmetric operators. Math. Pract. Theory **40**(8), 233–236 (2010)
105. Zagorodnyuk, S.M.: On a J-polar decomposition of a bounded operator and matrix representations of J-symmetric, J-skew-symmetric operators. Banach J. Math. Anal. **4**(2), 11–36 (2010)
106. Zhu, S., Li, C.G.: Complex symmetric weighted shifts. Trans. Am. Math. Soc., to appear
107. Zhu, S., Li, C., Ji, Y.: The class of complex symmetric operators is not norm closed. Proc. Am. Math. Soc. **140**, 1705–1708 (2012)

GPSR Compliance

The European Union's (EU) General Product Safety Regulation (GPSR) is a set of rules that requires consumer products to be safe and our obligations to ensure this.

If you have any concerns about our products, you can contact us on ProductSafety@springernature.com

In case Publisher is established outside the EU, the EU authorized representative is:

Springer Nature Customer Service Center GmbH
Europaplatz 3
69115 Heidelberg, Germany

Batch number: 10614741

Printed by Printforce, the Netherlands